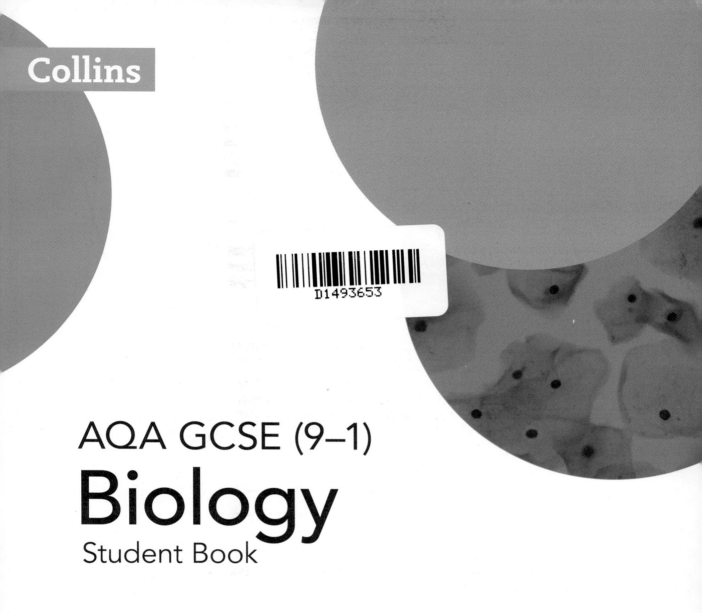

Collins

AQA GCSE (9–1)

Biology
Student Book

John Beeby
Anne Pilling
Series editor: Ed Walsh

D1493653

William Collins' dream of knowledge for all began with the publication of his first book in 1819.

A self-educated mill worker, he not only enriched millions of lives, but also founded a flourishing publishing house. Today, staying true to this spirit, Collins books are packed with inspiration, innovation and practical expertise. They place you at the centre of a world of possibility and give you exactly what you need to explore it.

Collins. Freedom to teach

HarperCollins Publishers
1 London Bridge Street
London SE1 9GF

Browse the complete Collins catalogue at
www.collins.co.uk

First edition 2016

10 9 8 7 6 5 4

© HarperCollins*Publishers* 2016

ISBN 978-0-00-815875-0

Collins® is a registered trademark of HarperCollins Publishers Limited

www.collins.co.uk

A catalogue record for this book is available from the British Library

Commissioned by Lucy Rowland and Lizzie Catford
Edited by Hamish Baxter
Project managed by Elektra Media Ltd
Copy edited by Sophia Ktori and Anna Clark
Proofread by Judith Shaw
Typeset by Jouve India and Ken Vail Graphic Design
Cover design by We are Laura
Printed by Grafica Veneta, S.p.A., Italy
Cover images © Shutterstock/Komsan Loonprom, Shutterstock/Everett Historical

MIX
Paper from responsible sources
FSC www.fsc.org **FSC™ C007454**

FSC™ is a non-profit international organisation established to promote the responsible management of the world's forests. Products carrying the FSC label are independently certified to assure consumers that they come from forests that are managed to meet the social, economic and ecological needs of present and future generations, and other controlled sources.

Find out more about HarperCollins and the environment at
www.harpercollins.co.uk/green

Approval message from AQA

This textbook has been approved by AQA for use with our qualification. This means that we have checked that it broadly covers the specification and we are satisfied with the overall quality. Full details of our approval process can be found on our website.

We approve textbooks because we know how important it is for teachers and students to have the right resources to support their teaching and learning. However, the publisher is ultimately responsible for the editorial control and quality of this book.

Please note that when teaching the GCSE Biology course, you must refer to AQA's specification as your definitive source of information. While this book has been written to match the specification, it cannot provide complete coverage of every aspect of the course.

A wide range of other useful resources can be found on the relevant subject pages of our website: aqa.org.uk

ACKNOWLEDGEMENTS

The publishers gratefully acknowledge the permissions granted to reproduce copyright material in this book. Every effort has been made to contact the holders of copyright material, but if any have been inadvertently overlooked, the Publisher will be pleased to make the necessary arrangements at the first opportunity.

Chapter 1
p. 12 Claudio Divizia/Shutterstock, Valeriy Velikov/Shutterstock fusebulb/Shutterstock; p. 13 royaltystockphoto.com/Shutterstock, royaltystockphoto.com/Shutterstock, Digieva/Shutterstock, Knorre/Shutterstock; p. 14 Jose Luis Calvo/Shutterstock Dimarion/Shutterstock, Alexander Raths/Shutterstock; p. 16 Africa Studio/Shutterstock, Science Photo Library; p. 17 Jubal Harshaw/Shutterstock; p. 18 Pan Xunbin/Shutterstock, royaltystockphoto.com/Shutterstock; p. 19 toeytoey/Shutterstock, Dlumen/Shutterstock, CNRI/SCIENCE PHOTO LIBRARY, Dr Jeremy Burgess/Science Photo Library, AMMRF, UNIVERSITY OF SYDNEY/SCIENCE PHOTO LIBRARY; p. 20 anyaivanova/Shutterstock, Ed Reschke/Getty, Visuals Unlimited, Inc./Dr. Gladden Willis/Getty; p. 20 Beeby Education Limited; p. 21 Lebendkulturen.de/Shutterstock, D. Kucharski K. Kucharska/Shutterstock, Power_J/Shutterstock, toeytoey/Shutterstock, Lebendkulturen.de/Shutterstock; p. 22 DR. MARLI MILLER/VISUALS UNLIMITED, INC. /SCIENCE PHOTO LIBRARY; p. 23 Zack Frank/Shutterstock; p. 25 Power_J/Shutterstock; p. 26 Leptospira/Shutterstock; p. 28 EPSTOCK/Shutterstock; p. 30 Lisa S./Shutterstock; p. 31 royaltystockphoto/Shutterstock; p. 32 Elena Pavlovich/Shutterstock; p. 33 memorisz/Shutterstock; p. 36 Ralf Herschbach/Shutterstock; p. 37 BMJ/Shutterstock, Dave Massey/Shutterstock; p. 38 wavebreakmedia/Shutterstock, Knorre/Shutterstock; p. 39 Africa Studio/Shutterstock; p. 40 borzywoj/Shutterstock; p. 43 Zaharia Bogdan Rares/Shutterstock; p. 46 StudyBlue inc.

Chapter 2
p.54 Heiti Paves/Shutterstock, Ethan Daniels/Shutterstock; p. 55 Jubal Harshaw/Shutterstock, MARCELODLT/Shutterstock, Brian Maudsley/Shutterstock; p. 61 Claudio Divizia/Shutterstock; p. 64 Andreas Altenburger/Shutterstock, bjul/Shutterstock, Anton_Ivanov/Shutterstock; p. 66 Lizard/Shutterstock; p. 67 Gemenacom/Shutterstock; p. 71 Jubal Harshaw/Shutterstock; p. 73 Stephen VanHorn/Shutterstock; p. 76 Brian Maudsley/Shutterstock; p. 78 JONATHAN PLEDGER/Shutterstock; p. 84 Marketa Mark/Shutterstock, YuG/Shutterstock, PearlNecklace/Shutterstock

Chapter 3
p. 86 Jubal Harshaw/Shutterstock; p. 87 eAlisa/Shutterstock, SUSUMU NISHINAGA/SCIENCE PHOTO LIBRARY, Jubal Harshaw/Shutterstock; p. 93 Mopic/Shutterstock; p.94 SCIENCE SOURCE/SCIENCE PHOTO LIBRARY, Sakarin Sawasdinaka/Shutterstock; p.95 dreamerb/Shutterstock, Lebendkulturen.de/Shutterstock, Abel Tumik/Shutterstock, Monika Vosahlova/Shutterstock, Yann hubert/Shutterstock; p. 102 Collins Separate Sciences B OCR Gateway ISBN-13-978-0-00-741534-2, p28; p. 107 unpict/Shutterstock; p. 116 PROFESSORS P.M. MOTTA & S. CORRER/SCIENCE PHOTO LIBRARY; p. 127 Steve Boice/Shutterstock, Bildagentur Zoonar GmbH

Chapter 4
p. 128 Stefano Carnevali/Shutterstock, royaltystockphoto.com/Shutterstock, fusebulb/Shutterstock, Maxim Tupikov/Shutterstock; p. 129 WitthayaP/Shutterstock, JPC-PROD/Shutterstock, Kazakov Maksim/Shutterstock; p. 131 Antonio Guillem/Shutterstock; p. 132 WitthayaP/Shutterstock; p. 133 Edyta Pawlowska/Shutterstock; p. 134 Maksym Bondarchuk/Shutterstock, solkanar/Shutterstock, Australis Photography/Shutterstock; p. 135 Ehab Edward/Shutterstock; p. 138 IAN GOWLAND/SCIENCE PHOTO LIBRARY; p. 139 Chaikom/Shutterstock; p. 140 NORM THOMAS/SCIENCE PHOTO LIBRARY; p. 141 Jina K/Shutterstock; p. 142 MichaelTaylor3d/Shutterstock, royaltystockphoto.com/Shutterstock, Thailand Travel and Stock/Shutterstock; p. 144 © Anne Pilling, Nigel Cattlin/Visuals Unlimited/Corbis; p. 145 gorillaimages/Shutterstock; p. 150 Roberto Piras/Shutterstock; p. 152 EM Karuna/Shutterstock, Zaharia Bogdan Rares/Shutterstock, P. FERGUSON, ISM/SCIENCE PHOTO LIBRARY; p. 154 JPC-PROD/Shutterstock; p. 156 Sarah Marchant/Shutterstock, Hellen Sergeyeva/Shutterstock, TwilightArtPictures/Shutterstock; p. 160 NORM THOMAS/SCIENCE PHOTO LIBRARY, popular business/Shutterstock, kay roxby/Shutterstock p. 161 Christian Musat/Shutterstock, PHOTO FUN/Shutterstock; p. 162 anyaivanova/Shutterstock, GEOFF KIDD/SCIENCE PHOTO LIBRARY; p. 163 vladimir salman/Shutterstock

Chapter 5
p. 170 Alexilusmedical/Shutterstock, BlueRingMedia/Shutterstock, Pavel Chagochkin/Shutterstock, Christophe Baudot/Shutterstock; p. 172 lzf/Shutterstock; p. 173 Andrii Muzyka/Shutterstock; p. 174 3Dme Creative Studio/Shutterstock; p. 176 bikeriderlondon/Shutterstock; p. 178 Steve Buckley/Shutterstock, UNIVERSITY OF DURHAM/SIMON FRASER/SCIENCE PHOTO LIBRARY; p.179 EPSTOCK/Shutterstock, LaKirr/Shutterstock; p. 180 India Picture/Shutterstock; p. 183 royaltystockphoto.com/Shutterstock; p. 188 Joseph Giacomin/Cultura/SCIENCE PHOTO LIBRARY; p. 187 Robert Przybysz/Shutterstock; p. 190 NYPL/SCIENCE PHOTO LIBRARY; p. 194 Santibhavank P/Shutterstock; p. 196 Andrey_Popov/Shutterstock; P. 197 AnglianArt/Shutterstock; p. 200 Shipov Oleg/Shutterstock; p. 202 ZEPHYR/SCIENCE PHOTO LIBRARY; p.205 Nigel Paul Monckton/Shutterstock, Lisa-Blue/iStock; p. 206 gopixa/Shutterstock; p. 210 Keystone/getty, originalpunkt/Shutterstock; p. 211 nevodka/Shutterstock; p. 212 ALAIN POL, ISM/SCIENCE PHOTO LIBRARY, fotografixx/istock; p. 214 Pamela Moore/istock; p. 217 Roman Prishenko/Shutterstock, Ray Ellis/SCIENCE PHOTO LIBRARY, JPC-PROD/Shutterstock, Michael Kraus/Shutterstock, Image Point Fr/Shutterstock, areeya_ann/Shutterstock; p.220 Daxiao Productions/ Shutterstock, MARTIN SHIELDS/SCIENCE PHOTO LIBRARY; p. 223 Barbol/Shutterstock; p. 225 Pat_Hastings/Shutterstock; p. 227 Jamroen Jaiman/Shutterstock, ElecImagery/Alamy, Norman Pogson/Shutterstock; p. 233 Anest/Shutterstock

Chapter 6
p. 236 A. BARRINGTON BROWN, GONVILLE AND CAIUS COLLEGE/SCIENCE PHOTO LIBRARY, STEVE GSCHMEISSNER/SCIENCE PHOTO LIBRARY, itsmejust/Shutterstock; p. 237 hxdbzxy/Shutterstock, Giovanni Cancemi/Shutterstock; p. 238 PHILIPPE PSAILA/SCIENCE PHOTO LIBRARY; p. 239 alybaba/Shutterstock; p. 240 Gio.tto/Shutterstock; p. 241 isak55/Shutterstock; p. 242 Leah-Anne Thompson/Shutterstock, JEAN-FRANCOIS PODEVIN/SCIENCE PHOTO LIBRARY; p. 244 A. BARRINGTON BROWN, GONVILLE AND CAIUS COLLEGE/SCIENCE PHOTO LIBRARY, adike/Shutterstock; p. 248 Steve McWilliam/Shutterstock; p. 252 khemporn tongphay/Shutterstock; p. 253 Anest/Shutterstock; p. 254 Levent Konuk/Shutterstock; p. 255 Africa Studio/Shutterstock; p. 256 Puwadol Jaturawutthichai/Shutterstock; p. 258 WILL & DENI MCINTYRE/SCIENCE PHOTO LIBRARY, SCIENCE PHOTO LIBRARY p.260 catwalker/Shutterstock; p. 263 Kazakov Maksim/Shutterstock, MARK THOMAS/SCIENCE PHOTO LIBRARY

Chapter 7
p. 272 Sergey Novikov/Shutterstock, Kjersti Joergensen/Shutterstock, Esteban De Armas/Shutterstock; p. 273 Kenneth Keifer/Shutterstock, Everett Historical/Shutterstock, Surrphoto/Shutterstock; p. 274 Charlotte Purdy/Shutterstock; p. 275 BMJ/Shutterstock, outdoorsman/Shutterstock, photos2013/Shutterstock; p. 276 Tanor/Shutterstock; p. 279 Ryan M. Bolton/Shutterstock, Claude Huot/Shutterstock, Guido Vermeulen-Perdaen/Shutterstock, Ben Queenborough/Shutterstock, NATURAL HISTORY MUSEUM, LONDON/SCIENCE PHOTO LIBRARY, Joe Ravi/Shutterstock; p. 280 Olga Popova/Shutterstock; p. 282 PAUL TAFFOREAU/ESRF/PASCAL GOETGHELUCK/ SCIENCE PHOTO LIBRARY; p. 283 Natursports/Shutterstock; p. 284 NATURAL HISTORY MUSEUM, LONDON/SCIENCE PHOTO LIBRARY; p. 285 Alexandr79/Shutterstock, chris2766/Shutterstock, Essua; p. 286 Chantelle Bosch/Shutterstock; p. 287 Patricia Chumillas/Shutterstock, Ben Queenborough/Shutterstock, Stubblefield Photography/Shutterstock, MIGUEL CASTRO/SCIENCE PHOTO LIBRARY; p. 288 manfredxy/Shutterstock; p. 289 Martin Fowler/Shutterstock, Steve McWilliam/Shutterstock; p. 290 SCIENCE PHOTO LIBRARY, NATIONAL LIBRARY OF MEDICINE/SCIENCE PHOTO LIBRARY; p. 291 Studiotouch/Shutterstock, HERMANN EISENBEISS/SCIENCE PHOTO LIBRARY; p. 292 Everett Historical/Shutterstock; p. 296 dwphotos/Shutterstock, Nate Allred/Shutterstock; p. 297 Henk Vrieselaar/Shutterstock, Sergey Fatin/Shutterstock, Eric Isselee/Shutterstock; p. 298 Volodymyr Stakhiv/Shutterstock; p. 299 ur Zoonar GmbH/Shutterstock, Jorge Salcedo/Shutterstock, Gordana Sermek/Shutterstock, Texturis/Shutterstock; p. 300 VOLKER STEGER/SCIENCE PHOTO LIBRARY; p. 301 Denton Rumsey/Shutterstock; p. 302 sundetman/Shutterstock; p. 303 Lightspring/Shutterstock, Geoffrey Budesa/Shutterstock; p. 304 science photo/Shutterstock; p. 305 stocksolutions/Shutterstock, a katz/Shutterstock; p. 306 www.GloFish.com; p. 307 PETER MENZEL/SCIENCE PHOTO LIBRARY; p. 318 Kiselev Andrey Valerevich/Shutterstock

Chapter 8
p. 322 apigude/Shutterstock, patostudio/Shutterstock, Taiga/Shutterstock, Peshkova/Shutterstock; p. 323 Kirsanov Valeriy Vladimirovich/Shutterstock, Ethan Daniels/Shutterstock, Critterbiz/Shutterstock, FotoJoost/Shutterstock; p. 324 Tischenko Irina/Shutterstock; p. 325 Zack Frank/Shutterstock, Martin Fowler/Shutterstock, Ron Zmiri/Shutterstock; p. 326 Vaclav Volrab/Shutterstock, Joanne Weston/Shutterstock, SJ Travel Photo and Video/Shutterstock; p. 328 so51hk/Shutterstock; p. 330 Onsuda/Shutterstock, Aquapix/Shutterstock, Leonardo Gonzalez/Shutterstock, Masahiro Suzuki/Shutterstock, Ugo Montaldo/Shutterstock; p. 334 Bjul/Shutterstock, Paula French/Shutterstock, Tom Roche/Shutterstock; p. 335 Polarpx/Shutterstock, Pichai Tunsuphon/Shutterstock, 7382489561/Shutterstock; p. 336 Yellowj/Shutterstock, richpav/Shutterstock; p. 337 Gajic Dragan/Shutterstock; p. 338 Darren Foard/Shutterstock, Pal Teravagimov/Shutterstock; p. 339 kamon_saejueng/Shutterstock, f9photos/Shutterstock; p. 340 Vladimir Melnik/Shutterstock, Joanna Zaleska/Shutterstock; p. 341 Straystone/Shutterstock, Tanor/Shutterstock; p. 342 topten22photo/Shutterstock; p. 343 Marek Velechovsky/Shutterstock; p. 344 Carlos Caetano/Shutterstock; p. 345 DrObjektiff/Shutterstock; p. 346 Joanna Stankiewicz-Witek/Shutterstock, Bildagentur Zoonar GmbH/Shutterstock; p. 350 Anton Gorlin/Shutterstock, Lukas Gojda/Shutterstock, ChiccoDodiFC/Shutterstock; p. 351 Konstantin Stepanenko/Shutterstock; p. 352 Richard Thornton/Shutterstock; p. 353 photoneye/Shutterstock, Lodimup/Shutterstock; p. 354 Gary Andrews/Shutterstock; p. 355 Bildagentur Zoonar GmbH/Shutterstock, Nathape/Shutterstock; p. 357 bikeriderlondon/Shutterstock, BOONCHUAY PROMJIAM/Shutterstock; p. 358 Huguette Roe/Shutterstock; p. 359 Anticiclo/Shutterstock; p. 360 Randimal/Shutterstock, gubernat/Shutterstock, Orla/Shutterstock; p. 362 Stephen Lavery/Shutterstock, sunsetman/Shutterstock; p. 363 Ilona Ignatova/Shutterstock, Vilainecrevette/Shutterstock; p. 364 Sergei Butorin/Shutterstock; p. 365 montree hanlue/Shutterstock; p. 366 Vladislav Gajic/Shutterstock; p. 367 Randimal/Shutterstock, Egon Zitter/Shutterstock; p. 368 Golden Rice Humanitarian Board www.goldenrice.org; p. 373 Sarah Pettegree/Shutterstock

Contents

You can use this book if you are studying Combined Science: Trilogy

 you will need to master all of the ideas and concepts on these pages

 you will need to master some of the ideas and concepts on these pages.

How to use this book

Remember! to cover all the content of the AQA Biology Specification you should study the text and attempt the End of chapter questions.

These tell you what you will be learning about in the lesson and are linked to the AQA specification.

This introduces the topic and puts the science into an interesting context.

Each topic is divided into three sections. The level of challenge gets harder with each section.

Each section has level-appropriate questions, so you can check and apply your knowledge.

Biology

Genetic engineering

Learning objectives:

- give examples of how plant crops have been genetically engineered to improve products and describe how fungus cells are engineered to produce human insulin.

- describe the process of genetic engineering

KEY WORDS

genetic engineering
GM crops
vector

Genetic engineering involves taking specific genes from one organism and introducing them into the genome of another. Scientists can now, more or less, transfer genes from any organism, including plants, animals, bacteria or viruses.

Producing human insulin

Patients with Type 1 diabetes need regular injections of the hormone insulin. Since the early 1920s, insulin was extracted from the pancreas of pigs or cattle. But these types of insulin differ *slightly* from human insulin in the amino acids they contain. They had some side effects.

With **genetic engineering** it became possible to genetically engineer the bacterium, *Escherichia coli*, and the fungus, yeast, to produce 'human' insulin. This is *identical* to the insulin produced by the human body.

Yeast produces a more complete version of the insulin molecule. Less processing is required, so this method is often preferred.

Figure 7.39 Human insulin production in India. This photograph shows the purification process

1. What is genetic engineering?

2. Name two organisms that can be genetically engineered to produce insulin.

Genetically engineered plants

Genetic engineering has transformed crop production. Genes from many organisms, often not even plants, are cut out of their chromosomes and inserted into the cells of crop plants. Such crop plants and other organisms are called genetically modified, **GM crops** or GM organisms (GMOs).

Plants have been engineered to be resistant to disease, and to increase yields, such as producing bigger, better fruit. Several types of crop plant have been produced that are resistant to diseases caused by viruses.

In the wet summer of 2012, potato plants became exposed to the potato blight fungus. In 2014 British scientists produced a GM potato that is resistant to potato blight. Genes from two wild relatives of the potato were inserted into the Desiree potato variety.

300 AQA GCSE Biology: Student Book

7.14

Figure 7.40 Pesticides are sprayed over crops to protect them from diseases. Disease resistant GM plants don't need the pesticides.

3 Give two reasons for the genetic modification of plant crops.

4 What types of organism cause disease in plants?

HIGHER TIER ONLY

The genetic engineering technique

Enzymes are used to remove the required gene, or genes, from the organism that carries the gene(s).

The gene is transferred, using a **vector**, to the organism that is to be modified. The vector is often a plasmid. The gene is inserted and sealed into the plasmid DNA using another enzyme. Bacteria have plasmids, and so do some eukaryotes, such as yeast. Viruses may also be used as vectors, including tobacco mosaic virus. Viruses that have had other genes modified, so that they are not infective, have been used in vaccine production.

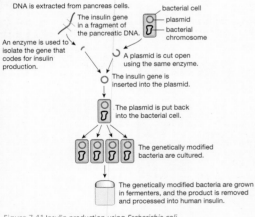

Figure 7.41 Insulin production using *Escherichia coli*

5 How is the required gene removed from the donor organism?

6 How is the gene transferred to the organism that is to be genetically modified?

DID YOU KNOW?

Genes for human insulin were cloned and transferred to *E. coli* in 1978. The first drug produced using genetic engineering was human insulin.

KEY INFORMATION

Remember: if genes are transferred to plants, this needs to be at an early stage of their development. Older organisms have too many cells that would need to be modified.

Google search: 'genetic engineering, genetic modification, GM insulin' 301

Biology

The first page of a chapter has links to ideas you have met before, which you can now build on.

This page gives a summary of the exciting new ideas you will be learning about in the chapter.

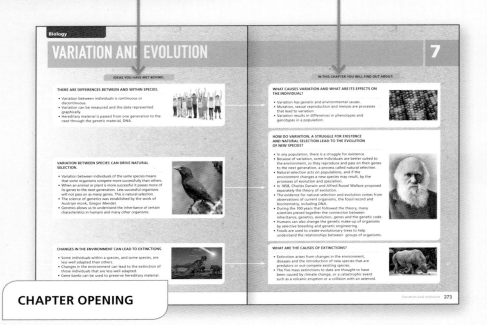

CHAPTER OPENING

The Key Concept pages focus on core ideas. Once you have understood the key concept in a chapter, it should develop your understanding of the whole topic.

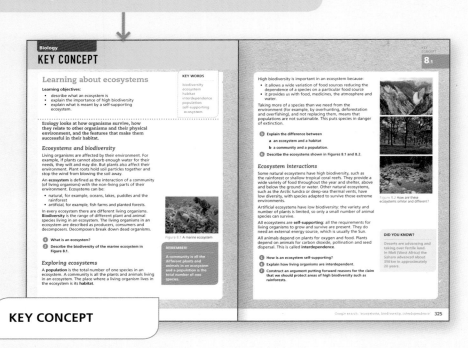

KEY CONCEPT

There is a dedicated page for every Required Practical in the AQA specification. They help you to analyse the practical and to answer questions about it.

The tasks – which get a bit more difficult as you go through – challenge you to apply your science skills and knowledge to the new context.

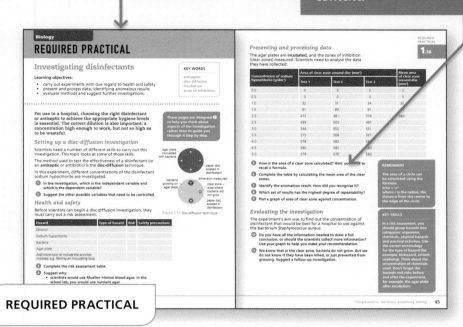

REQUIRED PRACTICAL

The Maths Skills pages focus on the maths requirements in the AQA specification, explaining concepts and providing opportunities to practise.

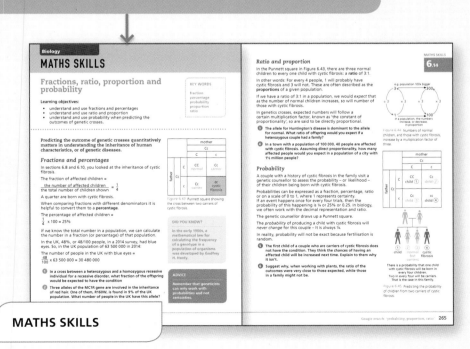

MATHS SKILLS

Biology

These lists at the end of a chapter act as a checklist of the key ideas of the chapter. In each row, the green box gives the ideas or skills that you should master first. Then you can aim to master the ideas and skills in the blue box. Once you have achieved those you can move on to those in the red box.

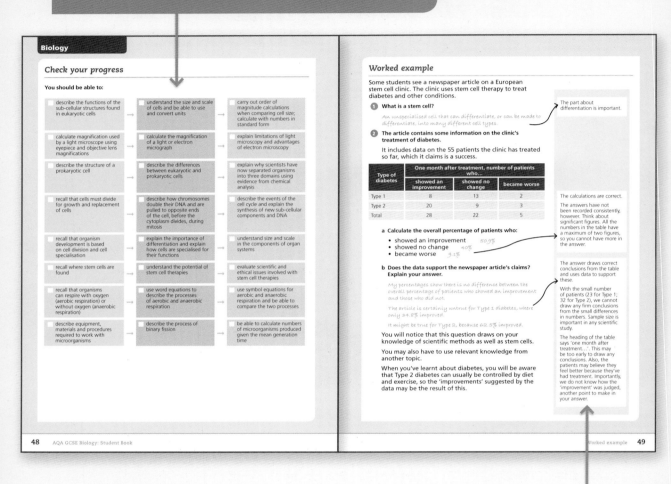

Use the comments to help you understand how to answer questions. Read each question and answer. Try to decide if, and how, the answer can be improved. Finally, read the comments and try to answer the questions yourself.

END OF CHAPTER

The End of chapter questions allow you and your teacher to check that you have understood the ideas in the chapter, can apply these to new situations, and can explain new science using the skills and knowledge you have gained. The questions start off easier and get harder. If you are taking Foundation tier, try to answer all the questions in the Getting started and Going further sections. If you are taking Higher tier, try to answer all the questions in the Going further, More challenging and Most demanding sections.

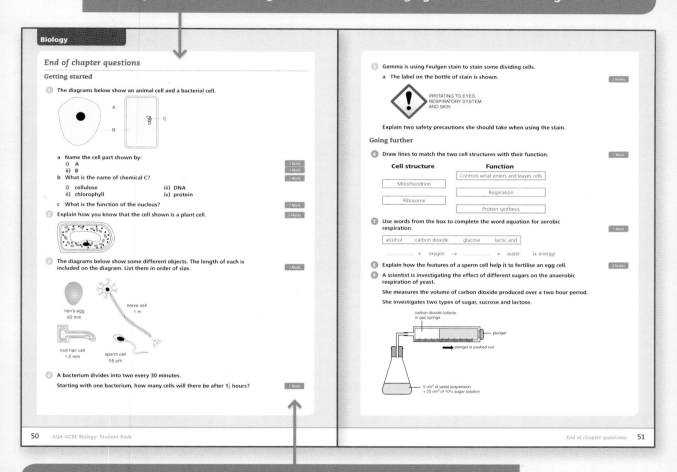

Biology

End of chapter questions

Getting started

1. The diagrams below show an animal cell and a bacterial cell.

   a Name the cell part shown by:
      i) A   1 Mark
      ii) B   1 Mark
   b What is the name of chemical C?   1 Mark
      i) cellulose   iii) DNA
      ii) chlorophyll   iv) protein
   c What is the function of the nucleus?   1 Mark

2. Explain how you know that the cell shown is a plant cell.   2 Marks

3. The diagrams below show some different objects. The length of each is included on the diagram. List them in order of size.   1 Mark

   hen's egg
   60 mm

   nerve cell
   1 m

   root hair cell
   1.5 mm

   sperm cell
   55 µm

4. A bacterium divides into two every 30 minutes.

   Starting with one bacterium, how many cells will there be after 1½ hours?   1 Mark

5. Gemma is using Feulgen stain to stain some dividing cells.

   a The label on the bottle of stain is shown.   2 Marks

   IRRITATING TO EYES,
   RESPIRATORY SYSTEM
   AND SKIN

   Explain two safety precautions she should take when using the stain.

Going further

6. Draw lines to match the two cell structures with their function.   1 Mark

Cell structure	Function
	Controls what enters and leaves cells
Mitochondrion	
	Respiration
Ribosome	
	Protein synthesis

7. Use words from the box to complete the word equation for aerobic respiration.   1 Mark

   | alcohol | carbon dioxide | glucose | lactic acid |

   .... + oxygen → .... + water   (+ energy)

8. Explain how the features of a sperm cell help it to fertilise an egg cell.   2 Marks

9. A scientist is investigating the effect of different sugars on the anaerobic respiration of yeast.

   She measures the volume of carbon dioxide produced over a two hour period.

   She investigates two types of sugar, sucrose and lactose.

   carbon dioxide collects
   in gas syringe

   plunger

   plunger is pushed out

   5 cm³ of yeast suspension
   + 25 cm³ of 10% sugar solution

There are questions for each assessment objective (AO) from the final exams. This will help you develop the thinking skills you need to answer each type of question:

AO1 – to answer these questions you should aim to **demonstrate** your knowledge and understanding of scientific ideas, techniques and procedures.

AO2 – to answer these questions you should aim to **apply** your knowledge and understanding of scientific ideas and scientific enquiry, techniques and procedures.

AO3 – to answer these questions you should aim to **analyse** information and ideas to: interpret and evaluate, make judgements and draw conclusions, develop and improve experimental procedures.

CELL BIOLOGY

ALL LIVING ORGANISMS ARE MADE OF CELLS.

- Cells are the building blocks of life.
- Cells contain specialised structures.
- Organisms such as bacteria are unicellular.
- All plants and animals are multicellular.

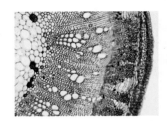

IN MULTICELLULAR ORGANISMS CELLS BECOME SPECIALISED.

- Specialised cells have a particular job to do.
- Specialised cells are organised into tissues, tissues into organs, and organs into body systems.

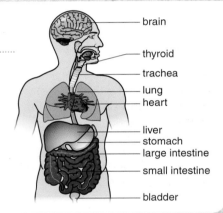

brain
thyroid
trachea
lung
heart
liver
stomach
large intestine
small intestine
bladder

ORGANISMS OBTAIN ENERGY BY THE PROCESS OF RESPIRATION.

- The energy that is released drives all the processes necessary for life.
- Most organisms respire by aerobic respiration, using oxygen.
- Some cells or organisms can survive without oxygen. They respire anaerobically.

MICROORGANISMS CAN HELP TO KEEP US HEALTHY AND PROVIDE US WITH FOOD.

- Microorganisms produce important food products by fermentation.
- Bacteria in the gut are important in keeping us healthy.

IN THIS CHAPTER YOU WILL FIND OUT ABOUT:

HOW HAVE SCIENTISTS DEVELOPED THEIR UNDERSTANDING OF CELL STRUCTURE AND FUNCTION?

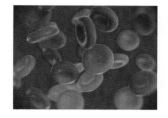

- The structures inside cells do different jobs within the cell.
- Cells can be studied using different types of microscopes.
- The cells of bacteria are different from the cells of plants and animals.

HOW DO WE DEVELOP INTO A COMPLEX ORGANISM FROM JUST A FERTILISED EGG CELL?

- The body's cells divide and the newly formed cells are identical to the existing cells.
- Cells differentiate to become specialised, and specialised cells are organised.
- When cell division accelerates out of control, cancer develops.
- Cells that are unspecialised in the embryo, and cells that remain unspecialised in us as adults, are called stem cells.
- Stem cells could be used to treat certain conditions and diseases that are currently untreatable.

HOW DO ORGANISMS OBTAIN THEIR ENERGY FROM FOOD?

- Anaerobic respiration: when some organisms run out of oxygen, they can respire without it.
- Many microorganisms can respire anaerobically, as can the muscles of mammals for short periods.

WHY IS IT IMPORTANT TO STUDY MICROORGANISMS, AND HOW DO WE GROW THEM IN THE LAB AND COMMERCIALLY?

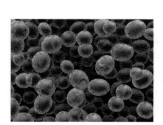

- The biochemistry of fermentation is involved in the production of alcoholic drinks and bread.
- Lab techniques are used to grow, or culture, microorganisms.
- Microorganisms reproduce, and the number of bacteria produced can be estimated.
- Tests can show how effective antibiotics, antiseptics and disinfectants are at inhibiting the growth of bacteria.

Looking at cells

Learning objectives:

- describe the structure of eukaryotic cells
- explain how the main sub-cellular structures are related to their functions.

KEY WORDS

DNA
chloroplast
chlorophyll
chromosome
eukaryotic
order of magnitude

Cell biology helps us to understand how parts of the cell function and interact with each other. It also helps us to learn how we develop, and about our relationships with other organisms.

Biomedical scientists use cells to look for signs of disease and in new drug development.

Plant and animal cells

Almost all organisms are made up of cells. Plant and animal cells have a basic structure.

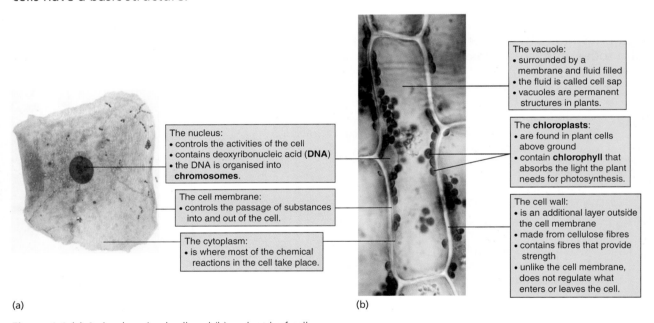

The nucleus:
- controls the activities of the cell
- contains deoxyribonucleic acid (**DNA**)
- the DNA is organised into **chromosomes**.

The cell membrane:
- controls the passage of substances into and out of the cell.

The cytoplasm:
- is where most of the chemical reactions in the cell take place.

The vacuole:
- surrounded by a membrane and fluid filled
- the fluid is called cell sap
- vacuoles are permanent structures in plants.

The **chloroplasts**:
- are found in plant cells above ground
- contain **chlorophyll** that absorbs the light the plant needs for photosynthesis.

The cell wall:
- is an additional layer outside the cell membrane
- made from cellulose fibres
- contains fibres that provide strength
- unlike the cell membrane, does not regulate what enters or leaves the cell.

(a)

(b)

Figure 1.1 (a) A simple animal cell and (b) a plant leaf cell

This type of cell, containing a true nucleus in the cytoplasm, is called a **eukaryotic** cell.

1 **List the sub-cellular structures found in both plant and animal cells.**

2 **Which sub-cellular structures are found only in plant cells?**

What is the function of:

- **the nucleus**
- **the cell membrane?**

Figure 1.2 Growing cells in a laboratory

3 What structure gives strength to a plant cell?

Cell size

The smallest thing we can see is about 0.04 mm, so you can see some of the largest cells with the naked eye. For all cells, however, we need a microscope to see them in any detail.

Most animal and plant cells are 0.01–0.10 mm in size. The unit we use to measure most cells is the micrometre, symbol µm. For some sub-cellular structures, or organisms such as viruses, it is best to use a smaller unit: the nanometre, symbol nm.

$$1 \text{ millimetre (mm)} = \frac{1}{1000} \text{ m} \quad \text{or } 10^{-3} \text{ m}$$

$$1 \text{ micrometre (µm)} = \frac{1}{1000} \text{ mm} \quad \text{or } 10^{-3} \text{ mm} \quad \text{or } 10^{-6} \text{ m}$$

$$1 \text{ nanometre (nm)} = \frac{1}{1000} \text{ µm} \quad \text{or } 10^{-3} \text{ µm} \quad \text{or } 10^{-9} \text{ m}$$

4 What size is the smallest thing our eye can see, in m?

5 What is the range in size of most animal and plant cells, in µm?

Order of magnitude

Figure 1.3 shows the size of plant and animal cells compared with some other structures.

| ant length 3 mm | hair diameter 100 µm | leaf cell length 70 µm | red blood cell diameter 7 µm | bacterium length 1 µm | virus 100 nm | DNA diameter 2.5 nm | carbon atom 0.34 nm |

Figure 1.3 Size and scale

When comparing the sizes of cells, scientists often refer to differences in **order of magnitude**. That's the difference calculated in factors of 10.

So, the difference in order of magnitude for the HIV and the plant cell:

The plant cell in Figure 1.1b is $100 \, \text{µm} = 0.1 \text{ mm} = 10^{-4} \text{ m}$.

The human immunodeficiency virus (HIV) is $100 \, \text{nm} = 0.1 \, \text{µm} = 10^{-4} \text{ mm} = 10^{-7} \text{ m}$.

The difference in order of magnitude is 10^3, expressed as 3.

6 A cell membrane measures 7 nm across. Convert this to micrometres.

7 A white blood cell measures 1.2×10^{-5} m. An egg cell measures 1.2×10^{-4} m. Calculate the difference in order of magnitude.

8 Suggest what substances might pass in or out of a muscle cell and explain why.

> **REMEMBER!**
>
> You'll notice that this system of units uses, and gives names to, multiples and sub-multiples of units at intervals of thousands (10^3) or thousandths (10^{-3}). A common exception is the centimetre, $\frac{1}{100}$ or 10^{-2} of a metre. But it is often convenient to use centimetres, particularly in everyday life.

The light microscope

Learning objectives:

- observe plant and animal cells with a light microscope
- understand the limitations of light microscopy.

The type of microscope you have used in the school laboratory is called a light microscope. Microscopes produce a magnified image of the specimen you are looking at, making them look bigger than they are.

Some early microscopes had just a single lens. The compound microscope has two.

As lens-making techniques improved, microscopes were developed with higher magnifications and resolutions.

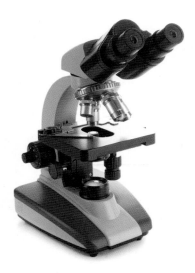

Figure 1.4 A light microscope

Magnification

The magnified image is produced by two lenses, an eyepiece and an objective lens. There is usually a choice of objective lenses.

Total magnification = magnification of eyepiece × magnification of objective lens

For instance, if the eyepiece has a **magnification** of ten, which is written × 10, and the objective lens has a magnification of × 40, the total magnification is × 400.

1. Calculate the total magnification with an eyepiece magnification of × 15 and an objective lens magnification of × 40.

2. What magnification would the objective lens need to be to give a total magnification of × 300 with an eyepiece of × 15?

DID YOU KNOW?

British scientist Robert Hooke first used the term 'cell'. He recorded the first drawings of cells using a compound microscope in his book *Micrographia*, which was 350 years old in 2015.

You may also have heard of Hooke for his law of elasticity, Hooke's law, in physics.

Magnification of images

The magnification described on the previous page is the magnification used to *view* an image. Microscope images, or **micrographs**, books or scientific papers must show the magnification in order to be meaningful.

$$\text{magnification of the image} = \frac{\text{size of the image}}{\text{size of real object}}$$

The cell in Figure 1.5 is 50 mm across on the page. In real life, it measures 40 μm.

To calculate the magnification, first convert the 50 mm into micrometres (or convert 40 μm to millimetres).

50 mm = 50 000 μm

The cell measures 40 μm

Therefore, the magnification of the image $= \dfrac{50\ 000}{40} = \times 1250$.

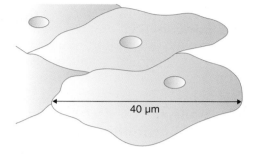

Figure 1.5 A drawing of a micrograph of a cell

3 A micrograph of a plant cell in a book is 150 mm long. The plant cell measures 120 μm long. Calculate the magnification.

4 Why is it essential to state the magnification of an image of a cell in a book but of little value on a website?

The limits of the light microscope

Very high magnifications are not possible with the light microscope. This is because of the light-gathering ability of the microscope and the short working distances of high-power lenses. The highest magnification possible is around × 1500.

Using higher magnification does not always mean that you can see greater detail in an image. This depends on the **resolving power**, or resolution. This is the ability to distinguish between two points. In other words, whether you see them as two points, or one.

The resolving power of a light microscope is around 0.2 μm, or 200 nm. This means that you could not separately pick out two points closer than 200 nm apart.

5 What is the maximum resolving power of the light microscope?

6 What is the maximum magnification possible with a light microscope?

7 Make a table to show the pros and cons of using a light microscope.

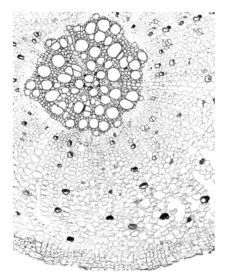

Figure 1.6 A micrograph of the cross section of a root. Magnification ×100

COMMON MISCONCEPTIONS

Do not confuse magnification, which is how much bigger we can make something appear, with resolving power, which is the level of detail we can see.

Think about a digital photo. You can make it as big as you like, but at a certain point you will not be able to see any more detail.

Looking at cells in more detail

Learning objectives:

- identify the differences in the magnification and resolving power of light and electron microscopes
- explain how electron microscopy has increased our understanding of sub-cellular structures.

The *transmission electron microscope (TEM)* uses an electron beam instead of light rays.

Some of the electrons are scattered as they pass through the specimen. Those able to pass through it are focused in TEMs using electromagnetic coils instead of lenses.

Electron microscopes

TEMs are used for looking at extremely thin sections of cells. The highest magnification that can be obtained from a transmission electron microscope is around × 1 000 000, but images can also be enlarged photographically.

The limit of resolution of the transmission electron microscope is now less than 1 nm.

The **scanning electron microscope (SEM)** works by bouncing electrons off the surface of a specimen that has had an ultra-thin coating of a heavy metal, usually gold, applied. A narrow electron beam scans the specimen. Images are formed by these scattered electrons.

SEMs are used to reveal the surface shape of structures such as small organisms and cells. Because of this, resolution is lower and magnifications used are often lower than for TEM.

Electrons do not have a colour spectrum like the visible light used to illuminate a light microscope. They can only be 'viewed' in black and white. Here, false colours have been added.

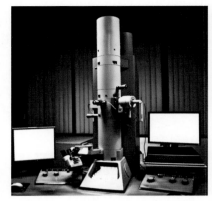

Figure 1.7 A transmission electron microscope. The electrons are displayed as an image on a fluorescent screen

1. What is the maximum resolution of an electron microscope?
2. What types of samples would a TEM and an SEM be used to view?
3. How has electron microscopy improved our understanding of cells?

Figure 1.8 A scanning electron micrograph of a cancer cell

Cell ultrastructure

The TEM reveals tiny sub-cellular structures that are not visible with the light microscope. It also shows fine detail in those structures.

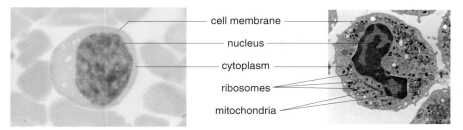

cell membrane
nucleus
cytoplasm
ribosomes
mitochondria

Figure 1.9 A white blood cell, as seen with a light microscope and a transmission electron microscope

We can see mitochondria and chloroplasts with the light microscope, but the electron microscope reveals their internal structure.

(a)

Mitochondria are where aerobic respiration takes place in the cell. A mitochondrion has a double membrane. The internal membrane is folded.

(b)

Chloroplasts are the structures in the plant cell where photosynthesis takes place. Like mitochondria, they also have a complex internal membrane structure.

(c)

Ribosomes are tiny structures where protein synthesis takes place. You can see them as dots in the micrograph. They can either lie free in the cytoplasm or may be attached to an internal network of channels within the cytoplasm.

Figure 1.10 Viewing (a) mitochondria, (b) chloroplasts and (c) ribosomes by transmission electron microscopy

The size of sub-cellular structures is important. Mitochondria and chloroplasts vary in size and shape. The complexity of a mitochondrion indicates how active a cell is. Chloroplast size varies from one species to another. Scientists sometimes investigate the ratio of the area of the cytoplasm to that of the nucleus in micrographs. A high ratio of cytoplasmic:nuclear volume can indicate that the cell is about to divide. A low one can be characteristic of a cancer cell.

4 Name one structure visible to the electron microscope, but not the light microscope.

5 What process happens in ribosomes?

6 Which type of microscope would be best suited to viewing the 3D structure of a cell? Explain why.

COMMON MISCONCEPTIONS

Don't assume that we always use electron microscopes in preference to light microscopes, or that electron microscopes are always used at high magnifications. Confocal microscopy is used in a lot of biomedical research. It can give high resolution images of live cells. And SEM is often used at low magnifications.

DID YOU KNOW?

Three scientists won the Nobel Prize in 2014 for the development of super-resolved fluorescence microscopy. It allows a much higher resolution than normal light microscopy. And, unlike electron microscopy, it has the advantage of allowing scientists to look at living cells.

REQUIRED PRACTICAL

Using a light microscope to observe and record animal and plant cells

Learning objectives:

- apply knowledge to select techniques, instruments, apparatus and materials to observe cells
- make and record observations and measurements
- present observations and other data using appropriate methods.

These pages are designed ❶ to help you think about aspects of the investigation rather than to guide you through it step by step.

Many scientists use electron microscopes to observe fine detail in cells. But much of the microscope work carried out – including in hospital and forensic science labs – is done with the light microscope.

Preparing cells for microscopy

Live cells can be mounted in a drop of water or dilute salt solution (saline) on a microscope slide.

Most cells are colourless. We must stain them to add colour and contrast. In the school laboratory, you may have used methylene blue to stain animal cells or iodine solution to stain plant cells.

1 **Write an equipment list for looking at cheek cells with a microscope. State why each piece of equipment is used.**

2 **Suggest why it's better to mount the cells in saline than in water.**

3 **The micrograph of the frog's blood (Figure 1.12) shows red blood cells (the lower micrograph) and two types of white blood cell.**

  a **Label the different types of cell and the cell structures that are visible. Hint: use a photocopy or printout of the page.**

  b **How is the structure of the frog's red blood cells different from that of human red blood cells?**

High and low power

The slide is first viewed with low power. This is because:

- the **field of view** with high power is small. It would be difficult to locate cells if starting with the high power objective.
- it enables you to see the layout of cells within the tissue.
- it's useful when estimating the numbers of different types

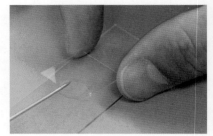

Figure 1.11 A glass coverslip is carefully lowered onto the cells or tissue, taking care to avoid trapping air bubbles. The coverslip keeps the specimen flat, and retains the liquid under it

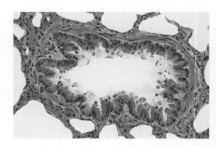

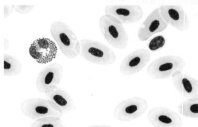

Figure 1.12 Cell biologists use other chemical stains. These are used to reveal or identify specific cell structures.

of cell on the slide or in a tissue (though here, high power may be needed).

A low power digital image (or drawing) can be used to show the arrangement of cells in a tissue. This includes regions of the tissue but not individual cells.

If required, the cells or tissue can then viewed with high power to produce a detailed image of a part of the slide.

  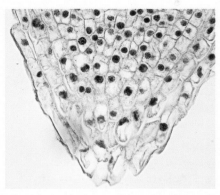

Figure 1.13 Low and high power micrographs, and a student diagram, of a plant root.

4 **Why is a slide viewed with low power first?**

5 **On a printout of a low power plan of the root (Figure 1.13), label the root cap, meristem (the region of cell division) and the region of cell elongation.**

Recording images

As you have seen in topic 1.3, a microscope drawing or micrograph is of little value if it gives no indication of size.

It's usual to add a magnification to the image. We can then envisage, or work out, the true size of a specimen.

Alternatively, we can use a **scale bar**. Any scale bar must be:

- drawn for an appropriate dimension
- a sensible size in relation to the image.

Look at Figure 1.14. For the top micrograph, the magnification of × 1000, means that a 10 *millimetre* scale bar can be drawn to represent 10 *micrometres*.

You will find out how scientists measure, or sometimes estimate, the size of cells in topic 1.17.

6 **Complete the scale bar for the bottom micrograph.**

7 **Calculate the length of the protists in Figure 1.14.**

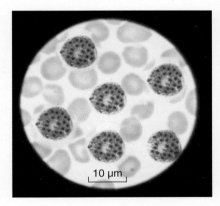

10 μm

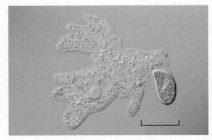

Figure 1.14 Light microscopy is also used to examine small organisms such as protists. The top image shows six blood cells infected with the malarial parasite. The bottom image shows two protists found in pond water - *Amoeba* on the left; *Paramecium* on the right (at × 200 magnification).

Primitive cells

Learning objectives:

- describe the differences between prokaryotic cells and eukaryotic cells
- explain how the main sub-cellular structures of prokaryotic and eukaryotic cells are related to their functions.

The oldest fossil evidence of life on Earth comes from Australia. It confirms that there were bacteria living around 3.5 billion years ago.

The bacteria probably formed thin purple and green mats on shorelines. The bacteria would have photosynthesised, but produced sulfur as waste instead of oxygen.

Prokaryotic cells

The cells of most types of organisms – such as all animals and plants – are **eukaryotic**. These have a cell membrane, cytoplasm containing sub-cellular structures called organelles and a nucleus containing DNA.

Bacteria are among the simplest of organisms. Along with bacteria-like organisms called archaeans, they belong to a group of organisms called the **Prokaryota**. These are single cells with a **prokaryotic cell** structure.

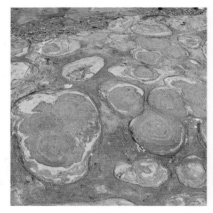

Figure 1.15 The organisms in this fossil are similar to purple bacteria that are living today.

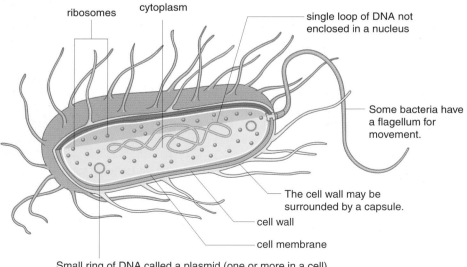

ribosomes
cytoplasm
single loop of DNA not enclosed in a nucleus
Some bacteria have a flagellum for movement.
The cell wall may be surrounded by a capsule.
cell wall
cell membrane
Small ring of DNA called a plasmid (one or more in a cell). Genes in the plasmids can give the bacterium advantages such as antibiotic resistance.

Figure 1.16 The structure of a prokaryotic cell

Prokaryotic cells are much smaller than eukaryotic cells, around 1 μm across. Their DNA is not enclosed in a nucleus. It is found as a single molecule in a loop. They may also have one or more small rings of DNA called **plasmids**.

1 List the differences between prokaryotic and eukaryotic cells.

2 Where is DNA found in prokaryotic cells?

A new classification system

By the 1970s, biologists had classified living organisms into five **kingdoms**.

Very small, microscopic organisms called archaeans were originally grouped in a kingdom with bacteria. But in 1977, American microbiologist Carl Woese suggested that certain types of organisms that lived in extreme environments or produced methane gas should be placed in a separate group.

Woese suggested that living things should be divided into three groups called **domains**: Bacteria, Archaea and Eukaryota.

3 What are the three domains of living things?

4 In which domain are plants and animals?

Chemical characteristics of archaeans

Acceptance of Woese's theory was a slow process. Even today, not everyone agrees with it, but chemical analyses have supported the idea that archaeans should be in a separate domain.

The ribosomes of archaeans are similar in size and structure to those of bacteria, but the **nucleic acid** in these structures is closer to that of eukaryotes.

When American biochemist Craig Ventner, one of the first scientists involved in the sequencing of the human **genome**, looked at the DNA of Archaea he was astounded to find that 'two-thirds of the genes [in Archaea] do not look like anything we've ever seen in biology before'.

5 Suggest 3 different environments where you might find Archaea.

6 What evidence suggests that archaeans should be placed in a separate domain to bacteria?

7 Suggest why scientists have only discovered Archaea quite recently.

DID YOU KNOW?

A 'superfood' called Spirulina is the dried cells of a blue-green bacterium. The cells contain high concentrations of protein, and are rich in essential fatty acids, vitamins and minerals.

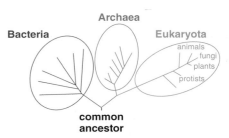

Figure 1.17 The three-domain classification system

Figure 1.18 Archaeans live in extreme environments, such as hot springs and salt lakes. Some produce methane and are important in the carbon cycle

REMEMBER!

You should aim to be able to discuss how information on Archaea, Bacteria and Eukaryota allows them to be placed in separate domains.

Cell division

KEY WORDS

mitosis
stem cells
daughter cell

Learning objectives:

- describe the process of mitosis in growth, and mitosis as part of the cell cycle
- describe how the process of mitosis produces cells that are genetically identical to the parent cell.

As an adult, we are made up of 37 trillion (3.7×10^{13}) cells. To produce these cells, the fertilised egg needs to undergo many cell divisions.

Chromosomes

As we grow, the cells produced by cell division must all contain the same genetic information.

The genetic information of all organisms is contained in the nucleus, in chromosomes, made of DNA. The DNA in resting cells is found in the nucleus as long, thin strands. For cell division, these strands form condensed chromosomes.

Human body cells have 46 chromosomes, or 23 pairs. Each chromosome in a pair has the same type of genes along its length.

1. **How many chromosomes are found in each human body cell?**

2. **How are the chromosomes arranged in a karyotype?**

Mitosis

New cells have to be produced for growth and development, and to replace worn out and damaged body cells.

When new cells are produced they must be identical to the parent cell. Cells divide to produce two new ones. This type of cell division is called **mitosis**. Two **daughter cells** are produced from the parent cell.

For some cell types, new cells are produced by the division of **stem cells** (discussed later in the chapter).

3. **When are new cells produced?**

4. **In this type of cell division:**
   - **how many chromosomes do daughter cells have?**
   - **how many daughter cells are produced?**

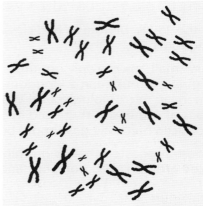

A photograph is taken of a dividing cell.

XX XX XX XX XX
1   2   3   4   5

XX XXXX XX XX XX XX
6   7 8 9 10 11 12

XX XX XX   XX XX XX
13 14 15   16 17 18

XX XX   XX XX XX
19 20   21 22 23

The chromosomes in the photograph are cut out and arranged into pairs. The pairs are arranged so that Pair 1 has the longest chromosomes; Pair 22 the shortest Pair 23 is the sex chromosomes.

Figure 1.19 A profile of a set of chromosomes, called a karyotype

MAKING CONNECTIONS

To come up with a figure for how many cells there are in the human body, scientists must *estimate* the number by adding up cell counts from different organs.

So that the daughter cells produced are identical to the parent cell, the DNA must first copy itself. Each of the 46 chromosomes then consists of two molecules of DNA.

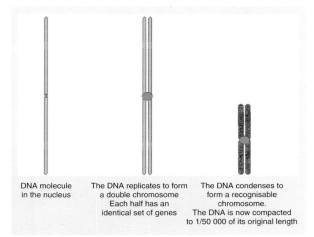

DNA molecule in the nucleus

The DNA replicates to form a double chromosome Each half has an identical set of genes

The DNA condenses to form a recognisable chromosome. The DNA is now compacted to 1/50 000 of its original length

Figure 1.20 It is these 'double' chromosomes that we always see in micrographs or illustrations of chromosomes

During mitosis, the double chromosomes are pulled apart as each new set of 46 chromosomes moves to opposite ends of the cell (Figure 1.21). Two nuclei then form. The cytoplasm and cell membrane then divides and two identical cells are produced.

5 **Why do chromosomes appear double, or X-shaped, in micrographs?**

The cell cycle

A cell that is actively dividing goes through a series of stages called the cell cycle. The cycle involves the growth of the cell and the production of new cell components and division.

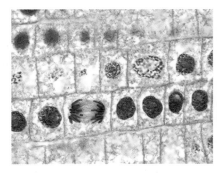

Figure 1.21 Mitosis in an onion cell

1. The cell grows. The number of sub-cellular structures, e.g. mitochondria and ribosomes, increases.

2. The DNA replicates to form two copies of each chromosome.

3. Further growth occurs and the DNA is checked for errors and any repairs made.

4. Mitosis – the chromosomes move apart and two nuclei form.

5. The cytoplasm divides into two and the new cell membrane separates off to give two new, identical cells.

6. Temporary cell resting period, or the cell no longer divides, e.g. a nerve cell.

Figure 1.22 The cell cycle

In actively dividing human cells, the whole cell cycle lasts 1 hour

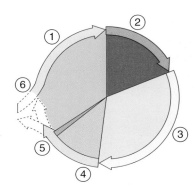

6 **Using Figure 1.22, calculate the proportion of the cell cycle spent in mitosis. You will need a protractor.**

7 **If the cell cycle lasts 2 hours, estimate the time spent in mitosis.**

8 **Mitosis occurs rapidly in a newly formed fertilised egg. Suggest another situation in the body where you might expect cells to be actively dividing by mitosis.**

Cell differentiation

KEY WORDS

differentiation
organ
organ system
specialised
tissue

Learning objectives:

- explain the importance of cell differentiation
- describe how cells, tissues, organs and organ systems are organised to make up an organism
- understand size and scale in relation to cells, tissues, organs and organ systems.

Cell division makes up only part of our growth and development.

For the first four or five days of our lives, the cells produced as the fertilised egg divides are identical. Then, some of our cells start to become specialised **to do a particular job.**

Cell adaptations

In a multicellular organism, many different types of cell take on different roles to ensure that the organism functions properly and as a whole.

As cells divide, new cells acquire certain features required for their specific function. This is **differentiation**. A cell's size, shape and internal structure are adapted for its role. Most animal cells differentiate at an early stage.

Figure 1.23 By this stage in its development, this human embryo has developed many of the 200 different cell types in the human body

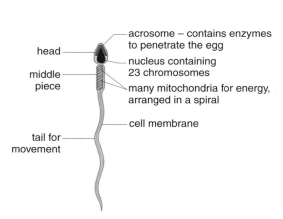

Figure 1.24 The function of a sperm cell is to swim in the female reproductive system with the aim of fertilising an egg. (Cell length 55 μm; width at widest point 3 μm)

head
acrosome – contains enzymes to penetrate the egg
nucleus containing 23 chromosomes
middle piece
many mitochondria for energy, arranged in a spiral
cell membrane
tail for movement

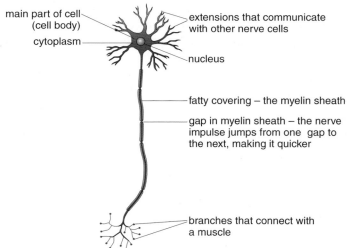

main part of cell (cell body)
cytoplasm
extensions that communicate with other nerve cells
nucleus
fatty covering – the myelin sheath
gap in myelin sheath – the nerve impulse jumps from one gap to the next, making it quicker
branches that connect with a muscle

Figure 1.25 Nerve cells carry messages, or electrical impulses, from one part of the body to another. This type of cell brings about movement of the skeleton. (Motor nerve cell length: up to 1 m or more; diameter 1–20 μm)

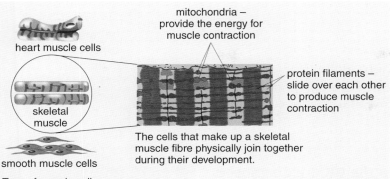

heart muscle cells

mitochondria – provide the energy for muscle contraction

skeletal muscle

protein filaments – slide over each other to produce muscle contraction

smooth muscle cells

The cells that make up a skeletal muscle fibre physically join together during their development.

Type of muscle cell

Protein filaments give the cells of heart and skeletal muscle a striped appearance. In smooth muscle, found, for instance, in the circulatory system, there are fewer filaments, which are thinner and less well-organised.

Figure 1.26 There are three types of muscle cell. Muscle cells contain protein filaments which move to make the muscle contract. Muscle cells link with each other so a muscle contracts as a whole.

KEY INFORMATION

You can work out how the structure of a cell is related to its function, even if you are not familiar with the type of cell in the question. Look at the size, shape and surface area of the cell, and what it contains, for example mitochondria, ribosomes or a food store.

1 **How are the following cells adapted to their functions:**
   - **a sperm cell**
   - **a muscle cell**
   - **a nerve cell?**

2 **Cells of the pancreas produce the hormone insulin. Insulin is a protein. Suggest how pancreatic cells are adapted for their function.**

Cells, tissues and organs

Some cells work in isolation, like sperm cells. Others are grouped as **tissues** and work together.

A tissue is a group of cells with a particular function. Many tissues have a number of similar types of cell to enable the tissue to function.

Tissues are grouped into **organs**. Organs carry out a specific function.

Different organs are arranged into **organ systems**, for example the circulatory system, digestive system, respiratory system, and reproductive system.

3 **Arrange the following in ascending order of size:**

   system   cell   human body   organ   tissue

4 **Name two other types of cell and one other type of tissue in the circulatory system.**

5 **Red blood cells have a biconcave shape which gives them a large surface area. How is this shape related to its function?**

cell – heart muscle cell

tissue – heart muscle

organ – the heart

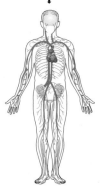

organ system – the circulatory system

Figure 1.27 The organisation of the human circulatory system

Cancer

Learning objectives:

- describe cancer as a condition resulting from changes in cells that lead to their uncontrolled growth, division and spread
- understand some of the risk factors that trigger cells to become cancerous.

Every year, over 300 000 people in the UK are diagnosed with cancer. It is estimated, however, that four in ten cases of cancer could be prevented by lifestyle changes.

What is cancer?

Normally, cells grow and divide by mitosis when the body needs new cells to replace old or damaged cells. When a cell becomes cancerous, it begins to divide uncontrollably. New cells are produced even though the body does not need them.

The extra cells produced form growths called tumours. Most tumours are solid, but cancers of the blood, for instance leukaemia, are an exception.

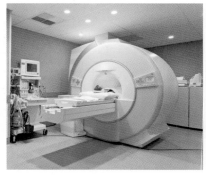
Figure 1.28 A CT scanner, used to detect cancer

1 **What is cancer?**

2 **Name one type of cancer that does not form a solid tumour.**

Types of tumour

Type of tumour	Characteristics
Benign	• slow growing • often have a capsule around them, so can be removed easily • not cancerous and rarely spread to other parts of the body • they can press on other body organs and look unsightly.
Malignant	• grow faster • can spread throughout other body tissues • as the tumour grows, cancer cells detach and can form **secondary tumours** in other parts of the body.

Malignant cells develop.

The malignant cells divide and can invade normal tissues.

Malignant cells can detach from the tumour and spread to other parts of the body.

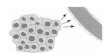
The tumour secretes hormone-like chemicals.

Blood vessels are stimulated to grow around the tumour; the blood vessels supply the tumour with food and oxygen.

Malignant cells detach from the tumour and are transported away in the blood.

The malignant cell squeezes through the capillary wall.

The cell divides to produce a secondary tumour.

Figure 1.29 The growth and spread of a tumour

3 Name two types of tumour.

4 Explain why a tumour needs a blood supply.

5 What is the name of the type of tumour formed when a cancer spreads?

What triggers cancer?

Chemicals and other agents that cause cancer are called **carcinogens**.

Carcinogens cause cancer by damaging DNA. A change in the DNA of a cell is called a **mutation**. Mutations can also occur by chance as a cell is dividing.

There are natural checks for such errors during the cell cycle. Some of our genes suppress developing tumours.

Several mutations, not just one, are necessary to trigger cancer. This is why we are more likely to develop cancer as we get older.

Mutations that lead to cancer can be caused by several agents:

- viruses (see Figure 1.30).
- chemicals in the home, industry or environment
- ionising radiation
- ultraviolet radiation
- lifestyle choices, such as alcohol intake or diet (see Figure 1.31).

DID YOU KNOW?

Many treatments for cancer come from plants, such as the Pacific yew and the Madagascan periwinkle. These drugs work by interfering with mitosis.

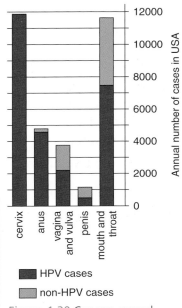

Figure 1.30 Cancers caused by the human papilloma virus (HPV)

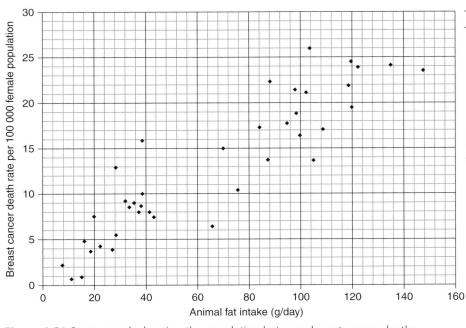

Figure 1.31 Scattergraph showing the correlation between breast cancer deaths and animal fat intake. Each data point represents a different country

6 Which type of cancer is caused *only* by HPV?

7 Describe the pattern shown by the scattergraph and draw a conclusion.

Stem cells

Learning objectives:

- describe the function of stem cells in embryonic and adult animals
- discuss potential benefits and risks associated with the use of stem cells in medicine.

The UK has a shortage of blood donors.

In the summer of 2015 the NHS announced that it planned to start giving people blood transfusions using artificial blood by 2017.

What are stem cells?

Stem cells are unspecialised cells that can produce many different types of cells.

Stem cells are found in the developing embryo and some remain, at certain locations in our bodies, as adults.

Figure 1.32 The red blood cells in the artificial blood will be produced using stem cells

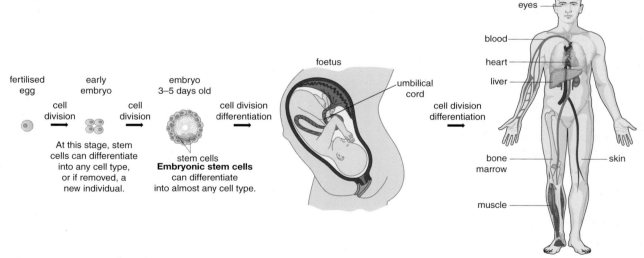

fertilised egg — early embryo — cell division — At this stage, stem cells can differentiate into any cell type, or if removed, a new individual.

embryo 3–5 days old — cell division differentiation — stem cells **Embryonic stem cells** can differentiate into almost any cell type.

foetus — umbilical cord — cell division differentiation

adult — brain — eyes — blood — heart — liver — bone marrow — skin — muscle

Adult stem cells are rare and found at certain locations only. Their role is to replace body cells that die through injury and disease. They can differentiate only into cells from the type of tissue where they are found, e.g. blood, muscle.

Figure 1.33 Stem cells in the human body

1. What is the function of adult stem cells?
2. Which type of stem cell can differentiate into more cell types?

Stem cell transplants

Transplanting stem cells, or transplants of specialised cells grown from stem cells, could help people with:

- injuries, e.g. spinal injuries leading to paralysis
- conditions in which certain body cells degenerate, e.g. Alzheimer's disease, diabetes and multiple sclerosis
- cancers, or following treatments for cancer such as chemotherapy or radiation, e.g. people with leukaemia.

Stem cell transplants also enable chemotherapy patients, who have had their bone marrow destroyed, to produce red blood cells.

The hope is that we will be able to **culture** stem cells in limitless numbers. Stem **cell lines** produced from patients with rare and complex diseases could transform the health service.

Figure 1.34 Embryonic stem cells

3 **Name two conditions that could be treated with stem cell transplants.**

4 **Why are stem cell transplants important for people who have had chemotherapy?**

Stem cell research and therapy is controversial

Stem cell research is necessary to find out more about stem cell development, and the best types to use in treatments.

The use of **embryonic stem cells**, which are removed from a living human embryo, is especially controversial.

Until recently, the embryos providing the stem cells were usually those left over from fertility treatments involving *in-vitro* fertilisation (IVF). Spare embryos would be destroyed if they had not been donated by the IVF couples for research.

British law now allows embryos to be created purely for scientific research. Some people object to this. Some religious beliefs argue that new life begins at the point of conception, so an embryo has rights. And who should decide when a human life ends?

These are moral and **ethical** questions. A moral question looks at whether something is right or wrong. An ethical question discusses the reasons why something might be right or wrong.

5 **Why do some people object to stem cell transplants?**

6 **Write down one ethical objection to stem cell research.**

7 **What are the potential benefits and drawbacks of using stem cells in medicine?**

DID YOU KNOW?

Stem cell transplants are not new. Transplants of bone marrow, which contain stem cells, have been carried out since 1968. But there are very few stem cells in bone marrow (only 1 in 10 000 bone marrow cells). We currently isolate these from blood, rather than bone marrow.

KEY INFORMATION

Current potential for adult stem cell use in therapies is restricted to certain cell lines, but it may be greater than once thought. Scientists are trying to induce them to differentiate into a wider range of tissues, a process called transdifferentiation.

Stem cell banks

Learning objectives:

- discuss potential benefits and risks associated with the use of stem cells in medicine.

KEY WORDS

donor
gene
mutation
therapeutic
 cloning
umbilical cord

Scientists predict that, in the future, vast banks of stored stem cells will be available to treat many medical conditions.

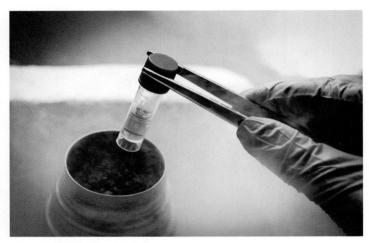

Figure 1.35 Stem cells can be stored in liquid nitrogen

Rejection of stem cell transplants

The stem cells from a bank originate from many different people. Rejection of stem cell transplants by a patient's immune system is, therefore, a problem.

One current solution is to find as close a match as possible between **donor** and patient cells. Another is to give the patient drugs to suppress their immune system. Scientists are looking for other ways to avoid transplant rejection.

One possible source of stem cells is blood left in the **umbilical cord** and placenta after a baby is born. Cord blood is easy to collect and store.

1 **Suggest sources of stem cells that would give the best match between donor and patient.**

2 **Suggest some possible advantages and disadvantages of having a baby's blood stored to treat possible disease or injury in later life.**

Therapeutic cloning

The idea of **therapeutic cloning** is to produce stem cells with the same **genes** as the patient. They would not be rejected by the patient's immune system.

The process involves nuclear transfer. The nucleus of a body cell from the patient is transferred to an egg cell that has had its nucleus removed.

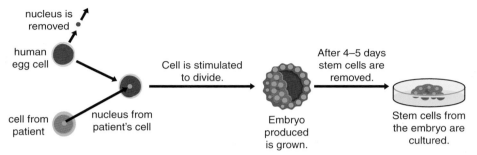

nucleus is removed

human egg cell

cell from patient

nucleus from patient's cell

Cell is stimulated to divide.

Embryo produced is grown.

After 4–5 days stem cells are removed.

Stem cells from the embryo are cultured.

Figure 1.36 Therapeutic cloning

The stem cells are used to treat the patient. The embryo is discarded.

3 **What is therapeutic cloning?**

4 **When are stem cells removed from the embryo?**

Scientific, ethical and social questions

Many questions arise from therapeutic cloning and stem cell therapy.

The first scientific question is how successful might these therapies be? Others consider safety: stem cells kept in culture can show similarities to cancer cells. After about 60 cell divisions, **mutations** have been observed. It is also possible for viruses to be transferred with stem cells, leading to infection.

There are also ethical questions:

- Is it morally right to create an embryo with the intent of destroying it?
- Could an embryo simply become a resource for researchers?

There are also important social questions. What are the potential benefits from successful stem cell treatment and do these outweigh the objections? And should patients be given false hope based on a currently unproven treatment? Public education on this issue is important.

5 **Give two questions scientists might have about therapeutic cloning.**

6 **Evaluate the risks and benefits as well as the ethical concerns associated with therapeutic cloning.**

DID YOU KNOW?

Scientists have succeeded in removing human skin cells and reprogramming them to become cells similar to embryonic stem cells. This removes some of the ethical concerns over stem cell transplants.

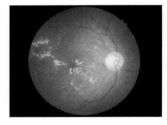

Figure 1.37 Blind patients have had their sight restored by stem cells. It has been possible to safely treat the part of the eye responsible for central vision

MAKING CONNECTIONS

You should be able to evaluate information from a variety of sources regarding practical, social and ethical issues relating to stem cell research and treatment.

KEY CONCEPT

Cell development

Learning objectives:

- give examples of where mitosis is necessary to produce identical daughter cells
- understand the need for the reduction division, meiosis
- describe the use and potential of cloned cells in biological research.

KEY WORDS

asexual reproduction
differentiation
gamete
meiosis
mitosis
placenta
zygote

Cell development involves the processes of cell growth, division and differentiation. These processes are closely linked, and are a key focus for current biological research.

Cells

The cell is the basic unit of life. You will have looked at cells with a microscope in school, probably cheek cells and onion skin cells. These illustrate the basic cell pattern, but most cells in all but the simplest of organisms are much more varied in their structure.

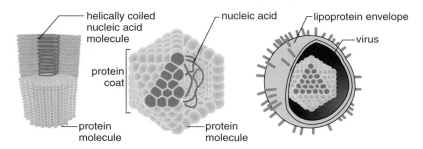

Figure 1.38 Viruses are not made up of cells. They consist simply of nucleic acid surrounded by a protein coat. Some have an outer envelope.

1. **Name one example of a human tissue where cells have merged.**

2. **Suggest why viruses can only live in other cells.**

Cell division

Human life begins as a fertilised egg cell, or **zygote.** This cell develops into an adult with trillions of cells. As new cell components are added, and the cell reaches a certain size, it divides by **mitosis**. Mitosis occurs in several other situations:

- to replace cells when they die or become damaged

- when single-celled, eukaryotic organisms reproduce by **asexual reproduction**, for example yeast
- when cancer cells divide
- when eukaryotic cells are cloned.

When an organism reproduces sexually, the sex cells, or **gametes**, cannot be produced by mitosis. If they were, the number of chromosomes in our cells would double every generation! We need another type of cell division, called **meiosis**.

3 **Give three examples of situations in which mitosis occurs.**

4 **Name one type of cell that does *not* divide by mitosis.**

Cell differentiation

Cells must become specialised for the development of complex, multicellular organisms.

Cells that can differentiate into other cell types are called stem cells. Embryonic stem cells, found after four to five days, can develop into *almost* any cell type. We can't say *'all'*, as they can't develop into cells of the **placenta**.

Stem cells have the potential to produce an unlimited amount of tissue for transplants. They are also important in medical research such as on how cells differentiate, and in the testing of drugs.

Stem cell research and treatments will require the cloning of cells. Some people object to the idea of these techniques for moral and ethical reasons.

Cancer cells divide uncontrollably by mitosis and do not differentiate into mature, specialised cells. Cancer cells early in the development of the disease can look almost normal, but in advanced cancers, differentiation in most cells is very limited.

5 **Why are some news articles that suggest that embryonic stem cells can differentiate into all cell types, strictly speaking, incorrect?**

6 **Stem cells are being used to test new drugs. What are the advantages of using human stem cells over using rats to test drugs?**

KEY SKILLS

For each chapter in the book, map out how different concepts you have learnt link with each other. Use a large sheet of paper or computer software.

Cells at work

Learning objectives:

- explain the need for energy
- describe aerobic respiration as an exothermic reaction.

KEY WORDS

active transport
aerobic
 respiration
exothermic
respiration

This runner is using energy to run a marathon. But we all need a continuous supply of energy – 24 hours a day – just to stay alive.

We need energy to live

Organisms need energy:

- to drive the chemical reactions needed to keep them alive, including building large molecules
- for movement.
- for keeping warm

Energy is needed to make our muscles contract and to keep our bodies warm. It's also needed to transport substances around the bodies of animals and plants.

In other sections of the book, you will also find out that energy is needed:

- for cell division
- to maintain a constant environment within our bodies
- for **active transport**. Plants use active transport to take up mineral ions from the soil, and to open and close their stomata
- to transmit nerve impulses.

Figure 1.39 An average runner uses around 13 000 kJ of energy for a marathon

1 List four uses of energy in animals.

2 List four uses of energy in plants.

Aerobic respiration

Respiration is the continuously occurring process used by all organisms to release the energy they need from food.

Respiration using oxygen is called **aerobic respiration**. This type of respiration takes place in animal and plant cells, and in many microorganisms.

Glucose is a simple sugar. It is the starting point of respiration in most organisms. The food that organisms take in is, therefore, converted into glucose.

This chemical reaction is **exothermic**. A reaction is described as exothermic when it releases energy. Some of the energy transferred is released as thermal energy.

Figure 1.40 Birds and mammals maintain a constant body temperature

3 **What is the purpose of respiration?**

4 **How do birds and mammals make use of the waste thermal energy?**

Bioenergetics

This is the equation for aerobic respiration:

glucose + oxygen → carbon dioxide + water (energy released)

$$C_6H_{12}O_6 + 6O_2 \rightarrow 6CO_2 + 6H_2O$$

This equation describes the overall change brought about through each of a series of chemical reactions. A small amount of energy is actually released at each stage in the series.

The first group of steps occurs in the cytoplasm of cells, but most of the energy is transferred by chemical reactions in mitochondria.

5 **When and where does respiration occur?**

6 **Give one characteristic feature of actively respiring cells.**

7 **Why do we often get hot when we exercise?**

Figure 1.41 Insect flight muscles have huge numbers of well-developed mitochondria

DID YOU KNOW?

The muscle an insect uses to fly is the most active tissue found in nature.

COMMON MISCONCEPTIONS

Don't forget that *all* organisms respire. The equation is the reverse of photosynthesis, but don't confuse the two. Photosynthesis is the way in which plants make their food.

Living without oxygen

Learning objectives:

- describe the process of anaerobic respiration
- compare the processes of aerobic and anaerobic respiration.
- explain how the body removes lactic acid produced during anaerobic respiration.

KEY WORDS

anaerobic
  respiration
fermentation
oxygen debt

Stewart is a brewer. He adds yeast to a mixture of malted barley and hops in water.

Figure 1.42 Yeast converts sugar into alcohol, or ethanol. The process is completed in around 3 days

Anaerobic respiration

The yeast respires using the sugary liquid. The yeast cells divide rapidly. After a few hours there are so many yeast cells that the oxygen runs out. The yeast is able to switch its respiration so that it can obtain energy *without* oxygen. Many microbes such as yeast can respire successfully without oxygen.

This is **anaerobic respiration** – respiration without oxygen.

Anaerobic respiration in yeast cells and certain other microorganisms is called **fermentation**.

Anaerobic respiration occurs in the cytoplasm of cells.

DID YOU KNOW?

Yeast is unable to use the starch in barley for respiration. Maltsters germinate the barley grains first to break down the starch into sugar.

1. **What is meant by anaerobic respiration?**

2. **Why do yeast cells switch from aerobic to anaerobic respiration in the process of making ethanol?**

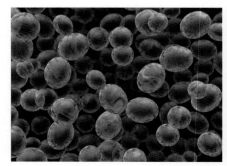

Figure 1.43 Yeast cells divide rapidly by mitosis. Many of the cells do not separate from each other

Baking

Yeast is also economically important in baking bread. Yeast is mixed with flour and some sugar. The ingredients are mixed together thoroughly and the dough is left to rise before baking it.

3 Explain why sugar is added to dough.

4 Why does the dough rise?

5 What happens to the alcohol made during bread production?

Figure 1.44 Dough is kneaded to mix the ingredients

The biochemistry of fermentation

The equation for fermentation is:

glucose → ethanol + carbon dioxide (energy released)

Anaerobic respiration is much less efficient than aerobic respiration. It produces only around a nineteenth as much energy. This is because the oxidation of glucose is incomplete. But in situations where there's little oxygen, it means that cells can stay alive, and the amount of energy produced is still enough to keep single cells running.

Certain plant cells can also use alcoholic fermentation to obtain their energy. These include plants that grow in marshes, where oxygen is in short supply. Pollen grains can also respire anaerobically.

6 Explain why it is helpful for pollen grains to respire anaerobically.

7 Write down the equation for fermentation.

Oxygen debt

Without oxygen, we would die. But when actively contracting, our muscles run short of oxygen. They are able to respire anaerobically for short periods of time. Lactic acid, and not ethanol, is produced.

glucose → lactic acid (energy released)

The incomplete oxidation of glucose causes a build up of lactic acid and creates an **oxygen debt**. As a result, our muscles become fatigued and stop contracting efficiently.

> **KEY SKILLS**
>
> You must be able to compare aerobic and anaerobic respiration: the need for oxygen, the products and the amount of energy transferred.

HIGHER TIER ONLY

When the period of exercise is over, lactic acid must be removed. The body's tolerance of lactic acid is limited. It is taken to the liver by the blood, and either oxidised to carbon dioxide and water, or converted to glucose, then glycogen. Glycogen levels in the liver and muscles are then restored.

These processes require oxygen. This is why, when the period of activity is over, a person's breathing rate does not return to normal straightaway. Oxygen debt is the amount of oxygen required to remove the lactic acid, and replace the body's reserves of oxygen. Paying back the oxygen debt can take from a few hours, to several days after a marathon.

8 For anaerobic respiration in muscle, write down the word equation, then work out the symbol equation.

9 Compare anaerobic with aerobic respiration.

Growing microorganisms

Learning objectives:

- describe the techniques used to produce uncontaminated cultures of microorganisms
- describe how bacteria reproduce by binary fission
- calculate the number of bacteria in a population.

KEY WORDS

agar plate
aseptic
 technique
autoclave
bacteria
bacterial
 growth curve
binary fission
colony

culture
culture
 medium
inoculating
 loop
nutrient
 broth
sterilise

We're most familiar with bacteria through the tiny minority of species that cause disease.

But harmless bacteria help us to live healthily. In our digestive system, they prevent harmful bacteria from gaining a foothold in our bodies and also produce essential nutrients.

Culturing bacteria

Owing to their size, it's best to grow bacteria in large numbers to study them. **Bacteria** are grown in **culture**, in or on a **culture medium**.

The culture medium is a liquid, such as **nutrient broth**, or a gel called agar. Different nutrients can be added to the agar. Because it's a gel, the agar contains the water required for the bacteria to grow.

All the equipment used must be **sterilised**. To make sure cultures and samples are *kept* uncontaminated by other microorganisms, and do not contaminate the environment:

- The inoculating loop must be sterilised by passing it through a Bunsen flame before and after use.
- The lid of the **agar plate** must be secured, but *not* sealed using adhesive tape.

After an investigation, agar plates are sterilised in an **autoclave** before disposal. The steps taken to ensure contamination is avoided are known as **aseptic technique**.

1. Explain why scientists need to work with uncontaminated cultures.

2. What piece of equipment is used to transfer bacteria from a culture to an agar plate?

Figure 1.45 Bacteria, or a sample under test, are transferred to an agar plate using a sterilised **inoculating loop**. After setting up the culture, the agar plates are incubated at a temperature appropriate for the bacteria to grow. Plates are incubated upside down

REMEMBER!

Cultures used for research must be kept pure. It's usual to work with one bacterium at a time. Samples must not be exposed to microorganisms in the environment, or valid conclusions cannot be drawn.

Quantitative studies with microorganisms

When supplied with nutrients and a suitable temperature, bacteria will multiply. They do this by dividing into two. The process is called **binary fission**. This is not the same as mitosis in eukaryotic cells. Binary fission involves prokaryotes with a single chromosome.

A live bacterium landing on the surface of agar will divide repeatedly to form a **colony**. A colony contains millions of bacteria.

DID YOU KNOW?

The Human Microbiome Project is cataloguing the genes of the microorganism population in our intestines. There are 100 times the number of species originally thought to be present.

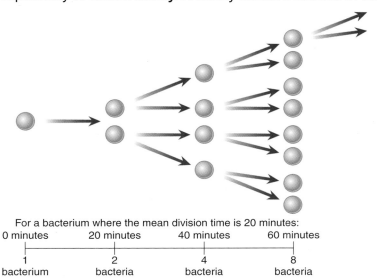

For a bacterium where the mean division time is 20 minutes:

0 minutes	20 minutes	40 minutes	60 minutes
1 bacterium	2 bacteria	4 bacteria	8 bacteria

Figure 1.46 Binary fission of a bacterium

3 **A bacterium has a mean division time of 20 minutes. Starting with one bacterium, how long would it take to produce a million bacteria?**

4 **If the mean mass of a bacterium is 1×10^{-12} g, estimate the mass of bacteria produced in Question 3. Suggest why the true mass is likely to be lower.**

Bacterial growth curves

With the optimum conditions, some bacteria can divide into two as often as every 20 minutes. Estimates of cells in culture are plotted to produce a **bacterial growth curve**.

Values on the *y*-axis are the *logarithms* of the numbers in the population. Otherwise, the range of numbers would be too large to fit appropriately onto the scale.

After a certain time, the culture may reach its stationary phase. Binary fission slows as food begins to run out and waste products build up.

5 **What is meant by exponential growth? Name a process which causes exponential growth.**

6 **Some cultures enter a *death phase*. Suggest the possible causes.**

7 **Predict what would happen if you introduced more food during the stationary phase.**

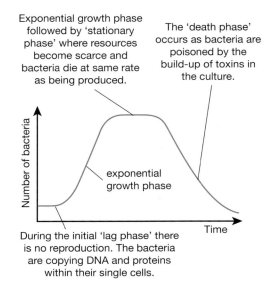

Exponential growth phase followed by 'stationary phase' where resources become scarce and bacteria die at same rate as being produced.

The 'death phase' occurs as bacteria are poisoned by the build-up of toxins in the culture.

During the initial 'lag phase' there is no reproduction. The bacteria are copying DNA and proteins within their single cells.

Figure 1.47 The growth curve of a bacterium

Testing new antibiotics

Learning objectives:

- use appropriate apparatus to investigate the effect of antibiotics on bacterial growth
- use microorganisms safely
- apply sampling techniques to ensure that samples are representative.

KEY WORDS

antibiotic
pathogen
sampling
 techniques

Bacteria are becoming resistant to antibiotics. A 2015 government report suggested that by 2050, 10 million people worldwide may die every year from diseases we can no longer cure.

Scientists are looking for new antibiotics to treat antibiotic-resistant bacteria.

Antibiotic sensitivity testing

The method used to test the effectiveness of an **antibiotic** is the disc-diffusion technique.

Figure 1.48 An agar plate is inoculated with the bacterium being tested and spread evenly across the plate. It is not incubated at this stage

A disc of filter paper is impregnated with the antibiotic. Several concentrations of the antibiotic are tested.

You can do this in the school lab by immersing the filter paper disc in a solution of the antibiotic, and allowing the antibiotic to drain off.

The disc is placed on the surface of an agar plate containing the bacterium being tested.

1 **The metal spreader is heated in a Bunsen flame before use and allowed to cool before spreading the bacteria. Suggest why.**

2 **When setting up the agar plate, explain why it is inoculated but not incubated.**

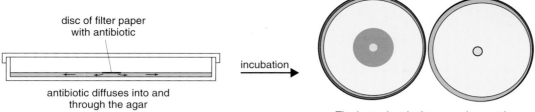

Figure 1.49 The larger the clear area, the more effective the antibiotic

The bacterium in the agar plate on the left is sensitive to the antibiotic.
The bacterium in the agar plate on the right shows resistance to the antibiotic.

disc of filter paper with antibiotic

antibiotic diffuses into and through the agar

incubation

Selecting the most appropriate apparatus and techniques

The apparatus and **sampling techniques** a professional scientist would use to carry out this investigation are almost identical to those you would use. However, there are two important differences:

- The standard medium for testing antibiotics is Mueller–Hinton agar, often with added blood. It contains beef and milk protein and is ideal for culturing human **pathogens**.
- Agar plates are incubated at 37°C – human body temperature. Samples in school must never be incubated above 25°C because of the risk of growing pathogens.

Figure 1.50 The sample of *Staphylococcus aureus* transferred for testing must be from a colony that looks identical to others on the plate

3 Name three ingredients of Mueller–Hinton agar.

4 Suggest why scientists use no more than 12 discs per plate.

Ensuring the investigation is valid

One of the most dangerous bacteria showing antibiotic resistance is methicillin-resistant *Staphylococcus aureus* (MRSA). A sample of bacteria for testing must be *representative* of the population of bacteria.

If the bacteria are spread appropriately, the clear zones will be uniformly circular and there will be continuous growth of bacteria across the remainder of the plate. Measurements are made with a ruler or callipers. Any plates where zones are not circular, or where there is poor growth of the bacterium, should be discarded.

5 What is meant by a representative sample? Why is it important that the sample is representative?

6 How is a representative sample of the bacterial culture taken?

7 Why is MRSA considered such a dangerous bacteria?

REMEMBER!

You should be able to describe the apparatus and techniques used when testing the effects of antibiotics, antiseptics and disinfectants.

DID YOU KNOW?

Two main strains of MRSA have caused problems in British hospitals since the 1990s. EMRSA16 is the most common form.

REQUIRED PRACTICAL

Investigating disinfectants

Learning objectives:

- carry out experiments with due regard to health and safety
- present and process data, identifying anomalous results
- evaluate methods and suggest further investigations.

KEY WORDS

antiseptic
disc-diffusion
incubation
zone of inhibition

For use in a hospital, choosing the right disinfectant or antiseptic to achieve the appropriate hygiene levels is essential. The correct dilution is also important: a concentration high enough to work, but not so high as to be wasteful.

Setting up a disc-diffusion investigation

Scientists need a number of different skills to carry out this investigation. This topic looks at some of those skills.

The method used to test the effectiveness of a disinfectant (or an **antiseptic** or antibiotic) is the **disc-diffusion** technique.

In this experiment, different concentrations of the disinfectant sodium hypochlorite are investigated.

① In the investigation, which is the independent variable and which is the dependent variable?

② Suggest the other possible variables that need to be controlled.

Health and safety

Before scientists can begin a disc-diffusion investigation, they must carry out a risk assessment.

These pages are designed to help you think about aspects of the investigation rather than to guide you through it step by step.

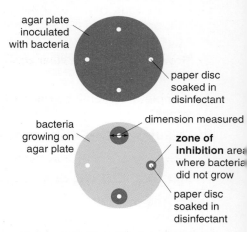

Figure 1.51 Disc-diffusion technique

Hazard	Type of hazard	Risk	Safety precautions
Ethanol			
Sodium hypochlorite			
Bacteria			
Agar plate			
Add more rows to include the activities involved, e.g. flaming an inoculating loop.			

③ Complete the risk assessment table.

④ Suggest why:
- scientists would use Mueller–Hinton blood agar; in the school lab, you would use nutrient agar
- you would incubate the plate at 25°C; the scientists at 37°C.

Presenting and processing data

The agar plates are **incubated**, and the zones of inhibition (clear zones) measured. Scientists need to analyse the data they have collected:

Concentration of sodium hypochlorite (g/dm³)	Area of clear zone around disc (mm²)			Mean area of clear zone around disc (mm²)
	Test 1	Test 2	Test 3	
0.0	0	0	0	0
0.5	0	0	0	0
1.0	32	31	34	32
1.5	91	89	91	90
2.0	470	381	379	380
2.5	499	505	497	
3.0	546	552	551	
3.5	575	568	567	
4.0	578	582	580	
4.5	580	580	580	
5.0	579	578	583	

5 How is the area of a clear zone calculated? Hint: you need to recall a formula.

6 Complete the table by calculating the mean area of the clear zones.

7 Identify the anomalous result. How did you recognise it?

8 Which set of results has the highest degree of repeatability?

9 Plot a graph of area of clear zone against concentration.

Evaluating the investigation

The experiment's aim was to find out the concentration of disinfectant that would be best for a hospital to use against the bacterium *Staphylococcus aureus*.

10 Do you have all the information needed to draw a full conclusion, or should the scientists collect more information? Use your graph to help you make your recommendation.

11 We know that in the clear zone, bacteria do not grow. But we do not know if they have been killed, or just prevented from growing. Suggest a follow-up investigation.

REMEMBER!

The area of a circle can be calculated using the formula:
area $= \pi r^2$
where r is the radius, the distance from the centre to the edge of the circle.

KEY SKILLS

In a risk assessment, you should group hazards into categories: organisms, chemicals, physical hazards and practical activities. Use the correct terminology for the type of hazard (for example, biohazard, irritant, oxidising). Think about the concentration of chemicals used. Don't forget the hazards and risks before and after the experiment, for example, the agar plate after incubation.

MATHS SKILLS

Size and number

Learning objectives:

- make estimates for simple calculations, without using a calculator
- be able to use ratio and proportion to calibrate a microscope
- recognise and use numbers in decimal and standard form.

KEY WORDS

calibrate
graticule
haemocytometer
standard form

The size of structures is important in biology, from whole organisms to molecules.

Estimating cell size

Accurate measurements are often essential. But estimating cell size or number is sometimes sufficient and may be quicker.

To estimate cell size, we can count the number of cells that fit across a microscope's field of view.

$$\text{Size of one cell} = \frac{\text{diameter of field of view}}{\text{number of cells that cross this diameter}}$$

If the field of view of this microscope, at this magnification, is 0.3 mm, or 300 μm, we can do a quick calculation without a calculator.

Each cell must be roughly (300 ÷ 5) μm, or 60 μm across. This is an approximation, but could be important.

1 Suggest how to estimate the field of view of a microscope.

2 State one advantage of estimating cell size over exact measurement.

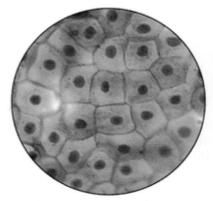

Figure 1.52: In this image, *approximately* five cells fit across the field of view. We round numbers up or down to make calculations straightforward.

DID YOU KNOW?

Scientists *estimate* cell or organism numbers when it is impossible or unnecessary to count them all.

Measuring cell size

To make accurate measurements of cell size a scientist **calibrates** their microscope. A **graticule** – piece of glass or plastic onto which a scale has been drawn – is placed into the eyepiece of the microscope.

A stage micrometer is placed on the microscope stage. This is simply a microscope slide onto which an accurate scale has been etched.

In Figure 1.53, 36 divisions on the eyepiece graticule are equivalent to 100 μm on the stage micrometer: 1 division is equivalent to $\frac{1}{36} \times 100$ μm = 2.8 μm

The cell highlighted in the right-hand diagram is 20 eyepiece divisions across: the width of the cell = (20 x 2.8) μm = 56 μm.

REMEMBER!

The digital point remains fixed. It is the digits that move as a number is multiplied or divided by powers of 10. So, as a number gets larger, the digits move to the left (and vice versa).

3 What would be the diameter of a cell that was 65 divisions on this graticule?

4 How many graticule divisions would a cell that was 35 μm across take up?

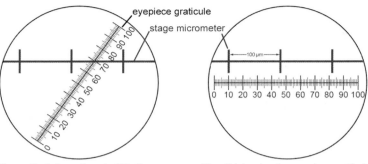

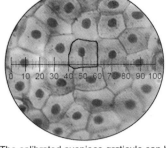

The graticule is enclosed within the eyepiece.
A stage micrometer is placed on the stage of the microscope.

The divisions on the eyepiece graticule and stage micrometer are lined up.

The calibrated eyepiece graticule can be used to make measurements of any cells or other structures viewed with that microscope.

Figure 1.53: Calibrating, then using an eyepiece graticule.

Numbers written in standard form

When writing and working with very large or very small numbers, it is convenient to use **standard form**. Standard form shows the magnitude of numbers as powers of ten.

Standard form numbers are written as: $A \times 10^n$

where: A is a number greater than 1 but less than 10. This could be decimal number such as A = 3.75, as well as an integer number such as A = 7. n is the index or power.

We use standard form with large numbers, small numbers and calculations. In standard form:

- when multiplying: multiply numbers and add powers (see example in Figure 1.54).
- when dividing: divide numbers and subtract powers.

Blood cell type	Width of an average cell (m)
Lymphocyte (small)	7.5×10^{-6}
Macrophage	5.0×10^{-5}
Megakaryocyte	1.5×10^{-4}
Neutrophil	1.2×10^{-5}

The sizes of different types of blood cell, written in standard form.

5 Look at the table of cell sizes. Arrange the cell types in descending order of size.

6 How many times larger is a megakaryocyte than a lymphocyte?

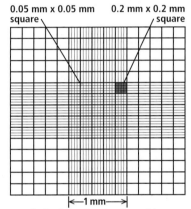

0.05 mm x 0.05 mm square   0.2 mm x 0.2 mm square

←—1 mm—→
depth of counting chamber = 0.1 mm

For a 0.2 × 0.2 mm counting chamber:
Dimensions:
top length: 0.2 mm = 2.0×10^{-1} mm
side length: 0.2 mm = 2.0×10^{-1} mm
depth: 0.1 mm = 1.0×10^{-1} mm
∴ volume of counting chamber =
add

$(2.0 \times 10^{-1}) \times (2.0 \times 10^{-1}) \times (1.0 \times 10t^{-1})$ mm³

multiply
= 4.0×10^{-3} mm³

Figure 1.54: Calculating the volume of a counting chamber. The counting chamber is a hollow on a microscope slide which holds a set volume of a fluid. It has a grid ruled onto it, and a depth of 0.1 mm. The number of cells in a given volume can be calculated.

Check your progress

You should be able to:

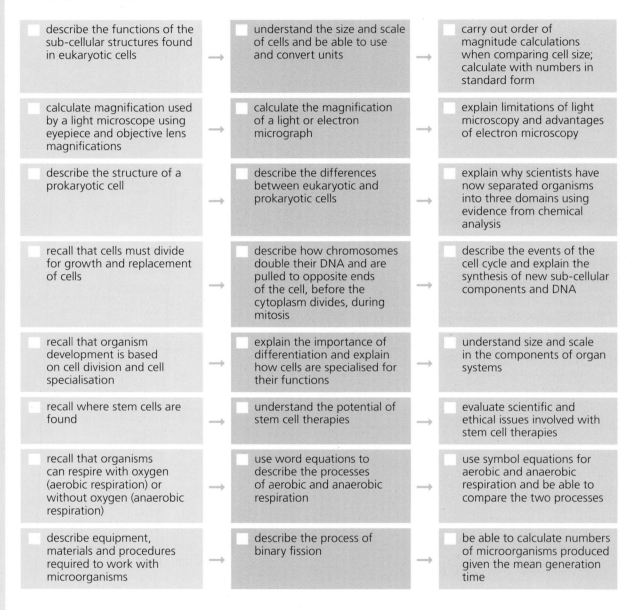

describe the functions of the sub-cellular structures found in eukaryotic cells →
understand the size and scale of cells and be able to use and convert units →
carry out order of magnitude calculations when comparing cell size; calculate with numbers in standard form

calculate magnification used by a light microscope using eyepiece and objective lens magnifications →
calculate the magnification of a light or electron micrograph →
explain limitations of light microscopy and advantages of electron microscopy

describe the structure of a prokaryotic cell →
describe the differences between eukaryotic and prokaryotic cells →
explain why scientists have now separated organisms into three domains using evidence from chemical analysis

recall that cells must divide for growth and replacement of cells →
describe how chromosomes double their DNA and are pulled to opposite ends of the cell, before the cytoplasm divides, during mitosis →
describe the events of the cell cycle and explain the synthesis of new sub-cellular components and DNA

recall that organism development is based on cell division and cell specialisation →
explain the importance of differentiation and explain how cells are specialised for their functions →
understand size and scale in the components of organ systems

recall where stem cells are found →
understand the potential of stem cell therapies →
evaluate scientific and ethical issues involved with stem cell therapies

recall that organisms can respire with oxygen (aerobic respiration) or without oxygen (anaerobic respiration) →
use word equations to describe the processes of aerobic and anaerobic respiration →
use symbol equations for aerobic and anaerobic respiration and be able to compare the two processes

describe equipment, materials and procedures required to work with microorganisms →
describe the process of binary fission →
be able to calculate numbers of microorganisms produced given the mean generation time

Worked example

Some students see a newspaper article on a European stem cell clinic. The clinic uses stem cell therapy to treat diabetes and other conditions.

1 **What is a stem cell?**

An unspecialised cell that can differentiate, or can be made to differentiate, into many different cell types.

> The part about differentiation is important.

2 **The article contains some information on the clinic's treatment of diabetes.**

It includes data on the 55 patients the clinic has treated so far, which it claims is a success.

Type of diabetes	One month after treatment, number of patients who...		
	showed an improvement	showed no change	became worse
Type 1	8	13	2
Type 2	20	9	3
Total	28	22	5

> The calculations are correct.
>
> The answers have not been recorded consistently, however. Think about significant figures. All the numbers in the table have a maximum of two figures, so you cannot have more in the answer.

a **Calculate the overall percentage of patients who:**

- showed an improvement   *50.9%*
- showed no change   *40%*
- became worse   *9.1%*

b **Does the data support the newspaper article's claims? Explain your answer.**

My percentages show there is no difference between the overall percentage of patients who showed an improvement and those who did not.

The article is certainly untrue for Type 1 diabetes, where only 34.8% improved.

It might be true for Type 2, because 62.5% improved.

> The answer draws correct conclusions from the table and uses data to support these.
>
> With the small number of patients (23 for Type 1; 32 for Type 2), we cannot draw any firm conclusions from the small differences in numbers. Sample size is important in any scientific study.
>
> The heading of the table says 'one month after treatment...'. This may be too early to draw any conclusions. Also, the patients may believe they feel better because they've had treatment. Importantly, we do not know how the 'improvement' was judged, another point to make in your answer.

You will notice that this question draws on your knowledge of scientific methods as well as stem cells.

You may also have to use relevant knowledge from another topic.

When you've learnt about diabetes, you will be aware that Type 2 diabetes can usually be controlled by diet and exercise, so the 'improvements' suggested by the data may be the result of this.

End of chapter questions

Getting started

1 The diagrams below show an animal cell and a bacterial cell.

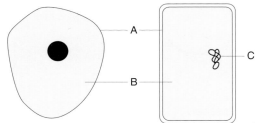

  a Name the cell part shown by:
    i) **A** `1 Mark`
    ii) **B** `1 Mark`
  b What is the name of chemical C? `1 Mark`

    i) cellulose   iii) DNA
    ii) chlorophyll   iv) protein

  c What is the function of the nucleus? `1 Mark`

2 Explain how you know that the cell shown is a plant cell. `2 Marks`

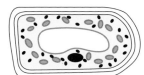

3 The diagrams below show some different objects. The length of each is included on the diagram. List them in order of size. `1 Mark`

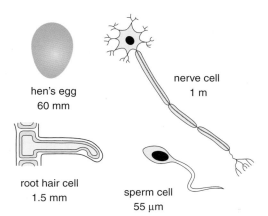

hen's egg
60 mm

nerve cell
1 m

root hair cell
1.5 mm

sperm cell
55 μm

4 A bacterium divides into two every 30 minutes.

Starting with one bacterium, how many cells will there be after $1\frac{1}{2}$ hours? `1 Mark`

5 Gemma is using Feulgen stain to stain some dividing cells.

The label on the bottle of stain is shown.

IRRITATING TO EYES,
RESPIRATORY SYSTEM
AND SKIN

`2 Marks`

Explain two safety precautions she should take when using the stain.

Going further

6 Draw lines to match the two cell structures with their function.

`1 Mark`

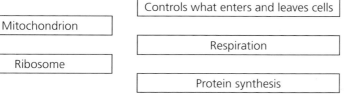

Cell structure	Function
	Controls what enters and leaves cells
Mitochondrion	
	Respiration
Ribosome	
	Protein synthesis

7 Use words from the box to complete the word equation for aerobic respiration.

`1 Mark`

alcohol	carbon dioxide	glucose	lactic acid

.... + oxygen → .... + water (+ energy)

8 Explain how the features of a sperm cell help it to fertilise an egg cell.

`2 Marks`

9 A scientist is investigating the effect of different sugars on the anaerobic respiration of yeast.

She measures the volume of carbon dioxide produced over a two hour period.

She investigates two types of sugar, sucrose and lactose.

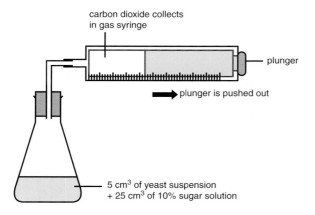

carbon dioxide collects
in gas syringe

plunger

plunger is pushed out

5 cm³ of yeast suspension
+ 25 cm³ of 10% sugar solution

a   Suggest ways in which she can be sure that the readings she takes are the result of using the two different types of sugar and not from other factors.   **4 Marks**

b   Her results are shown below.   **2 Marks**

Type of sugar	Volume of carbon dioxide produced (cm³)						
	Time (min.)						
	0	20	40	60	80	100	120
sucrose	0	0.2	1.8	5.6	7.6	7.9	7.9
lactose	0	0	0	0	0	0	0

Suggest reasons for the scientist's results.

More challenging

10   Give the *two* forms of tumour.   **1 Mark**

11   Write down the name of the technique used to produce embryonic stem cells from a person's body cells.   **1 Mark**

12   Explain why mitosis is important in plants and animals.   **2 Marks**

13   Below are some data on the link between viruses and neck cancer over a number of years.

Year	Percentage of neck cancer cases where virus DNA was detected in tumours
1965	20
1975	23
1985	28
1995	57
2005	68

Plot a graph of the data. Connect the points with a smooth curve.   **4 Marks**

14   Gemma is using Feulgen stain to stain some dividing cells. One of the images she sees is shown in the micrograph.   **2 Marks**

Describe what is happening in the micrograph.

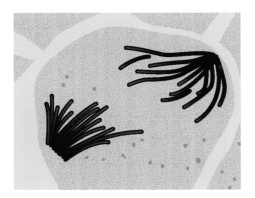

Most demanding

15 Describe how anaerobic respiration differs from aerobic respiration.

2 Marks

16 Some companies offer to store the stem cells found in the umbilical cord of a new born baby. These stem cells can be used for transplants or therapy for blood diseases. Suggest two possible advantages of using stem cells from an umbilical cord over using stem cells from bone marrow.

2 Marks

17 The graph below shows the correlation between colon cancer incidence and daily meat consumption in females. What conclusions can you draw from the graph? Evaluate the evidence for meat consumption affecting the incidence of colon cancer in females.

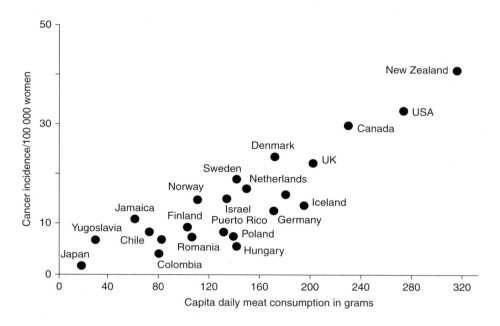

6 Marks

Total: 40 Marks

PHOTOSYNTHESIS

IDEAS YOU HAVE MET BEFORE:

PLANT CELLS ARE ADAPTED TO CARRY OUT THEIR FUNCTIONS.

- Plants have adaptations that allow them to survive and grow, such as air pores in the leaves.
- Cells have specific adaptations so that they can carry out a specialised function in the plant.
- Photosynthesis happens in the mesophyll cells in leaves. Water is absorbed through root hair cells.

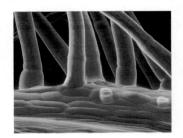

PLANTS MAKE THEIR FOOD WITH THE HELP OF SUNLIGHT.

- Green plants use water and minerals from the soil to grow. They use light, carbon dioxide and water to produce carbohydrates in their leaves by photosynthesis.
- The amount of photosynthesis can be affected by a range of different factors.
- The useful product of the photosynthesis reaction is glucose.

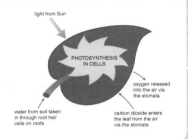

light from Sun

PHOTOSYNTHESIS IN CELLS

oxygen released into the air via the stomata

water from soil taken in through root hair cells on roots

carbon dioxide enters the leaf from the air via the stomata

PLANTS CAN MOVE NUTRIENTS AND WATER TO WHERE THEY ARE NEEDED.

- Plants have a network of vessels that transport water and minerals to their leaves and flowers.
- All cells contain watery cytoplasm. Plants need water so that the chemical reactions needed to sustain life can occur.

THE STRUCTURE OF A LEAF ALLOWS MOLECULES TO BE TAKEN IN FROM THE ENVIRONMENT, AND TO BE PASSED OUT.

- Diffusion is the movement of molecules from a higher concentration to a lower concentration until they are equally distributed.
- Substances that plants need diffuse into the plant. Other substances pass out of plants by diffusion.

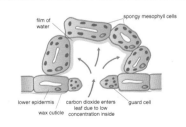

spongy mesophyll cells

film of water

lower epidermis

wax cuticle

carbon dioxide enters leaf due to low concentration inside

guard cell

IN THIS CHAPTER YOU WILL FIND OUT ABOUT:

HOW DO PLANTS' SPECIAL ADAPTATIONS HELP THEM SURVIVE AND GET ALL THEY NEED FROM THE ENVIRONMENT?

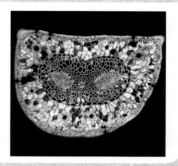

- Adaptations of cells and tissues in leaves allow them to photosynthesise efficiently.
- Stomata are adapted to control the exchange of gases.
- Cells and tissues in leaves, stems and roots are designed for the maximum exchange of substances in and out of the plant.

WHAT FACTORS AFFECT THE PHOTOSYNTHESIS REACTION? HOW DO THEY AFFECT IT?

- The useful products of photosynthesis are simple carbohydrates, for example glucose and sucrose.
- Different environmental factors interact to limit the rate of photosynthesis in different habitats at different times.
- The environment in which plants are grown can be artificially manipulated.

HOW IS THE SUPPLY OF WATER TO A PLANT AFFECTED BY ENVIRONMENTAL CONDITIONS?

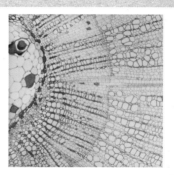

- There are two transport systems in plants: xylem transports water up the plant and phloem transports substances up and down the plant.
- Water movement through the plant is affected by different environmental factors.
- Water loss in plants is a consequence of adaptations for photosynthesis.

DIFFUSION ALSO ALLOWS SUBSTANCES TO PASS IN AND OUT OF CELLS. WHICH SUBSTANCES DIFFUSE INTO PLANTS?

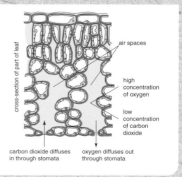

cross-section of part of leaf

air spaces

high concentration of oxygen

low concentration of carbon dioxide

carbon dioxide diffuses in through stomata

oxygen diffuses out through stomata

- Different factors affect the rate of diffusion in plant systems.
- Concentration gradients can affect the rate of photosynthesis.
- Substances move in and out of the leaf during different processes, for example, photosynthesis, respiration and transpiration.

Explaining photosynthesis

Learning objectives:

- identify the raw materials and products for photosynthesis
- describe photosynthesis by an equation
- explain gas exchange in leaves.

Plants and algae are amazing organisms. They transfer energy from outer space into chemical energy in glucose. Without plants, there would be no life! How do plants harness this energy?

Describing the process

Photosynthesis is the chemical reaction that plants use to produce glucose, so essentially they are able to make their own food. Photosynthesis needs two raw materials. These reactants are:

- carbon dioxide (absorbed through leaves)
- water (absorbed through roots and transported to leaves).

The reaction also needs light and **chlorophyll** in the leaves. The equation is:

$$\text{carbon dioxide} + \text{water} \xrightarrow{\text{light} + \text{chlorophyll}} \text{glucose} + \text{oxygen}$$

$$6CO_2 + 6H_2O \longrightarrow C_6H_{12}O_6 + 6O_2$$

This is an endothermic reaction. It requires energy in the form of light.

The products of photosynthesis are **carbohydrates**, for example glucose and starch. When iodine solution is added to a leaf, it will turn blue/black if starch is present. If starch is present, it shows that photosynthesis has happened.

DID YOU KNOW?

Some leaves have green parts and yellow/white parts. They are called variegated leaves. Variegated leaves only photosynthesise in the green parts where the chlorophyll is found.

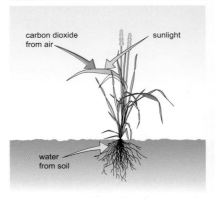

Figure 2.1 A plant needs light, water and carbon dioxide for photosynthesis

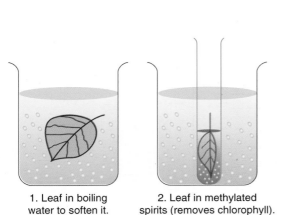

1. Leaf in boiling water to soften it.

2. Leaf in methylated spirits (removes chlorophyll).

3. Leaf washed. Few drops of iodine solution added.

4. Blue/black colour indicates starch produced by photosynthesis.

Figure 2.2 Testing a leaf for starch

Oxygen is made during photosynthesis and released into the air. It is a waste product on which animals depend.

1 Which raw materials are used in photosynthesis and what are the products of the reaction?

2 A leaf was tested for starch. The iodine solution stayed orange.

   a What does this tell you about the leaf?
   b What conditions was it kept in?

Gas exchange in leaves

Leaves take in carbon dioxide from the air and release oxygen when they photosynthesise. This is called gas exchange.

Some students investigated gas exchange in leaves. They used the same sized leaf in Tubes 1 and 2 and left them in a rack in bright light for an hour. Look at their method in the table.

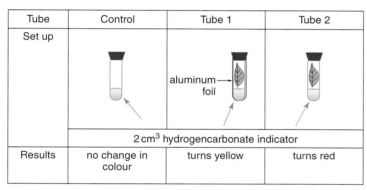

Tube	Control	Tube 1	Tube 2
Set up		aluminum foil	
	$2\,cm^3$ hydrogencarbonate indicator		
Results	no change in colour	turns yellow	turns red

Figure 2.3 Investigating gas exchange in leaves.

Hydrogencarbonate indicator is an orange solution.

- In more acidic conditions it turns yellow.
- In less acidic conditions it turns red.

Carbon dioxide dissolves in water to form an acidic solution.

3 Describe the results of the investigation and explain why a control was used.

4 Explain what the results tell you.

5 Why are light and chlorophyll needed for photosynthesis?

6 Predict what would happen to the hydrogencarbonate indicator if you removed the aluminium foil. Explain your answer.

COMMON MISCONCEPTION

Remember: light does not make food, but light is needed for the reactions to occur.

Looking at photosynthesis

Learning objectives:

- explain the importance of photosynthesis
- explain how plants use the glucose they produce.

KEY WORDS

endothermic
veins

Plants make carbohydrates (for example, sucrose and glucose) and oxygen when they photosynthesise. The plants release the oxygen that they do not need.

Learning about photosynthesis

Photosynthesis is a series of reactions that require energy. Reactions that need energy are called **endothermic** reactions. You will learn more about endothermic reactions in chemistry. In photosynthesis, energy is transferred from the environment to the chloroplast by light.

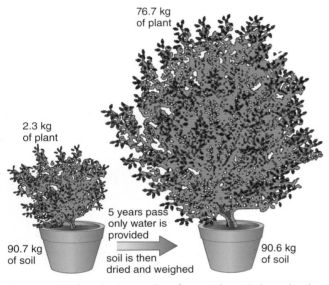

76.7 kg of plant

2.3 kg of plant

5 years pass only water is provided

90.7 kg of soil

soil is then dried and weighed

90.6 kg of soil

Figure 2.4 What do the results of van Helmont's investigation show?

Early scientists thought that plants grew just by using minerals in the soil.

In 1600, Jan Baptist van Helmont showed that the increase in the mass of a willow tree was not just due to the soil minerals.

In 1771, Joseph Priestley put a plant in a glass container with a lit candle (to use up the oxygen). He then left the investigation for 27 days. Priestley found the candle then burned again. This showed that oxygen was present again.

COMMON MISCONCEPTION

Leaves look green because they only reflect the green light. They do not absorb green light for photosynthesis.

1. **What did van Helmont's investigation show?**

2. **How could you investigate whether plants need light to photosynthesise?**

Using sugars

Glucose and sucrose molecules produced by photosynthesis are used in a number of ways. Sugars are soluble molecules that dissolve to be transported to wherever they are needed. **Veins** transport sugar to parts of plants shown in Figure 2.5 that do not photosynthesise.

Glucose is used by cells for respiration. Plants cannot make glucose at night so it is converted into insoluble starch for when it is needed. Starch is the main energy store in plants. It is found in every cell, where it is used for respiration in low light levels or the dark.

Plant cells respire sugars to provide energy, and produce carbon dioxide and water. The energy released is used for chemical reactions in cells, for example:

- building glucose into starch for storage
- building sugars into cellulose, which strengthens cell walls
- combining sugars with nitrate ions and other minerals to make amino acids for protein synthesis
- building fats and oils for storage

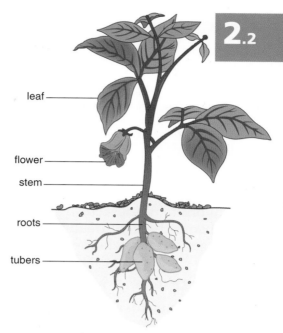

Figure 2.5 Why do plant organs need glucose?

3 When and why do plants respire?

4 List five ways that plants use glucose.

Plant respiration

Oxygen is a waste product of photosynthesis. When plant cells respire they use some of this oxygen. Plants respire continually because they need a constant supply of energy to keep them alive.

starch is insoluble and is made up of many glucose molecules joined together

glucose molecules are soluble

Figure 2.6 Starch is a large molecule and glucose is a small molecule

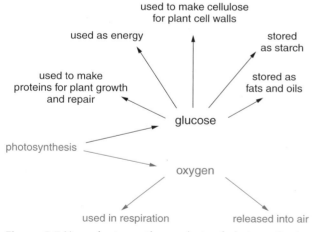

Figure 2.7 How plants use the products of photosynthesis

The rest of the oxygen is released through the stomata, back into the air.

5 Explain the relationship between photosynthesis and respiration in leaves.

6 Explain what is incorrect about the statement 'plants make oxygen for us to breathe'.

DID YOU KNOW?

Some sea slugs eat algae but do not digest it fully. The algae photosynthesise inside the slugs to give them more food and energy.

KEY INFORMATION

Stomata are tiny opening structures found on the outer leaf skin layer. Stomata consist of two specialised guard cells that surround the stoma, a tiny pore (hole).

Investigating leaves

KEY WORDS

epidermal tissue
palisade
  mesophyll
spongy
  mesophyll
vascular bundle

Learning objectives:

- identify the internal structures of a leaf
- explain how the structure of a leaf is adapted for photosynthesis
- recall that chloroplasts absorb energy from light for photosynthesis.

Leaves work to manufacture glucose. Chloroplasts in leaves absorb the light needed for photosynthesis. How do the organelles inside the leaf work together to do this?

The structure of a leaf

Leaves of green plants are plant organs that are adapted to allow them to photosynthesise efficiently. Most leaves:

- are broad, with a large surface for light to be absorbed
- are thin, so that carbon dioxide has a short distance to diffuse
- have **vascular bundles** (veins). These support the leaf and transport
  - water to the leaf
  - glucose away from the leaf.

1 **How does the shape of a leaf help photosynthesis to happen?**

2 **What is the function of vascular bundles?**

Inside the leaf

The diagram shows the internal structure of a leaf. Each part has a specific function.

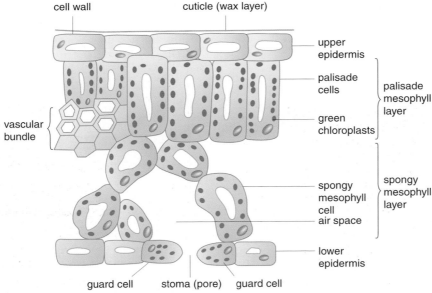

Figure 2.8 A diagram of a vertical section through a leaf

The upper surface is covered by a cuticle to protect it. Under the cuticle are layers of different cells:

- **epidermal tissues** cover the leaf, letting light penetrate
- **palisade mesophyll** carries out photosynthesis, absorbing most of the light
- **spongy mesophyll** has air spaces for diffusion of gases
- lower epidermal tissue protects the underside of the leaf.

Chloroplasts in the mesophyll are where light is absorbed for photosynthesis.

The lower epidermis has small pores called stomata. Each stoma allows gases to diffuse in and out of the leaf.

3 Describe the function of mesophyll cells.

4 Why do leaf cells contain chloroplasts but root cells do not?

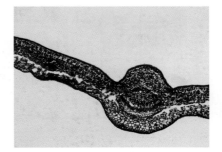

Figure 2.9 A micrograph of a section through a leaf. Compare what you can see here with the diagram in Figure 2.8

Adaptations for effective working

Using a microscope, look at a slide of a section through a leaf.

Identify the layers of cells. Calculate the magnification you used.

Each type of leaf cell has features that make it adapted for its function.

DID YOU KNOW?

Meristem tissue is found at the growing tips of roots and shoots. It differentiates into different types of plant cells.

Layer	Adaptation	Function
upper epidermal tissue	thin and transparent waxy cuticle	allows light to pass to the mesophyll to protect the leaf and stop water loss
palisade mesophyll	regular-shaped cells, arranged end-on, near upper surface; most chloroplasts at the top of the cells	absorb the maximum amount of light possible
spongy mesophyll	irregular-shaped cells; many air spaces	increases surface area for CO_2 absorption; allows gases to diffuse
lower epidermal tissue	many stomata; surrounded by guard cells	allow gases to diffuse; guard cells open and close stomata
vascular bundles	contain xylem and phloem tubes	transport substances around the plant

5 Suggest why stomata are found on the lower epidermal tissue.

6 Explain how the adaptations of the palisade mesophyll allow it to photosynthesise efficiently.

7 Explain the adaptations of the upper epidermal tissue.

REQUIRED PRACTICAL

Investigate the effect of light intensity on the rate of photosynthesis using an aquatic organism such as pondweed

KEY WORDS

photosynthesis
hypothesis
chloroplasts
mesophyll cells

Learning objectives:

*   use scientific ideas to develop a hypothesis
*   use the correct sampling techniques to ensure that readings are representative
*   present results in a graph.

Photosynthesis is the process by which green plants produce food. When plants photosynthesise they absorb light energy to power the reaction. The equation for photosynthesis is:

$$\text{carbon dioxide} + \text{water} \xrightarrow{\text{light}} \text{glucose} + \text{oxygen}$$
$$6CO_2 \quad + \quad 6H_2O \xrightarrow{\hspace{2cm}} C_6H_{12}O_6 \quad + \quad 6O_2$$

These pages are designed to help you think about aspects of the investigation rather than to guide you through it step by step.

Developing a hypothesis

Laura and Amy are investigating how light intensity affects the rate of **photosynthesis** in pondweed. They will measure how much gas is released by the pondweed in 1 minute when they put a lamp at 10 cm, 15 cm, 20 cm, 25 cm and 30 cm away from the beaker containing the pondweed.

Before they begin the investigation they are going to make a **hypothesis**. Hypotheses are developed using previous knowledge or observations. Laura and Amy know that:

*   photosynthesis produces glucose and oxygen
*   the oxygen will be released as bubbles of gas in the water
*   light energy is absorbed by chlorophyll found in the **chloroplasts** in the **mesophyll cells**
*   photosynthesis can be limited by different factors.

1. **When will the light intensity be greatest?**

2. **How will increasing the light intensity affect the rate of photosynthesis?**

3. **When the rate of photosynthesis increases what will happen to the amount of oxygen produced?**

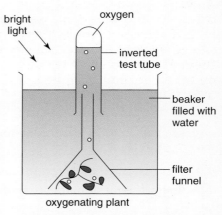

Figure 2.10 An oxygenating plant photosynthesising

4 What other variables could affect this investigation?

5 Suggest a hypothesis for the investigation that Laura and Amy are going to do.

Improving accuracy

Amy and Laura decided that they would check that their data were repeatable by taking repeat readings at each light intensity. Look at their results.

Distance of lamp from pondweed (cm)	Number of bubbles per minute		
	Test 1	Test 2	Test 3
10	105	121	124
15	58	50	54
20	26	30	32
25	15	17	16
30	12	13	14

6 What do you notice about the results in Test 1 compared with the results in tests 2 and 3?

7 Suggest why this happened.

8 Suggest what Laura and Amy should do with the Test 1 results.

9 Why is it important to take repeat readings when carrying out investigations?

Presenting results

Amy and Laura used a table to record their results because it was quick and the results were organised, but they found it hard to analyse them in this form. Their teacher said that they should think of a better way to present their results to help them. Laura thought that using a bar chart would be the best method but Amy disagreed. She presented her results as a line graph.

10 Which would be the correct way to present these results? Explain why.

11 Plot the graph of Amy and Laura's results using only Test 2 and 3 data.

> **REMEMBER!**
> ....
>
> When you make a hypothesis you must try to explain your prediction or observations using what you already know about science

Increasing photosynthesis

Learning objectives:

- identify factors that affect the rate of photosynthesis
- interpret data about the rate of photosynthesis
- explain the interaction of factors in limiting the rate of photosynthesis.

Plants grow faster in summer than in winter. This means that they must produce more food to allow them to grow in summer. Some factors can increase the rate of photosynthesis.

Plants in different habitats

Plants are found in every ecosystem, but their size, appearance and adaptations mean that they look very different. Tropical rainforests have dense plant life. In contrast, few plant species grow in **tundra** (Arctic regions with permanently frozen subsoil) and desert regions.

DID YOU KNOW?

Pine trees grow in a cone shape to expose more needles to the sun, increasing the rate of photosynthesis.

Figure 2.12 What are the environmental conditions of these habitats?

1. **How would you describe environmental conditions in tropical forest, tundra and desert ecosystems?**

2. **Suggest how the conditions you have described affect photosynthesis in each habitat.**

Limiting factors

Some students investigated the effect of light on the rate of photosynthesis. Look at their results (Figure 2.13). The students found that:

- Between A and B, the rate of photosynthesis increases as the light intensity increases. Because the rate depends on the light intensity, light intensity is called the **limiting factor**.
- Between B and C, increasing the light intensity has no effect on the rate of photosynthesis. Another factor is now the limiting factor.

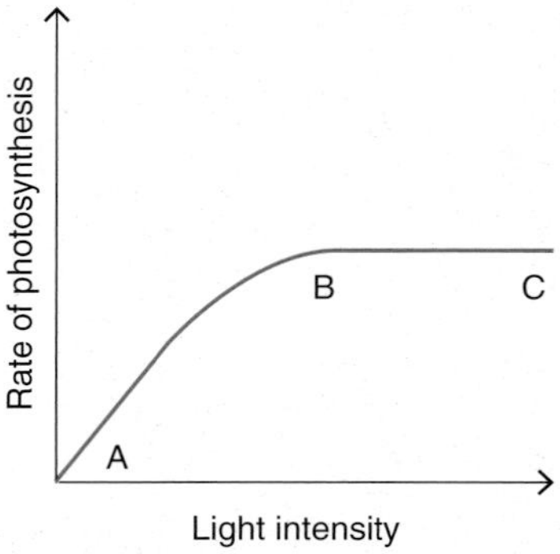

Figure 2.13 Can you explain the shape of the graph between B and C?

KEY INFORMATION

Remember that enzymes denature at temperatures greater than 40 °C.

3 **What other factors might limit the rate of photosynthesis between B and C?**

Plants need carbon dioxide to photosynthesise but there is only 0.04% in the atmosphere. It is often the limiting factor controlling the rate of photosynthesis.

Carbon dioxide levels around plants rise when there is no light. This is because the plants are respiring but not photosynthesising. As light levels increase the plants use the carbon dioxide up.

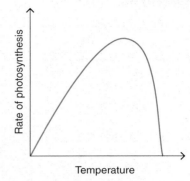

Figure 2.14 Explain what this graph shows

HIGHER TIER ONLY

Interacting limiting factors

Over one day, light, temperature and carbon dioxide levels change. Carbon dioxide may be the limiting factor when plants are crowded on a sunny day. Temperature may be the limiting factor in cooler months. Light may be the limiting factor at dawn.

Plants living in continual shade can adapt. They have a higher ratio of leaves to roots than other plants. The leaves are thinner, have a larger surface area and contain more chlorophyll to absorb light. A shortage of chlorophyll can limit the rate of photosynthesis.

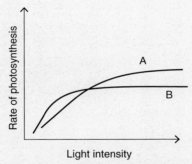

Figure 2.15 Photosynthesis in normal (A) and shade-adapted (B) leaves

The graph shows that shade-adapted leaves (B) are more efficient at absorbing low-intensity light than normal leaves (A).

4 **Suggest the limiting factors for photosynthesis over one complete warm summer's day.**

5 **Describe and explain adaptations for photosynthesis shown by some shade-tolerant plants.**

6 **Suggest one advantage and one disadvantage of a tree having needles rather than flat, broad leaves.**

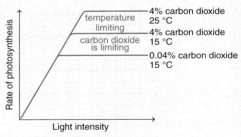

Figure 2.16 In the laboratory, we can see how different factors work in combination

Increasing food production

Learning objectives:

- explain how factors that increase food production can be controlled
- evaluate the benefits of manipulating the environment to increase food production
- understand and use the inverse square law in the context of light intensity and photosynthesis.

Farmers use their knowledge of the limiting factors of photosynthesis to increase crop *yields*. How can this be done? How do they change the factors?

HIGHER TIER ONLY

Greenhouses

Growing crops in greenhouses gives a large yield for a given area. This is intensive farming. Conditions in the greenhouse can be controlled to optimise the rate of photosynthesis. Greenhouses protect plants from weather conditions and from being damaged or eaten by animals.

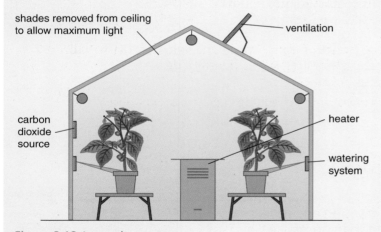

shades removed from ceiling to allow maximum light

ventilation

carbon dioxide source

heater

watering system

Figure 2.18 A greenhouse system

Figure 2.17 Why are dull cold days in winter not good for growing plants?

- Sunlight heats up the inside of greenhouses, causing the temperature to rise.
- A carbon dioxide source can be used to increase the concentration of the gas.
- Paraffin heaters are used in greenhouses. As the fuel inside them burns, they produce carbon dioxide. They can also be used to increase the temperature on cooler days and nights.
- Watering systems deliver a regular supply of water.
- Blinds can be used to control the amount of light.
- Humidifiers are used to add moisture to the air.

1 Why do many greenhouses have vents in the roof?

2 Explain why paraffin heaters are used in greenhouses.

3 How do greenhouses increase yield?

Optimising conditions

Farmers need to grow large fruits and vegetables without excessive leaf or root production. By optimising photosynthesis, and reducing effects of limiting factors, farmers can increase yields of crops. Using technology, commercial greenhouses increase yields by over twenty times, compared with traditional farming.

Glucose produced by photosynthesis that is not used for respiration, or stored, in the form of starch, is converted to other essential molecules such as cellulose, and amino acids and proteins. Mineral ions are required for these conversions. The mineral ions provided to plants can be optimised using a technique called **hydroponics**. Plants are grown in nutrient-rich liquid medium, which can be formulated for the type of plant, supported on a material such as rock wool.

Figure 2.19 Hydroponics is growing plants in mineral nutrient solutions, without soil

Computer systems control conditions.

- Nutrients are monitored and concentrations adjusted as needed.
- Temperature is controlled by sensors to within 0.1°C.
- Weather detection systems monitor external conditions and adjust vents and blinds to suit.
- Floors are covered in white plastic to reflect light.
- The glass used has a low iron content to ensure maximum light levels and even the metal supports are as thin as possible.
- Special lights are used to increase the hours of light. These are switched off for about 7 hours, allowing plants to transport the glucose that has been made.

> **KEY INFORMATION**
>
> Remember that greenhouses optimise carbon dioxide concentrations, light levels, water levels, temperature and nutrient availability.

4 What are the benefits and drawbacks of changing conditions inside a greenhouse?

Inverse square law

As the distance between the light source and a photosynthesising plant increases, the light intensity will decrease. This is described as being an inverse relationship. Light intensity obeys the **inverse square law**. The inverse square law states that if the light distance is doubled, the light intensity decreases by the square of the distance.

For example, if the distance between the light source and the plant is doubled, the light intensity is quartered (see Figure 2.20).

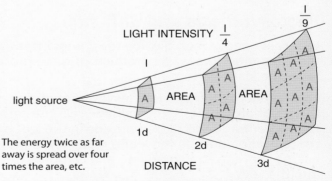

LIGHT INTENSITY $\frac{1}{4}$ $\frac{1}{9}$

light source

AREA   AREA

The energy twice as far away is spread over four times the area, etc.

1d   2d   3d

DISTANCE

Figure 2.20 Light intensity obeys the inverse square law: if the distance between the light source and the plant is doubled, the light intensity is quartered.

5 Look back at Amy and Laura's data on page 63. Draw a graph of rate of photosynthesis against $1/d^2$.

KEY CONCEPT

Diffusion in living systems

Learning objectives:

- describe the conditions needed for diffusion to occur
- calculate and compare surface area to volume ratios
- explain how materials pass in and out of cells.

KEY WORDS

passive transport
random
concentration gradient
equilibrium

In all living organisms, the transfer of many materials occurs by the process of diffusion. Diffusion is essential in living systems in the supply of important substances and in the control of certain processes.

Diffusion in living systems

Diffusion is sometimes called **passive transport**. This is because it happens due to the **random** motion of particles. No energy is required.

All that is needed for diffusion to happen is a **concentration gradient**. Diffusion can be defined as the net movement of particles from an area of high concentration to an area of lower concentration due to the random movement of particles, until **equilibrium** is reached. Equilibrium is the point at which the net movement of the particles stops changing.

1 **What effect do you think increasing the concentration gradient will have on the speed of diffusion?**

2 **What other factors will affect the speed of diffusion?**

Photosynthesis and diffusion

Leaf cells need carbon dioxide for photosynthesis and oxygen for respiration. The waste products of these processes must leave the leaf cells too.

Carbon dioxide diffuses into leaves through the stomata. During photosynthesis, the palisade cells use carbon dioxide; the concentration of carbon dioxide in the leaf is low but the concentration outside the leaf is relatively high. This results in a large concentration gradient, and carbon dioxide diffuses into the leaf.

The rate of diffusion in living systems can be increased by:

- increasing the surface area
- decreasing the distance the particles have to travel
- increasing the concentration gradient.

External conditions can also affect the rate of diffusion.

high concentration

low concentration

Figure 2.21 In diffusion, particles move from a higher concentration to a lower concentration

- Wind increases the concentration gradient of carbon dioxide between inside and outside, and the rate of diffusion increases.
- High humidity decreases the concentration gradient of water vapour between the inside and outside of the leaf, and diffusion slows down.

3 Explain how oxygen produced during photosynthesis passes out of the leaf.

4 Look back at section 2.3, 'Investigating leaves'. How are leaves adapted for maximum diffusion? Explain why this is important.

5 How does external temperature affect the rate of diffusion from a leaf?

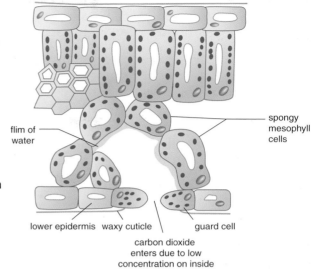

spongy
mesophyll
cells

flim of
water

lower epidermis   waxy cuticle   guard cell

carbon dioxide
enters due to low
concentration on inside

Figure 2.22 Why are the stomata spaced out?

Diffusion and cells

Cells are made largely of water containing many dissolved substances. These substances and water enter and leave the cells through cell membranes. Cell membranes allow some particles through, but block others. They are called *partially (or selectively) permeable membranes.*

Cells need continual supplies of dissolved substances, for example, oxygen and glucose, for cellular activities. Waste products, for example, carbon dioxide, need to be removed. To enter or leave a cell, dissolved substances have to be small enough to pass through the partially permeable cell membrane. Diffusion allows this to happen.

Cell membranes are similar to football nets. Large footballs cannot pass through the netting, but small golf balls easily pass through. In the same way, large particles, for example starch, cannot pass through the cell membrane. However, small glucose particles can easily pass through the membranes. Cell membranes are very thin to allow substances to easily diffuse through them.

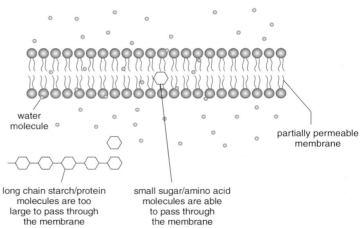

water
molecule

partially permeable
membrane

long chain starch/protein
molecules are too
large to pass through
the membrane

small sugar/amino acid
molecules are able
to pass through
the membrane

6 Explain how materials pass in and out of cells. Use a diagram to help you.

Figure 2.23 A cell membrane model. What is happening to the particles?

7 Will diffusion ever stop completely? Explain your answer.

Looking at stomata

Learning objectives:

- describe transpiration in plants
- describe the function of stomata
- explain the relationship between transpiration and leaf structure.

KEY WORDS

guard cell
transpiration
xylem

Gases continually pass in and out of leaves through tiny openings found on the lower surface of each leaf. The openings are called stomata. They control the exchange of gases in plants. Do all leaves have the same concentration of stomata?

Water movement in plants

Plants absorb water from soil through root hairs. The water passes into the root, up the stem and into the leaves. Water on the surface of the spongy mesophyll evaporates and diffuses out of leaves via the stomata. This movement of water through the plant and leaves is called **transpiration**.

Transpiration rate is affected by temperature, light availability, wind and humidity.

1 **What is transpiration?**

2 **Describe the path of water through a plant.**

3 **Predict what conditions will cause the rate of transpiration to increase.**

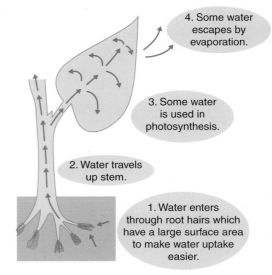

4. Some water escapes by evaporation.

3. Some water is used in photosynthesis.

2. Water travels up stem.

1. Water enters through root hairs which have a large surface area to make water uptake easier.

Figure 2.24 Where does the water in the leaf come from?

Transpiration

Plants need a constant flow of water. Water is transported in tubes called **xylem** in the veins. Evaporation of water from the leaf causes water to be drawn up the plant, just like a child sucking through a straw. Transpiration is important because:

- evaporation of water cools leaves
- water is a reactant in photosynthesis
- cells full of water support the plant
- water carries dissolved materials around the plant.

If too much water is lost through the stomata, plants may wilt and die. Surrounding each stoma are two **guard cells**. Guard cells open and close to control the exchange of gases and water loss.

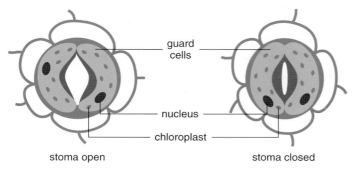

guard
cells

nucleus

chloroplast

stoma open   stoma closed

Figure 2.25 Open and closed stomata

4 **Predict why stomatal density may differ for different species.**

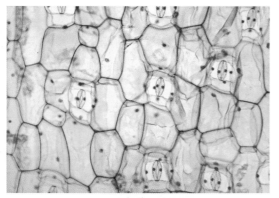

Figure 2.26 Make stomatal impressions of two different leaves using clear nail varnish. Use a microscope to calculate the stomatal density of each leaf

The effect of adaptations

A disadvantage of stomata is that while they are open for gas exchange, water loss occurs. Leaves are broad and thin to increase surface area. Water only has a short distance to diffuse out.

Marram grass grows on beaches. It has rolled leaves with thick cuticles and stomata are sunk into pits. Interlocking hairs hold water vapour. These features all reduce water loss.

5 **Explain how water loss in plants is a result of the adaptations for photosynthesis.**

6 **Explain how marram grass is adapted to prevent water loss.**

KEY INFORMATION

Water evaporates from the cell surfaces and the water vapour diffuses out of the leaf.

DID YOU KNOW?

Water lilies have stomata on the upper leaf surface so that they are not under water.

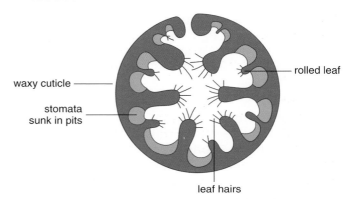

waxy cuticle

rolled leaf

stomata
sunk in pits

leaf hairs

Figure 2.27 Marram grass grows on exposed sand dunes. What conditions have the greatest effect on transpiration?

Moving water

Learning objectives:

- describe the structure and function of xylem and roots
- describe how xylem and roots are adapted to absorb water
- explain why plants in flooded or waterlogged soil die.

KEY WORDS

lumen
phloem
translocation
xylem
lignin

Plants do not have a heart or blood, but they still need a transport system to move food, water and minerals to every cell. The roots, stem and leaves form a plant organ system for the transport of substances.

Transport systems

Plants have two transport systems:

- **xylem** tissue carries water and minerals from the roots around the plant, especially to the leaves
- **phloem** tissue transports dissolved sugars made in the leaves around the plant. This is called **translocation**.

Xylem and phloem are found in continuous vascular bundles.

Water enters plants through the roots. Root hair cells near the root tip have hair-like projections that grow between soil particles to absorb water.

1 Why is water needed in the leaves?

2 What are the functions of xylem and phloem?

3 State what is meant by translocation.

COMMON MISCONCEPTION

Many students confuse xylem and phloem. Remember **phloem** carries food.

Adaptions for transport

Roots are narrow structures with a large surface area. Root hair cells also increase the surface area for water absorption. The water has only a short distance to travel to the xylem to be moved up and around the plant (Figure 2.28).

Xylem tubes are made from long cells with thick, reinforced walls containing **lignin**. This makes them strong. The cell walls are waterproof, which makes the cells die. The contents and end walls break down to form a vessel with a hollow centre or **lumen**. Water and minerals flow through the tubes. Xylem cells form the wood in a tree.

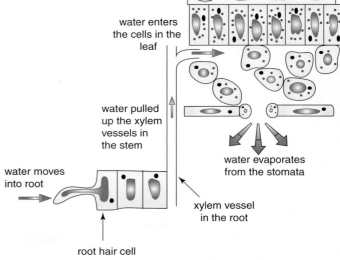

water enters the cells in the leaf

water pulled up the xylem vessels in the stem

water evaporates from the stomata

water moves into root

xylem vessel in the root

root hair cell

Figure 2.28 Use a microscope to observe root hair cells

Figure 2.29 shows the vascular bundles (xylem and phloem) in the different parts of the plant.

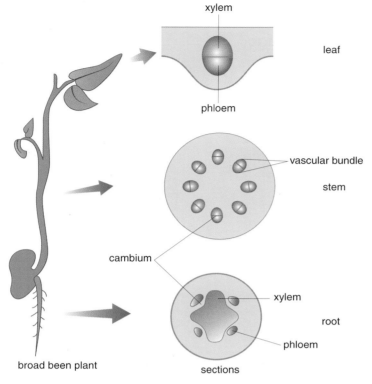

Figure 2.29 Use a microscope to draw sections through a root, stem and leaf to show xylem and phloem

4 **How are root hairs adapted to absorb water?**

5 **How is xylem adapted for its function?**

Water effects

Healthy plants balance water uptake and loss. If plants lose water faster than it is replaced, they wilt. Wilting protects the plant. Leaves droop and hang down to reduce the surface area for water loss by evaporation. The stomata close and photosynthesis stops to prevent water loss. The disadvantage is that the plant may overheat. Plants stay wilted until they get water, the temperature drops or the Sun goes off them.

Plants can have too much water. Waterlogged soil contains very little air. This means that the root cells do not receive the oxygen they need for respiration. They die and rot if the waterlogging lasts a long time.

6 **Describe the changes in a plant that help to prevent water loss.**

7 **A plant grown in a pot with no drainage is prone to dying. Explain why this might happen.**

Figure 2.30 Why has this plant wilted?

DID YOU KNOW?

Plant transport systems are found in the bark of trees. Some animals chew this away, causing the tree to die because it cannot transport water and glucose to where it is needed.

Investigating transpiration

KEY WORDS

potometer

Learning objectives:

- describe how transpiration is affected by different factors
- explain the movement of water in the xylem.

Water is an essential constituent of all plant tissues. Plants absorb large amounts of water but only use 2–5% of it for metabolic processes. Why do plants lose so much water?

Environmental factors and transpiration

Water for photosynthesis is transported to leaves in xylem vessels. Diffusion out of the spongy mesophyll causes a film of water to cover the cell surfaces. Water evaporates into the spaces between cells and the water vapour moves along the concentration gradient towards the stomata, and escapes from the leaf. When transpiration increases, water uptake by the plant increases.

Factors that increase the rate of transpiration are:

- an increase in temperature
- an increase in wind speed
- a decrease in humidity.

When plants photosynthesise, the stomata are open to allow carbon dioxide to diffuse into the leaf. Factors that increase photosynthesis will increase transpiration.

Some students investigated how light intensity affected the rate of transpiration (Figure 2.32).

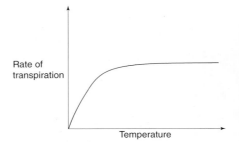

Figure 2.31 How does temperature affect transpiration rate?

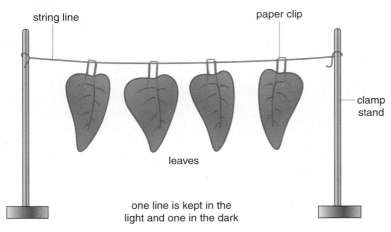

Figure 2.32 Experiment to show the effect of light on transpiration

1 In Figure 2.32, why were leaves put in the dark and the light?

2 What measurements could the students take?

3 Predict what you think will happen to the leaves that are put in the dark, and those put in the light. Give a reason for your prediction.

How plants control water loss

Potometers are used to measure the uptake of water by plants in different conditions. The movement of the air bubble in the tube shows how much water is taken up.

Different plants control water loss in different ways:

- waxy cuticle: in hot environments it can be very thick
- most stomata are on the underside of leaves, protected from environmental factors
- needle-like, spiny or rolled up leaves
- stomata in pits on the leaf surface.

4 Describe an experiment to show that transpiration rate is affected by air movement.

5 Describe what happens to transpiration rate as humidity decreases.

6 Describe how a leaf is adapted to reduce water loss.

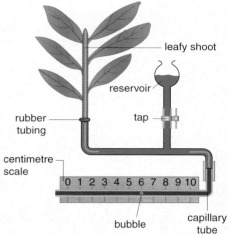

Figure 2.33 A potometer

Explaining changes in transpiration

On still days when water vapour diffuses out of the leaf, the diffusion and evaporation concentration gradients decrease until the water vapour concentration inside and outside the leaf are equal and transpiration stops.

On windy days, water vapour is blown away. The concentration gradients are maintained and the transpiration rate increases.

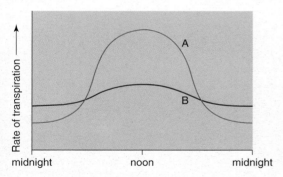

Figure 2.34 Explain how light intensity affects the rate of transpiration for plant A and plant B

KEY INFORMATION

Remember that water diffuses out of cells and evaporates to form water vapour that can escape from the leaf.

DID YOU KNOW?

Transpiration keeps plants cool. Some trees lose hundreds of thousands of litres of water in one day through transpiration.

7 Explain how the structure of a leaf is adapted to reduce water loss.

8 Why is water uptake rate not the same as transpiration rate?

9 Predict and draw graphs to describe and explain how transpiration rate is affected by:

a cold conditions   c dull cloudy conditions

b windy conditions   d still, humid conditions.

10 'Plants grow best in cooler, cloudy and humid conditions because they lose less water.' Discuss this statement. Do you agree with it? Explain why.

Moving sugar

Learning objectives:

- describe the movement of sugar in a plant as translocation
- explain how the structure of phloem is adapted to its function in the plant
- explain the movement of sugars around the plant.

KEY WORDS

sieve plates
translocation
companion cells

Sugars are needed by all cells for respiration. They are transported from the leaves for immediate use or storage. Amino acids are needed by all growth areas.

Phloem vessels

Phloem cells are elongated, thin-walled living cells that form columns or tubes.

The movement of sugars in plants is called **translocation.** Translocation happens in the phloem. Cell sap containing sugars and amino acids is moved from where these substances are produced to where they are needed.

water and food
cells have end walls with perforations
sieve plates
two-way flow
phloem sieve tube

Figure 2.35 How is phloem different from xylem? Make a model of phloem cells

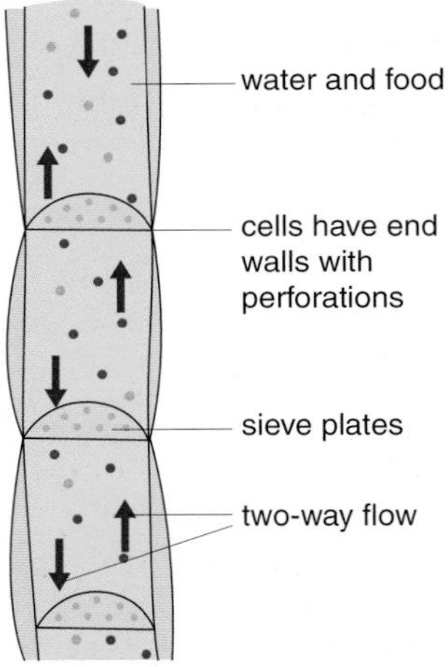

1. **Why are carbohydrates and proteins transported as sucrose and amino acids in the phloem vessels?**
2. **What is translocation?**

The structure of phloem vessels

The end cell walls of phloem cells do not break down, but they do have many small pores in them. These perforated walls are called **sieve plates**.

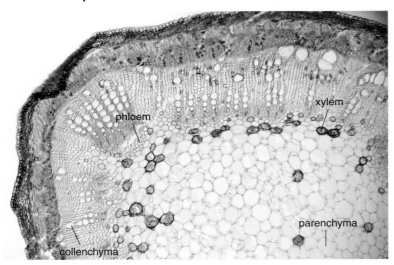

xylem
phloem
parenchyma
collenchyma

Figure 2.36 Use a microscope to observe and draw phloem tubes

Phloem cells contain cytoplasm but have no nucleus. They have a **companion cell** next to them. The companion cell controls the activities of the phloem but does not help with translocation. Substances in phloem vessels are moved by a process that requires energy.

Substances are transported in any direction in the phloem. Glucose made by photosynthesis is converted into a complex sugar called sucrose. Sucrose and amino acids are transported to all tissues in the plant. Different tissues use the substances in different ways.

3 Describe the structure of phloem cells.

4 Compare and contrast phloem and xylem cells.

How are phloem tubes adapted for their function?

Phloem tubes have:

- companion cells with a nucleus and many mitochondria, which provide the energy needed to move substances in the phloem
- limited amounts of cytoplasm and no nucleus to allow efficient movement of substances
- perforated sieve plates to allow the movement of substances through the phloem
- two-way flow of substances so that they are transported all over the plant.

5 Design a diagram to show how the plant uses the glucose made in photosynthesis.

6 Why do phloem tubes have companion cells?

7 Explain the function of the sieve plates.

KEY INFORMATION

Phloem is living, has thin walls and has sieve plates. Xylem is dead, has strengthened walls and no sieve plates.

DID YOU KNOW?

The strings that go up and down the length of bananas are phloem vessels.

MATHS SKILLS

Surface area to volume ratio

Learning objectives:

- be able to calculate surface area and volume
- be able to calculate surface area to volume ratio
- know how to apply ideas about surface area and volume.

In science, we use mathematical skills to help us understand what is happening. Surface area to volume ratio is very important in living things.

Finding the area of a surface

Alveoli provide a large surface area for gas exchange between the air and the bloodstream. If our lungs were smooth on the inside, like balloons, the surface area would be much less and materials would not be exchanged fast enough.

We can calculate surface area in different ways. For a rectangular shape, we multiply the length by the width. For example, an area of skin 4 cm long and 7 cm wide has a surface area of (4 cm × 7 cm) = 28 cm^2.

Living things are not generally made up of regular shapes, such as rectangles, so we have to use other ways of finding the area. One way is to use squared paper and to count how many squares are covered (or are largely covered) by the specimen.

Figure 2.37 Elephants have wrinkled skin to increase their surface area.

KEY INFORMATION

Note that both the length and the width have to be measured in the same units and that the answer is in those units, squared.

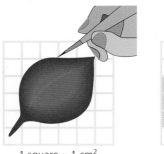

1 square = 1 cm^2 1 square = 1 cm^2

Figure 2.38 Counting squares that are largely covered is (approximately) balanced by not counting squares that are slightly covered.

1. **Estimate the surface area of the leaf in Figure 2.38.**

2. **Calculate the surface area of:**
   a) a piece of tree bark that is 30 cm long and 3 cm wide
   b) a razor shell that is 50 mm long and 8 mm wide.

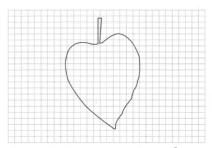

Figure 2.39 One square is 1 cm^2.

Working out the volume

It is easier for warm-blooded animals to keep warm on cold days if their volume is large, but then it is harder for these same animals to lose heat on a hot day. We calculate the volume of a cube by multiplying length by width by height. A die with a side of 2 cm has a volume of (2 cm × 2 cm × 2 cm) = 8 cm^3. Again, all the distances need to be in the same units and the volume is also in those units, cubed.

3 **What is the volume of:**
   a) a science laboratory 10 m wide, 15 m long and 3 m high?
   b) a block of wood 2 cm wide, 3 cm long and 4 cm high?

4 **Finding the volume of a tree branch is tricky, but one way is to immerse the branch in a tank full of water. Suggest how the volume is measured.**

Surface area to volume ratio

In science it is useful to compare the surface area with the volume. We do this by finding the ratio of one compared with the other. To find the **ratio**, divide the surface area by the volume. For example, for a cube with sides 2 cm long:

surface area = 2 cm × 2 cm × 6 = 24 cm^2;
volume = 2 cm × 2 cm × 2 cm = 8 cm^3;
surface area to volume ratio = 24:8 = 3:1

The shape of an organism also affects its surface area to volume ratio. **Spheres** have the smallest surface area compared with their volumes. Many small mammals have a shape that is almost spherical – for example, a mouse – and puppies and kittens curl up into a ball to sleep. As small animals, they want to minimise their surface area to volume ratio so as to minimise heat loss.

5 **Compare how surface area, volume, and surface area to volume ratio change as the size of a cube increases. Use cubes with sides of 1, 2, 3, 4, 6 and 8 cm.**

6 **Imagine the shapes below are animals. Look at A, B and C in Figure 2.40.**

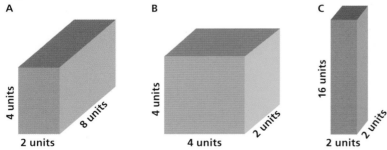

Figure 2.40 What do you notice about the surface area to volume ratios of these 'animals'?
   Which animal will have problems keeping:
   a) cool? Explain your answer.
   b) warm? Explain your answer.

7 **Devise a method to measure the volume of the air you breathe out in one breath.**

Check your progress

You should be able to:

use the word equation to describe photosynthesis	→	recall and use the symbol equation for photosynthesis	→	recall and use the balanced symbol equation for photosynthesis
know that chloroplasts absorb light and convert it to chemical energy	→	describe the use of light and chloroplasts in photosynthesis	→	explain that chloroplasts absorb energy to drive chemical reactions
understand that photosynthesis is an endothermic reaction	→	explain why photosynthesis is an endothermic reaction		
understand that plants respire and photosynthesise	→	explain why plants carry out respiration	→	describe the difference in gas exchange in plants during day and night
name the factors that affect photosynthesis	→	describe how the rate of photosynthesis can be increased	→	explain the effects of limiting factors on photosynthesis
identify the parts of a leaf and their function	→	describe how leaves are adapted for efficient photosynthesis	→	explain how the leaf's structure is adapted for photosynthesis
know the definition of diffusion	→	explain diffusion using the idea of particles	→	explain how substances pass in and out of cells
describe how water travels in plants	→	describe adaptations in xylem and phloem	→	explain adaptations of xylem and phloem
describe experiments on the rate of transpiration	→	describe how different factors affect transpiration	→	explain how different factors affect transpiration
recall that the movement of sugars is called translocation	→	describe how proteins and carbohydrates are transported in plants	→	explain how concentration gradients affect processes

Worked example

Laura is growing tomatoes in her greenhouse.

Laura knows that to increase her crop of tomatoes, she must increase photosynthesis.

1 **Complete the equation for photosynthesis**

$$CO_2 + H_2O \rightarrow C_6H_{12}O_6 + H_2O$$

Water (H_2O) has been correctly identified as a reactant, but has also been given as the product. The equation has not been balanced. The correct answer is:

$$6CO_2 + 6H_2O \rightarrow C_6H_{12}O_6 + 6O_2$$

2 **Laura heats her greenhouse. She wants to know the best temperature to grow the tomatoes in. Look at the graph.**

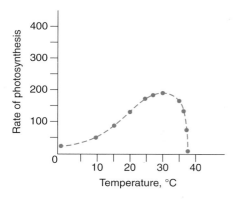

Explain what the graph shows

The rate increases from 5-30°C and then decreases to 0 at 37°C because the enzymes denature.

The pattern has been given (increases and then decreases) and the explanation for the decrease in rate is correct.

An alternative answer would be to explain that the rate increases from 5–30°C as the enzymes gain more energy or collide more often.

3 **What is the best temperature for Laura to grow the tomatoes?**

30°C

Correct answer given.

How does water for photosynthesis get to the leaves?

Through the xylem by osmosis.

One correct answer is given (xylem).

Another answer would be '…by transpiration'.

4 **Laura notices that when she turns the heaters up, she needs to water the plants more often. Explain why.**

More water evaporates.

One correct answer is given

A better answer would be: Transpiration increases because more water evaporates through the stomata.

End of chapter questions

Getting started

1

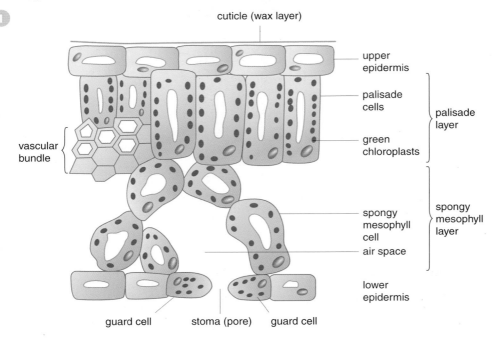

cuticle (wax layer)

upper epidermis

palisade cells

palisade layer

green chloroplasts

vascular bundle

spongy mesophyll cell

spongy mesophyll layer

air space

lower epidermis

guard cell   stoma (pore)   guard cell

a Why do plants photosynthesise?   1 Mark
b What structures inside plant cells absorb energy from light?   1 Mark
c What is the name of the solution used to test leaves for starch?   1 Mark
d What type of cells carry dissolved sugars around the plant?   1 Mark
e What are the products of photosynthesis?   2 Marks

2 Jenna is investigating the effect of light on leaf growth. To do this she needs to calculate the surface area of each leaf.
a Estimate the surface area of this leaf: 1 Mark

1 square = 1 cm²   1 square = 1 cm²

Jenna's results are shown below:

Leaf	South facing tree (most sun)	North facing tree (most shade)
	Area (cm³)	
1	14.0	29.0
2	18.0	19.5
3	13.5	28.0
4	21.5	29.5
5	15.5	23.0
Mean	16.5	

b   Calculate the mean area for the north facing tree.   `1 Mark`
c   Explain the difference in the results.   `2 Marks`

Going further

3   Give two uses of sugars in plants.   `2 Marks`

4   Describe how xylem vessels are adapted to their function.   `1 Mark`

5   Give a disadvantage of growing crops in greenhouses.   `1 Mark`

6   Why does Sally put her tomato plants in a greenhouse?   `4 Marks`

7   A scientist measured water uptake and transpiration in a tree over 18 hours.
The graph shows her data.

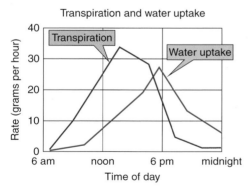

a   Transpiration is highest between 1 and 2pm. When is water uptake highest?   `1 Mark`
b   What conclusion can you make about water loss and uptake in this tree?   `1 Mark`

More challenging

8 a If you are investigating the effect of light intensity on the rate of photosynthesis, what other variables must you control? `2 Marks`

  b Draw a diagram to show how you could measure the effect of light intensity on photosynthesis using equipment you would find in a school laboratory. Show how you would set up the equipment and label each part. `2 Marks`

9 The images below show plants growing in the Sahara Desert in Africa.

  Explain the adaptations that these Saharan plants have that allow them to survive in this extreme environment. `6 Marks`

Most demanding

10 Write the balanced equation for photosynthesis. `2 Marks`

11 Suggest two reasons why loss of water is important for a plant. `2 Marks`

12 Water is transported from the roots to the rest of the plant by transpiration.

  An investigation was carried out over a 12-hour period to compare transpiration in the *Coleus* plant with transpiration in another type of plant called a *Begonia*.

  The mass of each plant was recorded before and after the 12-hour period to find out the effect of transpiration.

  The investigation was repeated five times with the same plants.

The table shows the change in mass of each plant over each 12-hour period.

| Trial | decrease in mass of plant in grams over 12-hour period after being watered | |
	Coleus	Begonia
1	3.7	1.1
2	4.5	1.3
3	2.8	0.8
4	1.6	0.6
5	3.2	1.0

Compare the results from the *Coleus* plant with the *Begonia* plant. Suggest an explanation for the difference in the results.

6 Marks

Total: 40 Marks

MOVING AND CHANGING MATERIALS

MOLECULES MOVE BY DIFFUSION.

- Molecules in gases and liquids move from a high concentration to a lower concentration until all areas have an equal concentration.
- Different factors can affect the rate of diffusion.
- The steepness of a concentration gradient affects the rate of diffusion.

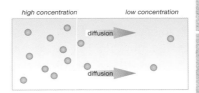

ANIMALS AND PLANTS TRANSPORT SUBSTANCES TO WHERE THEY ARE NEEDED.

- There are two transport systems in plants: xylem transports water up the plant from the roots to the leaves, and phloem transports substances up and down the plant.
- The circulatory system moves substances around the body in the blood.

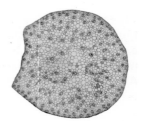

BLOOD IS USED TO TRANSPORT OXYGEN TO BODY TISSUES.

- Oxygen is taken from the lungs to the body. Carbon dioxide is returned from the body to the lungs.
- The function of the heart is to pump blood around the body.
- Blood is made up of red blood cells, white blood cells, plasma and platelets.

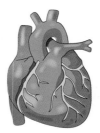

ENZYMES ARE IMPORTANT FOR THE REACTIONS THAT OCCUR IN OUR BODY.

- Enzymes speed up reactions inside the body.
- Different enzymes are used to speed up different reactions.
- The digestive system uses enzymes to digest food.

IN THIS CHAPTER YOU WILL FIND OUT ABOUT:

DO ALL MATERIALS MOVE BY DIFFUSION?

- Water moves from a high concentration of water to a lower concentration of water in living tissues by osmosis.
- The movement of water can affect the turgidity of living cells.
- Some substances that living cells need can be moved against a concentration gradient.

WHY DO SOME ORGANISMS NEED ORGAN SYSTEMS?

- Size affects the ability and efficiency of diffusion alone to supply cells with nutrients.
- Membrane surfaces and organ systems are specialised for exchanging materials to ensure that all body cells get the nutrients that they need.
- An example of an efficient exchange surface is root hair cells in plants.

DO ALL ORGANISMS MOVE MATERIALS IN THE SAME WAY?

- Small organisms do not have specialised organs for gaseous exchange or transport of some materials.
- Fish and mammals have evolved specialised exchange surfaces for gaseous exchange.
- Fish and mammals have specialised transport systems.
- The heart is an efficient pump for the transport system.

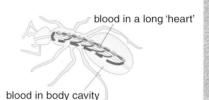

blood in a long 'heart'

blood in body cavity

HOW DO ENZYMES WORK?

- All reactions in cells are controlled by enzymes.
- The lock-and-key and collision theories are used to explain enzyme function and specificity.
- There are specific enzymes in the digestive system; their action is affected by different factors.
- The small intestine is adapted to make it an efficient exchange surface.

Explaining water movement

KEY WORDS

flaccid
osmosis
partially
  permeable
  membrane
turgid

Learning objectives:

- describe how water moves by osmosis in living tissues
- identify factors that affect the rate of osmosis
- explain what the term 'partially permeable membrane' means.

Some examples of substances that are transported in and out of cells by diffusion are carbon dioxide and oxygen. Living cells contain a lot of water. How does water move in and out of cells?

The diffusion of water

Osmosis is the diffusion of water through a **partially permeable membrane**. Partially permeable membranes have tiny holes in them.

A solution is dilute when there is a high water concentration. A dilute sugar solution has a high water concentration (the solvent) and a low sugar concentration (the solute). A solution is concentrated when there is a low water concentration. Most solute particles are too large to pass through the partially permeable membrane, but water molecules can pass through. Osmosis is the diffusion of water molecules from a dilute solution to a concentrated solution through a partially permeable membrane.

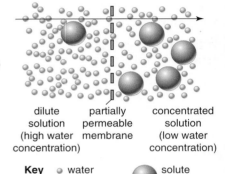

dilute solution (high water concentration)   partially permeable membrane   concentrated solution (low water concentration)

Key ○ water molecule   ● solute

Figure 3.1 Describe the movement of water particles

① How does water move in and out of living cells?

② Suggest why potatoes become soft when boiled.

Osmosis and cells

Look at the diagram.

In (a) the cell has a concentrated solution. Water molecules enter by osmosis from the surrounding dilute solution.

In (b) the cell has a dilute solution. Water molecules move out to the surrounding concentrated solution.

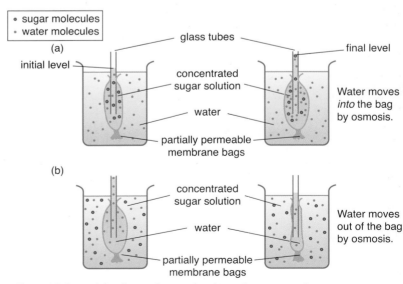

- sugar molecules
- water molecules

(a)
initial level
glass tubes
final level
concentrated sugar solution
water
Water moves *into* the bag by osmosis.
partially permeable membrane bags

(b)
concentrated sugar solution
water
Water moves out of the bag by osmosis.
partially permeable membrane bags

Figure 3.2 Model cells can be used to investigate osmosis

Living cells must balance their water content to work efficiently. Chemical reactions in cells use water. If the cytoplasm becomes concentrated, water enters by osmosis. If the cytoplasm becomes too dilute, water leaves the cell by osmosis.

Problems occur in animal cells when the external solution is more dilute than that inside the cell. Water enters; the cells swell and may burst.

When the external solution is more concentrated than that inside the cell, water moves out by osmosis. The cell shrinks and shrivels.

Plant cells have inelastic cell walls. Water enters the cell by osmosis and fills the vacuole. This pushes against the cell wall, making the cell **turgid.**

If water moves out the cell by osmosis, the vacuole shrinks and the cell becomes **flaccid.**

If too much water leaves the cell, the cytoplasm moves away from the cell wall.

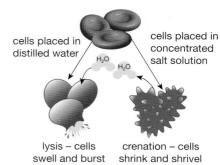

Figure 3.3 What will happen if red blood cells are placed in a very dilute or a very concentrated solution?

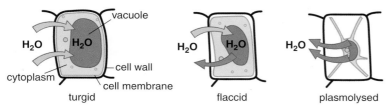

Figure 3.4 Explain the movement of water in these plant cells

3 **How are osmosis and diffusion similar and different?**

4 **Describe how osmosis affects animal cells.**

Explaining osmosis

Water molecules and sugar molecules in a solution move around randomly. When a sugar molecule hits the membrane, it bounces away. When a water molecule hits the membrane, it can pass through a hole to the other side.

In the Figure 3.5, there are more water molecules on the left, so more water molecules can pass through the membrane to the right-hand side than can pass in the opposite direction. The water molecules move both ways, but the net movement is from left to right.

5 **Explain why there are differences in the effects of water on plant and animal cells.**

6 **Explain osmosis.**

REMEMBER!

Osmosis is only the movement of water across a partially permeable membrane. No other molecules move by osmosis.

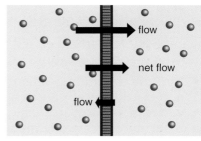

Figure 3.5 Water particles move both ways across the membrane

DID YOU KNOW?

Giant redwood trees grow over 90 m tall. The water pressure in the cells and the strong, lignified cell walls keep the plant upright.

Investigate the effect of a range of concentrations of salt or sugar solutions on the mass of plant tissue

KEY WORDS

osmosis
plasmolysed
partially permeable
 membrane

Learning objectives:

- use scientific ideas to develop a hypotheses
- plan experiments to test a hypotheses
- draw conclusions from data and compare these with hypotheses made.

Osmosis regulates the water content in living cells. Maintaining the internal concentration of a cell is important so that necessary reactions to keep the cell alive can take place.

These pages are designed ❗ to help you think about aspects of the investigation rather than to guide you through it step by step.

Developing a hypothesis

Gill and Aidan are going to investigate the effect of putting onion cells into water and then into salt solution. They will examine the cells through a microscope. If the cell is short of water the cytoplasm comes away from the cell wall. It is said to be **plasmolysed**.

Before they do the investigation they are going to produce a hypothesis. Hypotheses need to be developed using previous knowledge or observations. They know that:

- the concentration gradient between the solutions inside and outside a cell causes water to move by **osmosis**
- water moves towards a higher solute concentration through a **partially permeable membrane**
- the salt solution is more concentrated than the cytoplasm in the onion cells.

REMEMBER!

You choose how to change the independent variable and measure the changes in the dependent variable.

1 When the onion cells are put in water, how will the concentration inside the cell compare with that outside?

2 When the cells are put into salt solution, how will the concentration inside the cell compare with that outside?

3 When the cells are put into salt solution, in which direction will the water move?

4 Suggest a hypothesis for the experiment that Gill and Aidan are about to do.

Planning an investigation

Lily and Ahmed investigated the effect of a range of salt solutions on potato cylinders. They weighed the cylinders, placed them in salt solutions of various concentrations and then reweighed them. Before they reweighed the potato cylinders, they carefully dried them using a paper towel.

5 **What were the independent and dependent variables in the investigation?**

6 **Why did Lily and Ahmed dry the potato cylinders?**

7 **Lily and Ahmed collected these results:**

Concentration of NaCl solution (mol dm⁻³)	Starting mass (g)	Final mass (g)	Change in mass (g)	Percentage change in mass (%)
0.00	15.9	17.0		
0.15	19.2	20.1		
0.30	24.1	23.3		
0.45	20.7	19.2		
0.60	24.1	22.0		
0.75	14.9	13.5		

a   **Complete the table by calculating the change in mass and percentage change in mass of each cylinder**

b   **Are any of the results anomalous?**

c   **How should anomalous results be dealt with?**

d   **Why do Lily and Ahmed use percentage change in mass?**

e   **Suggest what conclusion Lily and Ahmed can make from this data.**

Evaluating the experiment

Lily and Ahmed had written a hypothesis before they did their experiment. They thought that 'the more concentrated the salt solution was that the potato was put in, the greater the loss of mass would be from the potato'.

8 **What conclusion would you draw from their data?**

9 **Was their hypothesis proved or disproved?**

10 **Explain why they got the result that they did.**

Learning about active transport

Learning objectives:

- describe active transport
- explain how active transport is different from diffusion and osmosis
- explain why active transport is important.

Diffusion and osmosis explain how gases and water move down a concentration gradient to enter (and sometimes exit) living things. Minerals are taken up into the root hair cells of a plant by another method.

Active transport

Cells can absorb substances that are at low concentration in their surroundings by **active transport**. These substances move against the concentration gradient, for example, when plants absorb nitrate ions through their root hairs from the soil water. The concentration of nitrate ions in soil water is usually less than the concentration of nitrate ions inside the root hair cells.

Nitrate ions naturally diffuse down their concentration gradient, out of the cell and into the soil, but plants transport the nitrates into their cells using active transport.

Active transport moves substances from a more dilute solution to a more concentrated solution (against a concentration gradient). This requires energy from respiration.

1 Describe how minerals are absorbed by plants.

2 How is active transport different from diffusion and osmosis?

More active transport

Investigations have shown that a plant can absorb different minerals in different amounts. The plant can select which minerals it needs. Look at the graph. Algae absorb a lot of chlorine, but only a little calcium in comparison.

In 1938, scientists discovered that an increase in mineral uptake by a plant happened at the same time as an increase in its respiration rate. This shows that the process needs energy – because the minerals are absorbed against the concentration gradient.

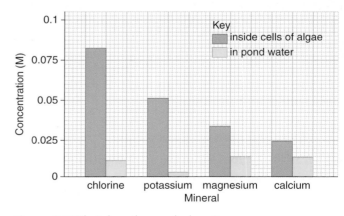

Figure 3.6 What does the graph show?

The greater the rate of cellular respiration, the more energy is available for active transport to happen.

Different cells use active transport for different purposes:

- Marine fish have cells in their gills that can pump salt back into the salty sea water.
- Cells in the thyroid gland take in iodine to use in the production of hormones.
- Cells in the kidney reabsorb sodium ions from urine.
- The villi in the small intestine absorb glucose from the gut (higher sugar concentration) into the blood (lower sugar concentration).
- Crocodiles have salt glands in their tongues that remove excess salt from their bodies.

Figure 3.7 Villi in the small intestine.

3 **Suggest how cells that carry out active transport are adapted to do this.**

4 **Explain why it is important for a plant to transport actively only the minerals that it needs.**

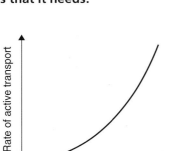

Figure 3.8 Explain the relationship between active transport and respiration

> **KEY INFORMATION**
>
> **Active transport needs energy from respiration to move substances from a low concentration to a high concentration.**

	Substances moved:	Conditions:	Requirements:	Energy needed:
Diffusion	any dissolved substance or any gas	high concentration to low concentration	down a concentration gradient	No
Osmosis	water	dilute solution to concentrated solution	through a partially permeable membrane	No
Active transport	dissolved substances	more dilute solution to more concentrated solution	against a concentration gradient through a partially permeable membrane	Yes

Investigating the need for transport systems

Learning objectives:

- describe how the size of an organism affects the rate of diffusion
- explain how changes in conditions affect the rate of diffusion
- explain the need for exchange surfaces and transport systems using surface area to volume ratio.

> **KEY WORDS**
>
> surface area
> exchange
>   surfaces

Many chemical reactions happen inside living cells. Substances must enter the cell to fuel these reactions and the waste products of the reactions need to be removed. Larger cells have greater chemical activity, so need more substances, for example, nutrients and oxygen, and more substances have to be removed, for example, carbon dioxide.

Size matters

Most cells are no more than 1 mm in diameter. This is because in small cells nutrients, oxygen and waste substances can diffuse quickly in and out. As the volume of a cell increases, the distance increases between the cytoplasm at the centre of the cell and the cell membrane. In cells bigger than 1 mm diameter, the rate of exchange with the surrounding environment may be too slow for diffusion to meet the cell's needs. The cell would probably not survive.

1. **How does the size of a cell affect the chemical activity inside the cell?**

2. **Describe one chemical activity that takes place in cells.**

Figure 3.9 Which will have the faster diffusion rate?

Looking at surface area to volume ratio

The **surface area** of a cell affects the rate at which particles can enter and leave the cell.

The volume of the cell affects the rate of chemical reactions within the cell (how quickly materials are used in reactions and how fast the waste products are made).

Look at the diagram. The cubes represent cells.

- cubes have 6 faces, so the surface area = length × width × 6
- the volume of a cube = length × width × height
- surface area to volume ratio = surface area ÷ volume

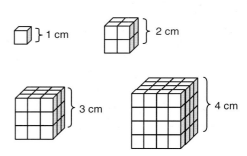

Figure 3.10 Cubes of increasing size

3 Calculate the surface area, volume and surface area to volume ratio (SA:V) for each of the cubes. Cube A has been done for you.

Cube	Number of internal cubes	Surface area (cm²)	Volume (cm³)	SA:V
A	1	6	1	6
B	8			
C	27			
D	64			

> **KEY INFORMATION**
>
> The feathery projections in a fish's gills are used to increase surface area. Water is continually taken in through the mouth and forced over the gills to maintain the concentration gradient.

4 What happens to the surface area and volume as the size increases?

5 What happens to the surface area to volume ratio as the size increases?

6 What does a decrease in surface area to volume ratio make it more difficult to do?

Transport systems and exchange surfaces

With a small surface area to volume ratio, multicellular organisms need surfaces and organ systems specialised for exchanging materials. Once an organism is multicellular and has several layers of cells, the oxygen and nutrients take longer to diffuse in and are all used up by the outer layers of cells. Exchange systems allow transport into and out of all cells for the organism's needs. As well as exchanging substances, cells also have to lose thermal energy fast enough to prevent overheating.

To function efficiently, **exchange surfaces** need the following features:

- a large surface area to maximise exchange
- a thin membrane to provide a short diffusion path
- a method of transporting substances to and from the exchange site, for example, a blood supply to carry nutrients around the body, and lungs to take oxygen into the body.

Organism		SA:V
bacterium		6 000 000
amoeba		60 000
fly		600
dog		6
whale		0.06

7 How are efficient exchange surfaces adapted to carry out their function?

8 Explain why large organisms need transport systems but small organisms do not.

Explaining enzymes

Learning objectives:

- describe what enzymes are and how they work
- explain the lock-and-key theory
- use the collision theory to explain enzyme action.

KEY WORDS

biological catalyst
collision theory
denatured
lock-and-key
    theory
metabolism
optimum

Chemical reactions happen all the time. *Metabolism* describes all the reactions in a cell or the body. The energy transferred by respiration in cells is used for the continual enzyme-controlled processes of metabolism that synthesise new molecules.

A special catalyst

A catalyst is a chemical that speeds up a reaction without being used up itself. This means it can be reused. Different reactions need a different catalyst.

Enzymes are **biological catalysts**. Enzymes catalyse most chemical reactions that happen in cells, for example, respiration, protein synthesis and photosynthesis. Enzymes help to:

- break down large molecules into smaller ones
- build large molecules from smaller ones
- change one molecule into another molecule.

1 **What is an enzyme?**

2 **Explain what enzymes do in a living organism.**

How do enzymes work?

Enzymes are large protein molecules made from folded, coiled chains of amino acids. Each enzyme has a unique sequence of amino acids. The active site, the area that attaches to the substrate (reactant), has a very specific shape that fits its substrate exactly. Enzymes catalse specific reactions due to the shape of the active site.

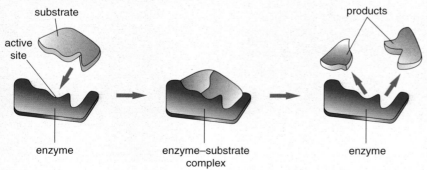

Figure 3.11 The **lock-and-key theory** is a simple explanation of how enzymes work

Only one type of substrate can fit into the active site of an enzyme, like a key fits into a lock. Once the substrate is attached to the active site, it is changed into a product.

If the shape of the active site changes, the enzyme is **denatured** and cannot catalyse the reaction. This is an irreversible change. The enzyme cannot work because the substrate cannot fit into the active site.

Enzyme-controlled reactions are affected by:

- pH
- temperature.

Every enzyme has an **optimum** pH and an optimum temperature. Above or below these levels, the rate of reaction will slow down. Extremes of pH or temperature can denature an enzyme.

Enzymes are chemicals. They are not living and, therefore, cannot be killed. They are denatured.

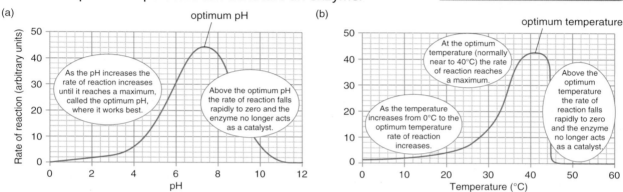

Figure 3.12 Graphs to show how (a) pH and (b) temperature affect the rate of an enzyme-catalysed reaction. What is the optimum pH for the reaction in (a)?

3 **Biological washing detergents contain enzymes. Explain why clothes washed at 60°C will not be as clean as those washed at 40°C.**

4 **How does the lock-and-key theory explain the specificity of enzymes?**

Explaining enzyme action

Look at the graph in Figure 3.13. It shows the amount of product produced during an enzymic reaction.

1 Reactants start to collide; as collisions increase more product is made.

2 Enzyme is added, holding the reactant in place.

3 There is an increase in collisions between the reactants: the rate of reaction also increases.

4 Reactants are being used up and there are fewer collisions, so the rate of reaction decreases and the amount of product made levels off.

This is known as **collision theory**.

5 **Use the collision theory (Figure 3.13) and particle diagrams (Figure 3.11) to explain how enzymes work.**

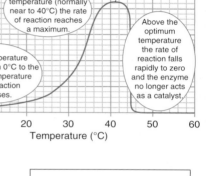

Figure 3.13 How do enzymes affect a reaction?

DID YOU KNOW?

Some bacteria live in hot springs. Their enzymes have an optimum temperature of over 80°C. Other bacteria live on the ocean floor. Their enzymes have optimum temperatures of below 0°C.

REQUIRED PRACTICAL

Investigate the effect of pH on the rate of reaction of amylase enzyme

KEY WORDS

amylase
starch
iodine solution

Learning objectives:

- describe how safety is managed, apparatus is used and accurate measurements are made
- explain how representative samples are taken
- make and record accurate observations
- draw and interpret a graph from secondary data using knowledge and observations.

Amylase is an enzyme that controls the breakdown of starch in our digestive systems. Starch turns a blue-black colour when iodine (an orange solution) is added.

Making accurate measurements and working safely

In this investigation digestion is modelled, using solutions of starch and **amylase** in test tubes, to find the optimum pH required for the reaction. Amylase and **starch** solution are each added to a different test tube and put in a water bath or a beaker of water at 25 °C and left for 5 minutes. The pH buffer solution is added to the amylase, before adding the starch. To carry out this method small volumes of chemicals and enzymes (less than 10 cm³) must be accurately measured. To do this a 10 cm³ measuring cylinder or a 10 cm³ calibrated dropping pipette could be used.

The iodine test is used to indicate the presence of starch. Iodine solution is added to the food being tested and if starch is present the result is a strong blue/purple colour. The investigation involves taking samples from the mixture at timed intervals and adding them to **iodine solution** in a spotting tile. To do this a glass rod or a dropping pipette could be used. The time at which the solution no longer turns a blue-black colour with iodine solution (the iodine solution remains orange) is recorded.

The procedure is repeated for buffer solutions of different pH values.

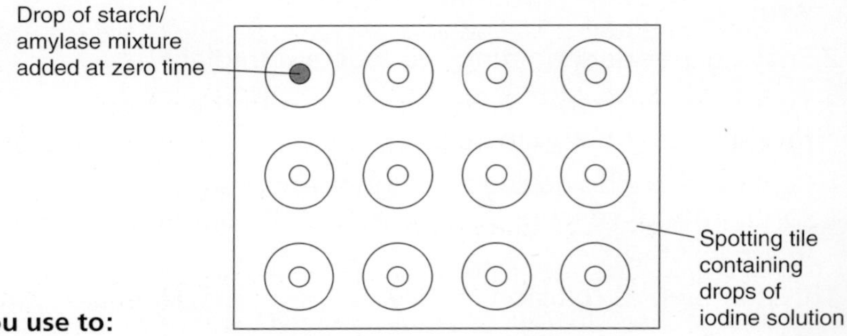

These pages are designed ❗ to help you think about aspects of the investigation rather than to guide you through it step by step.

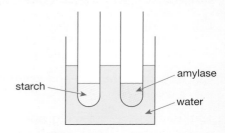

Figure 3.14

Drop of starch/amylase mixture added at zero time

Spotting tile containing drops of iodine solution

Figure 3.15

1 Which piece of apparatus would you use to:

  a   measure small volumes accurately?
  b   add drops of solution to the iodine?

2 To calculate the time taken for the reaction to occur, the teacher has told the students to count up the drops of iodine solution used. Explain how this is an accurate measurement of the time taken.

3 Give two safety precautions that should be taken when doing the investigation.

Planning your sampling

Tom carries out a trial run of the procedure before starting his investigation. Tom found that at pH7 the colour change took 40 seconds to happen. He then decided when to take samples of the solution.

4 What colour will Tom see in the spotting tile when all the starch has been broken down by the amylase?

5 Explain why Tom will need to take continuous samples throughout his investigation.

6 Why did Tom carry out a trial before starting the investigation?

7 From the result of the trial run, suggest suitable time intervals for Tom to take the continuous samples in his investigation.

> **REMEMBER!**
>
> A measurement is repeatable if you repeat the investigation using the same method and equipment and obtain the same results.
> A measurement is reproducible if the investigation is repeated by someone else or by using different equipment or methods, and the same results are obtained.

Using results to draw and interpret a graph

Tom completed the investigation. He sampled the test solutions every ten seconds. This is his data.

pH of solution	Time for colour change to occur (seconds)		
	Test 1	Test 2	Test 3
5	150	160	160
6	70	70	80
7	40	50	40
8	80	80	60
9	130	90	140

Tom decided to draw a graph of the data in his table to make it easier for him to see trends in the data.

8 Should Tom draw a bar chart or a line graph?

9 Tom repeated each pH test three times.

    a Why did he repeat the tests?
    b Tom thought he had an anomalous result. Which result do you think it was?
    c What should Tom do with the anomalous result?

10 Calculate the mean time for the colour change to happen at each pH.

11 Plot the graph of Tom's data.

12 Use the graph to calculate the rate of reaction at each pH.

13 Use your knowledge of enzymes to help you identify patterns, make inferences and draw conclusions.

Learning about the digestive system

Learning objectives:

- identify and locate the organs in the digestive system, and describe their functions
- describe how the products of digestion are absorbed into the body
- explain why the small intestine is an efficient exchange surface.

KEY WORDS

absorption
emulsify
lymphatic system
peristalsis

Our body breaks down the food we eat, so that all cells can receive the nutrients that they need to survive. How does this happen?

The digestive system

The digestive system is a long tube that runs from the mouth to the anus. It consists of several organs working together to digest and absorb food. Each organ is adapted to perform a different function.

Digestion is completed in the small intestine. The soluble food passes through the small intestine wall into the blood. This is called **absorption**. The blood transports the products of digestion to the body cells.

1 Why do we digest food?

2 What is 'absorption'?

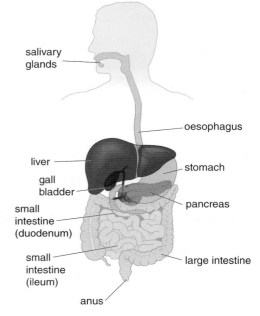

Figure 3.16 The human digestive system

Organs in the digestive system

Each part of the digestive system carries out a specific function.

Part	Adaptation	Function
salivary gland	produces saliva	moistens food; has enzymes to digest food
oesophagus	muscular walls	moves food to the stomach by **peristalsis**
stomach	strong muscles produces hydrochloric acid produces enzymes	mix food kills harmful microbes; provides optimum pH for stomach enzymes digest food
liver	produces bile (alkaline)	neutralises stomach acid stores carbohydrates (as glycogen) emulsifies fats
gall bladder	small bag-like structure	stores bile
pancreas	produces enzymes	provides enzymes to digest food in the small intestine
small intestine (duodenum)	produces enzymes large surface area	digestion of food absorption of soluble food
large intestine	special cells to absorb fluids	absorbs water; solidifies waste
anus	strong muscle	releases waste

Proteins are digested into amino acids. Carbohydrates are digested into glucose. These soluble products pass from the small intestine to the blood capillaries.

Bile **emulsifies** fats into smaller droplets to increase the surface area. The alkaline pH of bile and large surface area of fat droplets increase the rate of fat digestion into fatty acids and glycerol. Fatty acids and glycerol diffuse into the **lymphatic system**.

How the products of digestion are used:

- Glucose is used for respiration and to make new carbohydrates.
- Amino acids are synthesised (built) into proteins in the ribosomes. Protein synthesis is catalysed by enzymes.
- Fatty acids and glycerol are used:
  › for energy
  › to build cell membranes
  › to make hormones.

3 Explain the difference between emulsification and digestion.

4 Suggest why fatty acids are not absorbed into the blood.

5 Why is digestion important?

A special exchange surface

The small intestine is an effective exchange surface because:

- at about 7 m long, there is ample time for absorption of soluble molecules as food travels along
- it has a very thin, permeable membrane for easy diffusion
- there are many small projections called villi, with tiny projections called microvilli, increasing the surface area for absorption
- blood capillaries transport molecules away, maintaining the concentration gradient for diffusion.

Between meals, the concentration of dissolved food molecules in the blood can be higher than in the intestine. At this time, the molecules are moved into the blood using active transport.

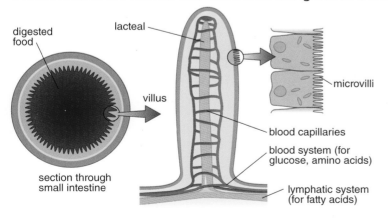

Figure 3.18 Villi help make the small intestine an effective exchange surface

6 Explain why the small intestine is an effective exchange surface.

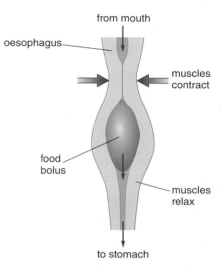

Figure 3.17 Explain how food is moved along the digestive system by peristalsis.

COMMON MISCONCEPTION

Remember that emulsification breaks fats into smaller drops but enzymes *digest* fats into fatty acids and glycerol.

DID YOU KNOW?

The flattened surface of the small intestine is about 250 m² (the size of a tennis court).

Explaining digestion

Learning objectives:

- describe how physical digestion helps to increase the rate of chemical digestion
- name the sites of production and action of specific enzymes
- interpret data about digestive enzymes.

In 1822, Alexis St Martin was shot. Dr William Beaumont saved his life, but the wound did not heal. Dr Beaumont used the hole to watch what happened in Alexis's stomach after he ate food!

Figure 3.19 Pieces of meat in the stomach were much smaller after a few hours. What had happened?

Physical or chemical digestion?

In the mouth, teeth are used to cut and grind food into smaller pieces. This is physical digestion. It allows food to pass through the digestive system more easily. It increases the surface area of the foods to speed up chemical digestion.

Muscles in the stomach wall also help to physically digest food by squeezing it. Muscles in other parts of the digestive system also squeeze the food to keep it moving by peristalsis.

Even small pieces of food cannot pass into the blood. Enzymes are produced in some parts of the digestive system. Enzymes break down food into very small soluble molecules, so they can be absorbed by the blood. This is chemical digestion.

1   **Why is physical digestion important?**

2   **How is chemical digestion different from physical digestion?**

Digestive enzymes

Most enzymes work inside cells, but the digestive enzymes work outside cells. They are produced by cells in glands and in the lining of the gut. The enzymes pass into the gut to mix with the food. There are three groups of enzymes in digestion:

- **Carbohydrases** break down carbohydrates into simple sugars. Amylase is a carbohydrase that breaks down starch.
- **Proteases** break down proteins into amino acids.
- **Lipases** break down fats into fatty acids and glycerol.

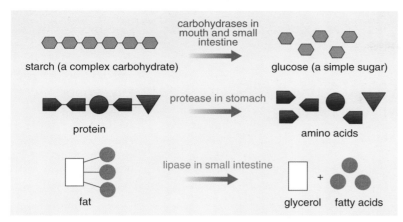

Figure 3.20 Why is chemical digestion important?

Different parts of the gut produce different enzymes, as shown in the table.

Enzyme	Site of production	Reaction
carbohydrase	salivary glands, pancreas, small intestine	carbohydrates → simple sugars
protease	stomach, pancreas, small intestine	proteins → amino acids
lipase	pancreas	lipids (fats) → fatty acids + glycerol

3 **Where are proteases produced and what reaction do they catalyse?**

4 **Suggest why digestive enzymes do not work inside cells.**

COMMON MISCONCEPTION

Remember that bile is not an enzyme and it does not *digest* fat molecules. Bile *emulsifies* fat droplets to increase their surface area to speed up their digestion by lipase enzymes.

DID YOU KNOW?

Your stomach produces about 3 litres of hydrochloric acid a day.

Speeding up digestion

Enzymes are affected by temperature and pH. Our body temperature is kept nearly constant at 37 °C. That is the optimum temperature of enzymes in our body.

Enzymes also have an optimum pH. Salivary amylase in the mouth works best at pH 6·7–7·0. The protease enzyme in your stomach works best in acidic conditions, so the stomach produces hydrochloric acid. The stomach also produces mucus which coats the stomach wall to protect it from the acid and enzymes. Protease enzymes that are made in the pancreas and small intestine need alkaline conditions. Bile, made in the liver, neutralises the stomach acid so that these enzymes can work effectively.

5 **Look at Figure 3.21a. What does this graph tell you about pepsin and trypsin enzymes?**

6 **Explain Figure 3.21b using the collision theory.**

(a)

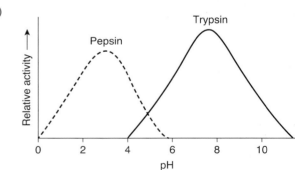

(b)

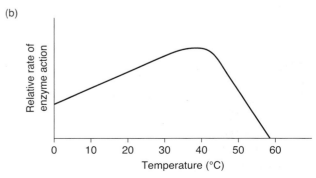

Figure 3.21

REQUIRED PRACTICAL

Use qualitative reagents to test for a range of carbohydrates, lipids and proteins

KEY WORDS

qualitative
reagents
carbohydrates
Benedict's test

Learning objectives:

* suggest appropriate apparatus for the procedures
* describe how safety is managed and apparatus is used
* describe how accurate measurements are made
* interpret observations and make conclusions.

Foods can contain carbohydrates (starch and sugars), protein, fat and small amounts of minerals and vitamins.

Describing how apparatus is used and working safely

Qualitative tests can be used to test for the presence of different food groups using ground up food. The food is added to distilled water, stirred and filtered.

The **Benedict's test** is used to test for sugars: the filtrate is transferred to a test tube, Benedict's **reagent** is added and the tube is placed in a water bath of boiling water for five minutes to see if a colour change occurs.

To test for lipids, the filtrate is added to a test tube and shaken gently with Sudan III stain.

The Biuret test is used to test for protein.

> These pages are designed to help you think about aspects of the investigation rather than to guide you through it step by step.

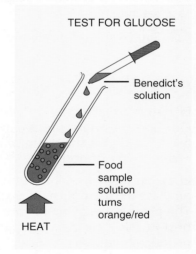

TEST FOR GLUCOSE

Benedict's solution

Food sample solution turns orange/red

HEAT

Figure 3.22

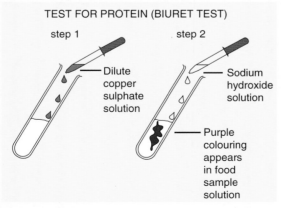

TEST FOR PROTEIN (BIURET TEST)

step 1

Dilute copper sulphate solution

step 2

Sodium hydroxide solution

Purple colouring appears in food sample solution

Figure 3.23

1 Which piece of apparatus could be used to:

a grind the food up?

b add drops of indicator solution?

c measure small volumes of liquids

2 Why is the ground up food stirred with the distilled water and then filtered?

3 Describe the test that could be carried out to show if protein was present in the food.

4 State two safety precautions that should be taken for each test.

Making and recording observations

Ravi is going to test three different foods to see which food groups, including starch, are present. She needs to design a table for the results that all the data will fit into. It needs to be fully labelled and include units. Ravi is going to test each food for each food group twice and is also going to use a control tube, using water.

5 Why is Ravi going to repeat the test for each food group?

6 Why is a control tube set up?

7 Construct a table that Ravi can use that will fit all the data she is going to collect.

Interpreting observations

Food samples A, B and C were tested for different food groups.

Complete this table to show the initial colour of the reagent and the colour change of a positive test.

Food group	Colour change of a positive test	
	Initial colour	Final colour
Glucose		
Protein		
Fat		

In the first test, the foods were boiled with Benedict's reagent, in the second test Biuret reagent was added and in the third test Sudan III reagent was added.

Food sample	Colour with Benedict's reagent	Colour with Biuret reagent	Colour with Sudan III reagent
A	brick red	blue	red layer at top of tube
B	blue	purple-pink	no red layer
C	brick red	purple-pink	no red layer

8 Which food/foods contained these different food groups? Explain your answers.

a   No protein
b   Glucose
c   Lipids

Looking at more exchange surfaces

KEY WORDS

gills
spiracles
trachea

Learning objectives:

- identify the structures responsible for gas exchange in fish, amphibians and insects
- describe the adaptations of different gas exchange surfaces
- explain the gas exchange surfaces in amphibians.

Gas and solute exchange surfaces in organisms are adapted to maximise their effectiveness. The more complex the organism, the more complex the exchange surface to supply the organism with what it needs to survive.

Examples of more exchange surfaces

Tadpoles and worms that live in water use their skin and external **gills** for gas exchange. The gills are 'feathery' projections that provide a large surface area for gas exchange with the water. Gas exchange in adult amphibians takes place mainly through their skin, but they also use lungs.

Insects have no transport system, so gases are taken direct to the respiring tissues in their body. An insect has tiny holes called **spiracles** along the side of its body, which open into small tubes called tracheae. The end of each **trachea** contains a small amount of water and connects to the body cells. Gases diffuse into the cells through the water.

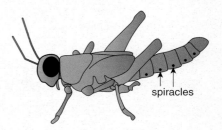

Figure 3.24 Insects use trachea for gas exchange

Fish have very thin gills that are covered with protective muscular flaps. Water is continually taken in through the mouth, forced over the gills and then leaves the body through the flaps.

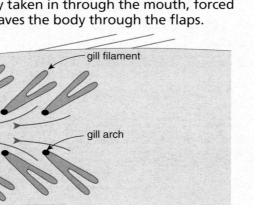

Figure 3.25 Fish use gills for gas exchange

1 What structures do fish, insects and amphibians use for gas exchange?

2 How does gas exchange in tadpoles differ from that in fish?

Adaptations in exchange surfaces

Gas exchange happens at a respiratory surface. This is a membrane that separates the external environment and the interior of the body. Unicellular organisms use their cell membrane for gas exchange, but for large organisms the exchange surface is part of specialised organs like lungs and gills, or leaves in plants.

All respiratory exchange surfaces have:

- a large surface area
- a thin permeable membrane
- a moist exchange surface.

Many exchange surfaces also have a ventilation system and an efficient transport system to keep the diffusion gradients as high as possible to maximise exchange.

Look at the exchange surfaces in the table.

> **KEY INFORMATION**
>
> All exchange surfaces have a large surface area, thin membranes and are moist. Many have an efficient transport system.

Exchange surface		Adaptation
skin		thin membrane for diffusion moist for dissolving gases organisms usually have a large SA:V
tracheae	 body wall — tracheole air → spiracle trachea — body cell (reinforced with rings of chitin which collapse)	spiracles can close to prevent evaporation, to keep exchange surfaces moist tracheae have many branches to increase surface area insects pump air in and out of tracheae tracheae are stiffened to prevent collapse short diffusion distance
gills	 gill cover — gill filament water in water out — gill bar **one gill**	feathery projections to increase the surface area very thin walls for diffusion water pumped over the gills
lungs	**gas exchange in an air sac** bronchiole oxygen diffuses in — carbon dioxide diffuses out alveolus (plural: alveoli) red blood cell — blood capillary	alveoli have thin membrane millions of alveoli to provide large surface area moist for dissolving gases constant ventilation to maintain the concentration gradient

3 Describe the adaptations of different gas exchange surfaces.

4 Suggest why adult amphibians sometimes use their lungs as well as their skin for gas exchange.

5 Explain how ventilation maintains the concentration gradient when you breathe out.

> **DID YOU KNOW?**
>
> Some aquatic insects can submerge for long periods. They carry a bubble of air with them and breathe from it.

Learning about plants and minerals

Learning objectives:

- describe how mineral ions from the soil help plants to grow
- explain how root hair cells are adapted for efficient osmosis
- describe the function of different mineral ions in a plant.

Like animals, plants also need to exchange materials. Leaves are exchange surfaces that are specialised for photosynthesis. Plants also need minerals and water. The root hair cells are adapted for the uptake of these materials.

Roots and minerals

Root hair cells are found on plant roots, just behind the root tip. Root hair cells help to anchor the plant, and absorb water and mineral ions from the soil.

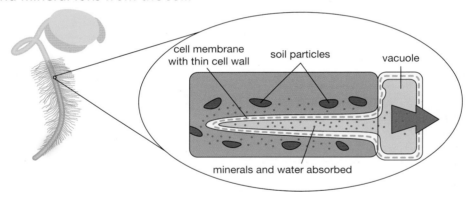

cell membrane with thin cell wall | soil particles | vacuole

minerals and water absorbed

Figure 3.26 Root hair cells are specialised exchange surfaces

Fertilisers containing minerals can be added to the soil to improve plant growth. Fertilisers contain **nitrates**, **phosphates**, potassium and magnesium. It is the root hairs that absorb the water and the minerals.

Each root hair cell has a long thin exchange surface reaching out between the soil particles. Water is absorbed by osmosis and mineral ions are absorbed by active transport.

1 **What are the functions of root hairs?**

2 **How do plants absorb the minerals they need?**

Root hair cells

Plants need water to maintain the shape of their cells and for photosynthesis. Minerals are needed to make proteins and other chemicals to help the plant grow.

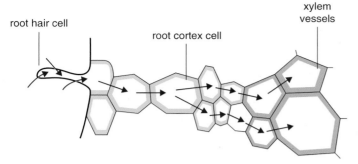

Figure 3.27 Roots are adapted for the transport of water and minerals

Root hair cells are an efficient exchange surface for osmosis because they:

- provide a large surface area for absorption of water
- have no cuticle, just a thin membrane to allow absorption
- have a thin cell wall to reduce the distance for osmosis
- have a large permanent vacuole to absorb as much water as possible
- are close to the xylem, so that materials can be moved around the plant.

These adaptations mean that a lot of water and mineral ions can get into the cell very quickly and then pass around the plant efficiently.

3 **How are root hair cells adapted for osmosis?**

4 **Why do plants need minerals?**

Looking at minerals

Mineral ions are only found in low concentrations in the soil, compared with the concentration inside the root hair cell. They need to be taken in by active transport. Root hair cells contain many mitochondria to supply the energy needed for this.

The table lists some of the minerals that you might hear about. You only need to remember about nitrates and magnesium for your examinations.

Mineral	Use in the plant
nitrates, containing nitrogen (N)	to make amino acids for protein synthesis
phosphates, containing phosphorus (P)	in respiration to make DNA and new cell membranes
potassium (K)	in respiration in photosynthesis to make enzymes
magnesium (Mg)	needed to make chlorophyll for photosynthesis

5 **Why are enzymes important in plants?**

6 **A farmer adds fertiliser to the soil to make his plants grow better. It is a warm, windy day. Describe how the plants take in and transport the fertilisers, and why this happens more quickly on a windy day. (Hint: look at Chapter 2, 'Photosynthesis'.).**

Investigating how plants use minerals

KEY WORDS
culture solution
deficiency

Learning objectives:

- describe why plants need different mineral ions
- explain the effects of mineral deficiencies on plant growth
- explain the importance of fertilisers.

Plants need minerals to grow and stay healthy. What happens if plants do not get the minerals they need?

Fertilisers and plant growth

Many years ago, people cut down trees and burned them so that they could plant crops. After a few years, the crops had used up all the minerals in the soil, and the plants did not grow well. The farmers would then cut down more trees to have new land to grow their food.

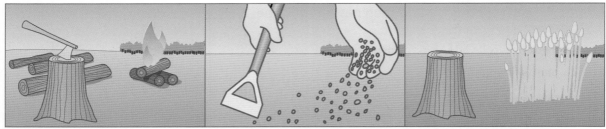

trees cut down and burned

wheat seeds planted in the warm ash and covered with soil

crops harvested after six weeks

Figure 3.28 This is the 'slash-and-burn' method of farming

Over the years, farmers found they could replace soil minerals using fertilisers. This means the same crop can be grown in the same field for several years. Fertilisers help increase the yield of crops. If a plant does not get enough of any of the minerals that it needs, its growth is poor.

1 Why are fertilisers important?

2 Why do gardeners add fertiliser to water before watering their plants?

Investigating minerals

Plants can be grown, without soil, in special aerated solutions to investigate the effect of a lack of a mineral. A control plant is grown with all the essential minerals a plant needs. Other plants are grown in different **culture solutions**, each one missing only one mineral at a time. The minerals investigated are nitrates, phosphates, potassium and magnesium.

A mineral **deficiency** happens when a plant cannot get enough of a mineral from the soil or culture solution for healthy

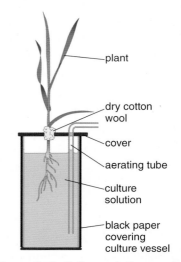

plant

dry cotton wool

cover

aerating tube

culture solution

black paper covering culture vessel

Figure 3.29 Culture solutions contain known amounts of specific minerals

growth. Plants only need minerals in small amounts. Mineral deficiencies are easy to analyse and treat.

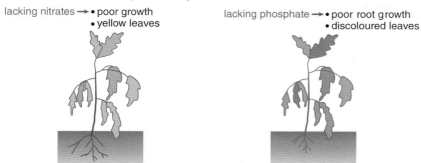

lacking nitrates → • poor growth
• yellow leaves

lacking phosphate → • poor root growth
• discoloured leaves

lacking potassium → • poor fruits and flowers
• discoloured leaves

lacking magnesium → • yellow leaves

Figure 3.30 Symptoms of mineral deficiencies in a plant

3 **How would you know when a plant has a mineral deficiency?**

4 **Describe the effect of:**

a a magnesium deficiency

b a phosphate deficiency.

Food production

Intensive farming methods use machines and chemicals to produce as much food as possible, quickly and cheaply. The use of fertilisers has increased crop yields but the area of land used has stayed the same or decreased.

Farmers use two types of fertilisers:

- inorganic fertiliser
- organic fertiliser.

Inorganic fertilisers are man-made. They come from concentrated sources of minerals, are used in small amounts and can be absorbed immediately. They do not smell, are easy to apply and store.

Organic fertilisers come from animal or plant matter. They take time to break down in the soil and slowly release the minerals. They can reduce soil erosion and improve water retention in the soil.

5 **Describe the similarities and differences between inorganic and organic fertilisers.**

6 **What are the advantages of using fertilisers in farming?**

KEY INFORMATION

Each mineral has different deficiency symptoms that can be used to identify why a plant is not growing well.

DID YOU KNOW?

In hydroponics systems, the plant roots are in water that contains the correct amounts of fertiliser and oxygen.

Learning about the circulatory system

KEY WORDS

double circulation
lumen

Learning objectives:

- identify the parts of the circulatory system
- describe the functions of the parts of the circulatory system
- explain how the structure of each part of the circulatory system relates to its function.

Very small animals have no need for a transport system. But as the distance from the internal cells to the animal's surface increases, a transport system becomes vital to supply cells with oxygen and nutrients, and to take away waste materials.

Transport systems

Insects have an open circulation system with no blood vessels. The blood flows slowly round the body cavity.

In closed circulatory systems, blood flows in vessels. There are two types of closed circulatory systems:

- a single circulation (for example, in fish); the blood flows in one circuit around the body
- a **double circulation** (for example, in humans); the blood flows in two circuits around the body:
    - › from the heart to the lungs
    - › from the heart to the rest of the body.

double circulatory system

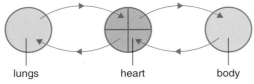

lungs   heart   body

Figure 3.31 Why does blood travel to the lungs?

Blood is pumped out of the heart into arteries under high pressure. As the blood flows from the arteries, through capillaries and then veins, the pressure decreases.

1. What is a double circulation system?
2. Why do closed circulation systems need a heart?

How does the system work?

Each type of blood vessel is adapted to carry out slightly different functions.

Arteries	Veins	Capillaries
thick, elastic wall / small lumen	thin wall / large lumen / valve	wall is one cell thick
carry blood from the heart	carry blood to the heart	carry blood from arteries to veins
blood under high pressure with a pulse	blood under low pressure, flows smoothly	pressure falls and pulse disappears
thick walls, not permeable	thinner walls, not permeable	walls are one cell thick and permeable
small **lumen**	large lumen	
no valves	valves along their length prevent backflow of blood	no valves
carry oxygenated blood (except pulmonary artery)	carry deoxygenated blood (except pulmonary vein)	blood slowly loses its oxygen

Blood flows from the heart to arteries, arterioles (small arteries), capillaries, venules (small veins), veins and then back to the heart.

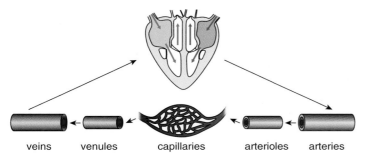

veins   venules   capillaries   arterioles   arteries

Figure 3.32 The sequence of blood flow

3 **Describe how blood is transported around the body.**

Explaining adaptations

The advantages of the double circulation system include:

- blood pressure is higher, especially to the body
- there is a higher blood flow to body tissues
- oxygenated blood is separate from deoxygenated blood.

Adaptations of the blood vessels include:

- the thick elastic walls of arteries withstand the high pressure of the blood
- capillary networks have a large exchange surface area
- the thin permeable walls of capillaries mean that substances have only a short distance to diffuse
- large lumen in the veins gives the least flow resistance
- valves in the veins prevent the backfow of blood.

4 **Explain how the structure of blood vessels is adapted to their function.**

5 **Why is a double circulation an advantage to an active animal?**

KEY INFORMATION

Arteries carry blood away from the heart.

DID YOU KNOW?

Every cell in your body is 0.05 mm away from a capillary, so substances have a very short distance to diffuse.

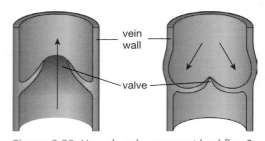

vein wall

valve

Figure 3.33 How do valves prevent backflow?

Exploring the heart

Learning objectives:

- describe the structure and function of the heart
- identify the functions and adaptations of the parts of the heart
- explain the movement of blood around the heart.

KEY WORDS

aorta
atrium (plural: atria)
coronary artery
pacemaker
vena cava
ventricle

Active or larger animals need a transport system with a pump. This is to maintain a constant supply of materials to meet the demand of all cells throughout the body, and to remove waste.

The heart

The heart is made of muscle. Heart muscle continually contracts and relaxes. It uses a lot of energy. Heart muscle receives oxygen and glucose for respiration from the blood brought by the **coronary artery**.

The heart has two pumps (a double circulation) that beat together about 70 times every minute of every day.

Each pump has an upper chamber (**atrium**) that receives blood and a lower chamber (**ventricle**) that pumps blood out. Both **atria** fill and pump blood out at the same time, as do both ventricles. The natural resting heart rate is controlled by a group of cells located in the right atrium that act as a **pacemaker**.

Blood from the lungs contains oxygen and enters the heart at the left atrium. It passes into the left ventricle and is pumped out to the body.

Blood from the body contains very little oxygen and enters the heart at the right atrium, passes into the right ventricle and is pumped to the lungs to be oxygenated.

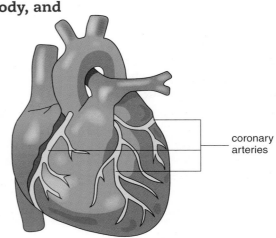

coronary arteries

Figure 3.34 Why does the heart need an oxygen supply?

1 What is the function of the heart?

2 Describe how blood flows through the heart.

The parts of the heart

The heart has four main blood vessels:

- The pulmonary vein transports oxygenated blood from the lungs to the left atrium.
- The **aorta** (main artery) transports oxygenated blood from the left ventricle to the body.
- The **vena cava** (main vein) transports blood from the body to the right atrium.

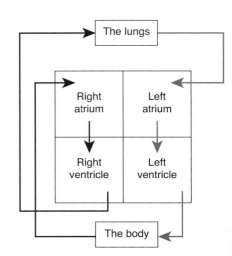

Figure 3.35 Blood flow through the heart

- The pulmonary artery transports deoxygenated blood from the right ventricle to the lungs.

Ventricles have thicker walls than atria because they pump blood further. The left ventricle pumps blood around the body. It has a thicker wall than the right ventricle, which only pumps blood to the lungs. Valves between the atria and the ventricles prevent the backflow of blood. They open to let blood through and then shut.

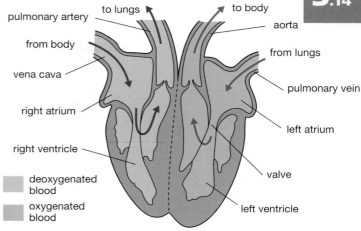

Figure 3.36 Why does the heart need two pumps?

3. **Describe the functions of the different chambers of the heart.**

4. **Describe how the atria and ventricles move blood through the heart.**

Explaining blood flow

In the second century, Galen thought blood was made in the liver. He said it flowed round the body, but was used as fuel for the muscles. Galen also thought there were holes in the septum, which allowed blood to flow from one side of the heart to the other. In the seventeenth century, Galen's ideas were disproved by William Harvey, who explained heart structure and described the blood vessels.

KEY INFORMATION

Atria receive blood, ventricles pump it out.

DID YOU KNOW?

The complete cardiac cycle normally takes 0.8 seconds.

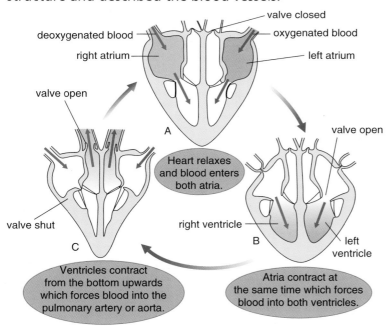

Figure 3.37 The cardiac cycle

5. **Explain the sequence of contractions and valve openings as blood passes through the heart.**

6. **If a coronary artery supplying the left ventricle becomes blocked, what effect does this have on the functioning of the heart? Explain your answer.**

Studying blood

Learning objectives:

- identify the parts of the blood and their functions
- explain the adaptations of red blood cells
- explain how red blood cells and haemoglobin transport oxygen efficiently.

Effective exchange systems in animals need a method of transporting substances to and from the exchange site, for example, mammals have an efficient circulatory system.

What is blood?

Blood is a tissue. It is a mixture of cells, solutes and a liquid. The liquid part of blood is **plasma**. It is straw coloured.

Red blood cells, white blood cells and platelets are suspended in the plasma. Each part of the blood has a specific function:

- Plasma transports substances around the body, for example, carbon dioxide.
- Red blood cells carry oxygen from the lungs to body cells.
- White blood cells help to protect the body against infection.
- Platelets are cell fragments which help the clotting process at wound sites.

There are millions of red blood cells in the plasma. This is why blood looks red.

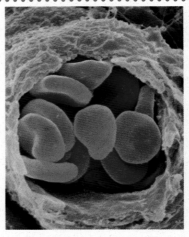

Figure 3.38 Why does blood look red?

1 Look at the scanning electron micrograph of blood (Figure 3.38). Identify the red blood cells, white blood cells and platelets.

2 What is the function of:

  a red blood cells

  b white blood cells?

Looking closer

The blood parts can be separated by spinning them very fast in a machine called a centrifuge. About 55% of blood is plasma; plasma consists of roughly 90% water and 10% solutes. It is very important because it transports many substances, for example:

- hormones
- antibodies
- nutrients, such as glucose, amino acids (proteins), minerals and vitamins
- waste substances, like carbon dioxide and urea.

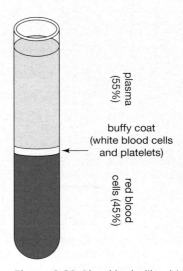

plasma (55%)

buffy coat (white blood cells and platelets)

red blood cells (45%)

Figure 3.39 Blood looks like this if it is separated in a centrifuge

Red blood cells transport oxygen from the lungs to tissues all over the body. Blood is able to transport oxygen efficiently because in 1 mm³ of blood there are about 5 million red blood cells. Red blood cells:

- are tiny, allowing them to pass through the narrow capillaries
- have a biconcave disc shape, giving them a large surface area to volume ratio which increases the efficiency of diffusion of oxygen into and out of the cell, and reduces the diffusion distance to the centre of the cell
- contain **haemoglobin** which binds to oxygen to transport it from the lungs to the body tissues
- have no nucleus, increasing the space available for haemoglobin.

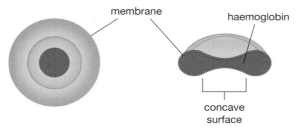

membrane   haemoglobin

concave surface

Figure 3.40 Red blood cells are adapted to carry oxygen

3 Describe the role of plasma.

4 How are red blood cells adapted to their function?

How the blood carries oxygen

Haemoglobin binds with oxygen at high concentration to form a bright red compound called **oxyhaemoglobin**. The bonds between the haemoglobin and oxygen are weak, and oxyhaemoglobin dissociates to haemoglobin and oxygen in low oxygen concentrations.

oxygen + haemoglobin $\rightleftharpoons$ oxyhaemoglobin

5 Describe how haemoglobin transports oxygen.

6 Use SA:V to explain how red blood cells are adapted to their function.

7 Sickle cell anaemia is a serious inherited blood disorder where the red blood cells develop abnormally and contain defective haemoglobin. Explain why somebody with sickle cell anaemia is likely to feel very tired and breathless while exercising.

KEY INFORMATION

Red blood cells transport oxygen; plasma transports many substances, including carbon dioxide.

DID YOU KNOW?

Crabs have blue blood, some worms have green blood and starfish have clear or pale yellow blood!

Investigating gas exchange

Learning objectives:

- identify the parts of the human gas exchange system and know their functions
- explain how gas exchange occurs in humans
- explain the adaptations of the gas exchange surfaces.

Our body needs a constant supply of oxygen for cellular respiration. Mammals have evolved a specialised exchange surface to provide a continuous supply of oxygen and to remove carbon dioxide.

The breathing system

The lungs are in the thorax. They are surrounded by the ribcage to protect them. Between the ribs are the intercostal muscles. These help to **ventilate** (move air into and out of) the lungs. Gas exchange happens in the **alveoli**.

Under the lungs is the muscular diaphragm. This also helps to ventilate the lungs.

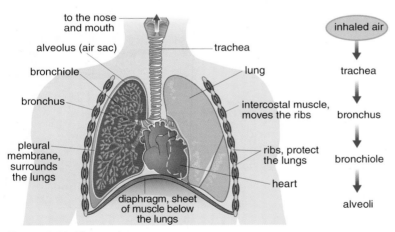

Figure 3.41 The respiratory system

Air is filtered, warmed and moistened in the mouth and nasal passages. It passes into the trachea, through one of the bronchi into the lungs. In the lungs, air passes through the many bronchioles until it reaches an **alveolus** (air sac). Oxygen then passes across the specialised cell membrane of the alveolus into the blood.

1 Describe how air reaches the alveoli.

2 What is gas exchange and where does it happen?

Gas exchange

Oxygen is used to release energy from food during respiration. Carbon dioxide is a waste product of the reaction. Gas exchange is:

- taking in oxygen
- releasing carbon dioxide.

Gas exchange happens in the alveoli. Blood transports the gases to and from the surface of each alveolus through the capillaries that cover the alveoli.

Air entering the alveoli has a greater oxygen concentration than the deoxygenated blood flowing through the lungs. This steep concentration gradient allows efficient diffusion to happen.

Deoxygenated blood contains a greater concentration of carbon dioxide, so it diffuses out of the blood and into the alveoli to be breathed out.

3 Explain how gas exchange happens in the alveoli.

4 What is the difference between breathing and respiration?

COMMON MISCONCEPTION

Remember that breathing is gas exchange and respiration is the release of energy from food.

DID YOU KNOW?

The lungs have over 300 million alveoli and the left lung is slightly smaller than the right, so there is room for your heart.

Why are alveoli good exchange surfaces?

Alveoli are efficient exchange surfaces. They have evolved adaptations to ensure maximum gas exchange:

- They are tiny spheres. Smaller spheres have a much larger surface area than larger ones, resulting in efficient diffusion of gases.
- The exchange surface at the alveoli wall is very thin (just one cell thick). This means that the diffusion distance for the gases is very short.
- Each alveolus is surrounded by blood capillaries to ensure a good blood supply. Oxygen is constantly moved into the blood and carbon dioxide is constantly taken to the lungs to be removed. This means that gas exchange happens at the steepest concentration gradients possible.
- The alveoli surfaces are moist. Gases dissolve to allow efficient diffusion across the exchange surface.

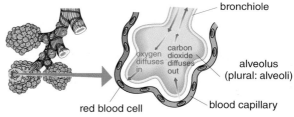

Figure 3.42 Gas exchange in the alveoli

5 Describe and explain how alveoli are adapted to their function.

6 Explain how cells located deep in our body can receive the oxygen they need and dispose of carbon dioxide efficiently.

Learning about coronary heart disease

Learning objectives:

- identify the causes and symptoms of coronary heart disease
- describe possible treatments of coronary heart disease
- evaluate the possible treatments of coronary heart disease.

Heart disease is one of the main causes of death in the UK. Cholesterol is made in the liver and transported in the blood. High levels of cholesterol are linked to heart disease.

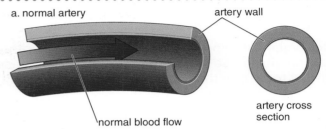

a. normal artery / artery wall / normal blood flow / artery cross section

Heart problems

In **coronary heart disease**, fatty material builds up inside the coronary arteries. Blood flow is reduced and less glucose and oxygen reach the heart for respiration. Less energy is available for the heart to contract. If cells are starved of nutrients, they can die and a heart attack may happen. Factors that contribute to coronary heart disease include genetic factors, gender, age, diet and if they smoke or not.

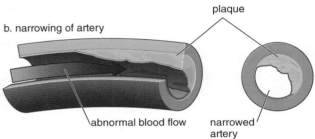

b. narrowing of artery / plaque / abnormal blood flow / narrowed artery

Heart valves prevent the backflow of blood. They can become faulty due to heart attack, infection or old age. Faulty valves may not open fully or can leak. Symptoms of leaky valves include:

Figure 3.43 How does coronary heart disease affect blood flow?

- tiredness and lack of energy
- breathlessness.

1. Stent pushed through catheter into position using X-rays.

1 Describe the symptoms of coronary heart disease.

2 What causes coronary heart disease?

Treatments for heart problems

- Some people cannot control their heart rate. **Artificial pacemakers** can be fitted under the skin. A wire is passed from a vein to the right atrium. This sends electrical impulses to the heart to control heartbeat.
- Faulty heart valves can be replaced using biological (from humans or other mammals) or mechanical valves.
- High cholesterol levels are treated with drugs called **statins** which stop the liver producing as much cholesterol. Patients can also change their diet to help reduce cholesterol.
- **Stents** are used to treat narrow coronary arteries. If the coronary artery is too damaged, bypass surgery is used. A vein is transplanted from the leg to bypass the blockage.

2. Balloon inflated, expanding stent.

3. Catheter taken out, leaving stent to open up artery.

Figure 3.44 Inserting a stent

When heart failure occurs, the only effective treatment is a donor heart transplant. An artificial heart can be used in the short term, while the patient is waiting for a heart transplant, or to allow the heart to rest to help recovery.

3 How is coronary heart disease treated?

4 What are artifical valves and what do they do?

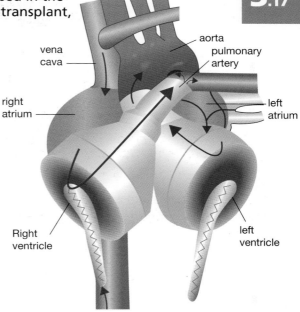

Figure 3.45 An artificial heart

Evaluating heart treatments

All treatments help keep patients alive.

Treatment	Advantage	Disadvantage
artificial valves	no rejection	can damage red blood cells patient needs anti-clotting drugs
biological valves	red blood cells not damaged	valves can harden and need replacing
stents	little risk	fatty deposits can rebuild
bypass surgery	no rejection	major surgery
statins	reduce cholesterol	possible side effects, e.g. liver damage
artificial pacemakers	major surgery not required	immune system can reject the pacemaker may need replacing
heart transplant	better quality of life	major surgery anti-rejection drugs needed (leading to greater infection risk) shortage of donors

5 Evaluate the advantages and disadvantages of treating cardiovascular diseases by drugs, mechanical devices or transplant.

6 Why are doctors increasingly prescribing statins for their patients?

KEY INFORMATION

The coronary artery takes oxygen and glucose to heart muscles for respiration.

DID YOU KNOW?

Over 300 000 people in the UK have a heart attack every year.

MATHS SKILLS

Extracting and interpreting information

Learning objectives:

- to extract and interpret information from tables, charts and graphs.

Tables, charts and graphs give a lot of information in a small space. Extracting and interpreting data from graphics is a vital skill for all scientists.

DID YOU KNOW?

Correlations prove a relationship but do not prove cause and effect.

Looking at tables

Tables allow us to **classify** and **compare** data. To read a table and extract information from it, you must first read the headings carefully.

Light intensity (units)	Rate of photosynthesis (number of bubbles per minute)
1	2
3	11
6	26
9	48
11	47
14	48

The headings of each column help us to read the table.

Column 1 tells us that this table is about light intensity, and column 2 tells us that the data collected was the rate of photosynthesis, which was measured by counting the number of bubbles released in a minute.

Look at the table.

1. **a** What was the light intensity when 26 bubbles were released in a minute?
   b what happened to the rate between 9 light intensity units and 14 light intensity units?

2. **a** At what light intensity was the rate of photosynthesis highest?
   b What trends do you notice in the data?

Looking at charts and graphs

The most commonly used statistical diagrams in science are line graphs, bar graphs and pie charts. Pie charts are used to show proportions and the relative importance of different data sets. The relative size of the angle at the centre of each sector represents the size of the data set. Different sectors can be compared without even knowing the numbers in the different groups.

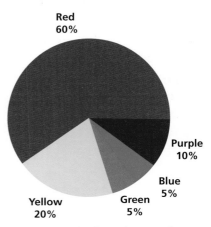

Figure 3.46 Pie chart showing the amount of time that hummingbirds spent at different coloured feeders.

Red 60%
Purple 10%
Blue 5%
Green 5%
Yellow 20%

In a bar graph each bar represents the number of a specific data set. The height of the bars allow us to compare the quantity of each sub-group within a data set.

Plot one data set against another to show whether or not there is a relationship, or correlation, between them. When the points are more or less in a straight line, it is said that there is a good **correlation** between the two data sets. Correlations can be positive, negative, or zero. Positive correlation shows that as one set of values increases, so does the second set (a 'more:more' relationship). Negative correlation shows that as one set of values increases, the other data set decreases (a 'more:less' relationship). Zero correlation shows that one data set increasing or decreasing has no effect on the other.

Another type of graph is the pictogram, which uses pictorial forms to represent data, for example, statistical data.

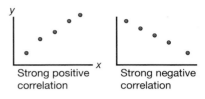

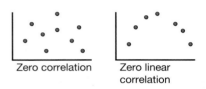

Figure 3.47 Correlations can be positive, negative, or zero.

3 **What can you deduce from this scatter graph?**

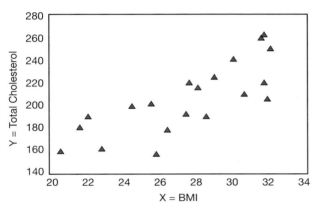

Figure 3.48 What can you deduce from this scatter graph?

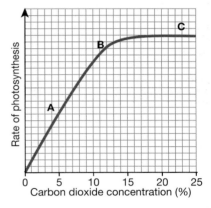

Figure 3.49 This line graph gives information about a factor that is relevant to plants

Line graphs

To interpret or read a line graph:

- Look at the labels, scales and units on each axis. What are they telling you?

- Look at the general **trend** or trends on the graph. As factor X increases, what happens to factor Y?

- Pick out specific points of interest. Follow the line along the graph until you find the data point that corresponds with a value on the other axis.

4 **What does the graph in Figure 3.49 tell you?**

Check your progress

You should be able to:

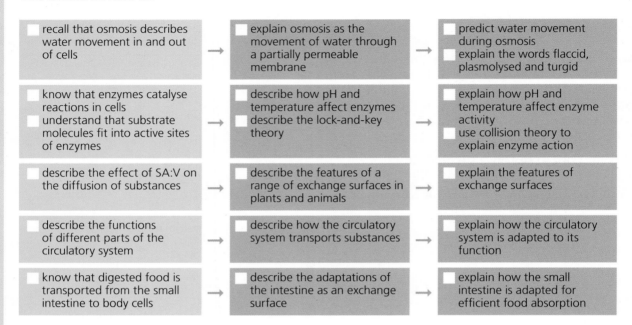

recall that osmosis describes water movement in and out of cells → explain osmosis as the movement of water through a partially permeable membrane → predict water movement during osmosis
explain the words flaccid, plasmolysed and turgid

know that enzymes catalyse reactions in cells
understand that substrate molecules fit into active sites of enzymes → describe how pH and temperature affect enzymes
describe the lock-and-key theory → explain how pH and temperature affect enzyme activity
use collision theory to explain enzyme action

describe the effect of SA:V on the diffusion of substances → describe the features of a range of exchange surfaces in plants and animals → explain the features of exchange surfaces

describe the functions of different parts of the circulatory system → describe how the circulatory system transports substances → explain how the circulatory system is adapted to its function

know that digested food is transported from the small intestine to body cells → describe the adaptations of the intestine as an exchange surface → explain how the small intestine is adapted for efficient food absorption

Worked examples

The diagram shows the human heart.

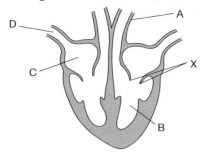

1. **Label parts A–D on the diagram.**

    A Artery

    B Left ventricle

    C Right atrium

    D Vena cava

2. a **What is the name of part X?**

    Valve

    b **Part X can become damaged or may leak. What effect will this have?**

    Person will have a lack of energy and breathlessness.

    c **How can a damaged or leaky part X be treated?**

    Transplant a valve from another person.

    d **Give an advantage and a disadvantage of the treatment.**

    It will not be rejected.

3. **The heart pumps blood to the lungs so that gas exchange can happen.**

    Why are alveoli an efficient exchange surface?

    They have a large surface area compared with volume to give a large area for gas exchange. Good blood supply to take oxygen to body cells where it is used for respiration for energy for growth.

B–D correctly identified. Giving 'artery' for A is insufficient.

'Aorta' is a an alternative answer.

Valve is correctly identified.

Two symptoms correctly identified.

One treatment has been identified.

Artificial valves are another treatment.

The student has correctly identified an advantage of one treatment but they have only answered half the question for this.

Two correct answers are identified. Then the student forgets the question being asked and moves away from the topic of alveoli.

Overall, this student shows a good understanding of the topic but does not give complete answers. Careful reading of the questions, and consideration of exactly what is being asked for, in addition to more detailed responses, might be helpful.

End of chapter questions

Getting started

1 In what part of the lung does gas exchange take place?　　`1 Mark`

2 Insects have tiny holes called spiracles along the side of the body. What is the function of the spiracles?　　`1 Mark`

3 How is glucose absorbed into the blood?　　`1 Mark`

4 What is the function of the heart?　　`1 Mark`

5 Describe the difference between respiration and breathing.　　`2 Marks`

6 Lindsay is investigating the effect of pH on the rate of reaction of amylase enzyme. She wants to find out the time taken to completely digest a starch solution at a range of pH values. She measures out 2 cm³ of amylase solution and puts it in a test tube before adding 1 cm³ of a pH solution.

Next she adds 2 cm³ of starch solution, starts the stop clock and then stirs the mixture using a pipette.

After 20 seconds she tests the mixture using Iodine reagent.

a　Why should Lindsay use a syringe to measure the solutions?　　`1 Mark`

b　Why is it important that the mixture is stirred?　　`1 Mark`

c　Why does Lindsay use iodine reagent?　　`1 Mark`

　The rate of enzyme activity can be calculated using the following equation:

d　Rate of enzyme activity = 1000/time taken

　Calculate the rate of enzyme activity if the time taken for amylase to break down starch was 125 seconds.　　`1 Mark`

Going further

7 What is the function of bile?　　`1 Mark`

8 What does it mean if an enzyme is denatured?　　`1 Mark`

9 Give two ways in which the small intestine is adapted for efficient absorption.　　`2 marks`

10 Amoeba is a single-celled organism that lives in fresh water. It has a vacuole that fills with water, moves to the outside of the cell and bursts. A new vacuole starts to form. Use osmosis to explain why amoeba needs a vacuole.　　`4 Marks`

11 Yeast was allowed to carry out anaerobic respiration at different temperatures. The carbon dioxide produced was collected for 5 minutes. The results are shown in the table:

Temperature °C	Amount of carbon dioxide collected in 5 minutes in cm³		
	Test 1	Test 2	Test 3
20	0.8	0.9	1.0
30	23.4	22.8	22.4
40	39.4	40.8	41.7
50	39.1	31.9	30.8

a Identify the anomalous result.

1 Mark

b Describe the effect of temperature on respiration in yeast shown by the data.

1 Mark

More challenging

12 Explain how plant cells become turgid.

2 Marks

13 How are plant roots adapted for the exchange of materials?

2 Marks

14 Active transport and diffusion are processes that are vital to a plant's survival. Explain how active transport and diffusion contribute to a plant's survival.

6 Marks

Most demanding

15 Define the term 'emulsification' and explain its purpose.

2 Marks

16 The table below shows the surface area and volumes of a bison and a prairie dog. Both are herbivores.

	Surface area	Volume	Surface area to volume ratio SA:V
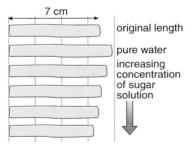	1128	2016	
	10	2	

Calculate the surface area to volume ratio for both and suggest why the prairie dog needs to eat more comparative to its size than the bison.

2 Marks

17 Potato cores were cut to the same length. Each core was placed in a different concentration of sugar solution. After 30 minutes they were measured. The result are shown in the diagram below.

7 cm

original length

pure water

increasing concentration of sugar solution

Using the diagram, describe and explain the results.

6 Marks

Total: 40 Marks

HEALTH MATTERS

HEALTH CAN BE AFFECTED BY DRUGS AND DISEASE.

- Smoking is damaging to health because cigarette smoke contains harmful chemicals.
- Drugs can be medicines and may help people suffering from pain or disease.
- Lack of exercise and a poor diet can lead to a higher BMI.
- Alcohol affects our response time and behaviour.

BACTERIA ARE SINGLE-CELLED LIVING ORGANISMS.

- Bacteria multiply by simple cell division every 20 minutes, if conditions are favourable.
- Temperatures greater than 25°C increase the likelihood that pathogens will grow.
- Antiseptics can kill bacteria.
- A specific antibiotic can kill a specific bacterial colony.

ORGANS WORK TOGETHER IN SYSTEMS TO PERFORM CERTAIN FUNCTIONS.

- Air enters the respiratory system through the nose.
- The trachea and bronchi are organs in the respiratory system.
- The stomach produces hydrochloric acid.
- The skin covers the outside of the body.
- Blood is the transport system in the body; red blood cells transport oxygen from the lungs to all cells.

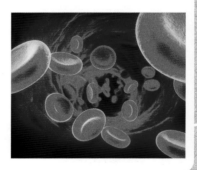

GREEN PLANTS ALL HAVE ORGAN SYSTEMS THAT HAVE SPECIFIC FUNCTIONS.

- Plants photosynthesise to increase biomass, which is tranferred to animals in higher trophic levels.
- Plants can be damaged by a range of ion deficiency conditions.
- In self-supporting ecosystems plant and animal populations are interdependent.

IN THIS CHAPTER YOU WILL FIND OUT ABOUT:

FACTORS THAT AFFECT OUR CHANCES OF CATCHING A NON-COMMUNICABLE DISEASE.

- Factors in our environment can increase our risk of disease.
- Our lifestyle can increase the chance of us developing a non-communicable disease.
- Sometimes a number of risk factors for developing a disease interact.
- Lifestyle factors can increase the risk of a person developing cancer.

HOW ARE COMMUNICABLE DISEASES SPREAD?

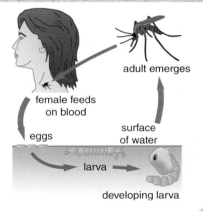

- Pathogens are microorganisms that cause disease in plants and animals.
- Bacteria, viruses, fungi and protists can cause disease.
- Toxins produced by bacteria make us feel ill.
- Understanding the lifecycles of some pathogens allows us to control the spread of disease.

adult emerges

female feeds on blood

eggs

surface of water

larva → developing larva

HOW DO WE CONTROL THE SPREAD OF DISEASE?

- The skin, nose, respiratory system and stomach all protect us from pathogens.
- The immune system is our major defence system against disease.
- White blood cells protect us from bacterial infections in a number of ways.
- Vaccination protects us from viral and bacterial pathogens.

HOW ARE PLANTS AFFECTED BY DISEASE AND PROTECTED FROM ATTACK?

- Plant diseases can usually be detected by visible symptoms.
- Viral, bacterial and fungal pathogens cause plant diseases.
- Plants have physical, mechanical and chemical defence systems to protect them from attack by herbivores.

...

Learning about health

Learning objectives:

- recall the difference between health and disease
- explain how some diseases interact
- evaluate data about lifestyle and health.

The word 'health' can mean many different things, depending on the people involved and their situation.

Health and disease

Good health is complete physical and mental well-being. It is about feeling good and having a positive frame of mind. **Mental health** is as important as physical health.

The major causes of physical and mental ill health include:

- disease
- diet
- stress
- life situations.

Diseases are disorders that affect part or all of an organism. They can be **communicable**, like measles and HIV, or **non-communicable**, such as cancer and cardiovascular disease.

Different types of disease may interact; for example, viruses living in cells can cause cancers. Cervical cancer is linked to infection with human papilloma virus (HPV), which causes genital warts.

Being overweight is strongly related to having high blood pressure. Having high blood pressure damages the body's arteries in the long term, making the walls thick and stiff, rather than flexible and elastic.

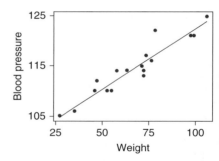

Figure 4.1 What do these data show?

1. **Describe what we mean by 'health'.**

2. **Name the major causes of ill health.**

3. **Name one communicable and one non-communicable disease.**

Diet, life situations and stress

Poor nutrition can contribute to the risk of developing some diseases:

- high fat or sugar-rich diets can cause high blood pressure, **depression**, heart disease and strokes, eating disorders and type-2 diabetes.
- low-calcium diets cause osteoporosis.
- red meat and processed meat increase the risk of bowel cancer.

Stress develops when life situations occur, for example, moving home, death in the family or divorce.

When you are stressed, hormones are released, blood vessels constrict and blood pressure rises. Stress often causes depression. It increases the risk of obesity, heart disease, Alzheimer's disease and diabetes; it can also trigger asthma in people who already have it.

Different diseases can interact:

- **immune system** diseases, such as AIDS, can cause an increased risk of contracting infectious diseases.
- immune reactions initially caused by a pathogen can cause allergies, such as skin rashes. Severe viral respiratory infections in early childhood can trigger asthma as children grow.
- severe physical ill health, for example obesity or cancer, can cause depression and other mental illnesses.

Figure 4.2 Name a cause of asthma

4 Explain how lifestyle factors affect mental health.

5 Explain one example of how different diseases interact.

HPV and cervical cancer

HPV is found in most cervical cancers. People with weak immune systems are less able to resist HPV. Not everyone who gets HPV develops cancer. Worldwide, cervical cancer is the second most common female cancer. It often takes many years to develop after having HPV and does not usually show symptoms until it is quite advanced. It is hard to treat.

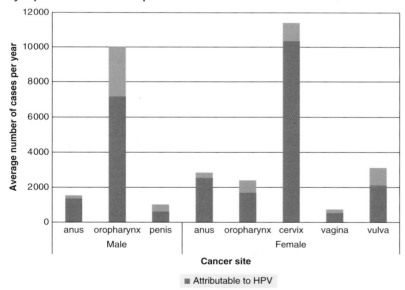

Figure 4.3 What do these data show?

6 Explain the graph in Figure 4.3.

7 Discuss the human and financial cost of cancer on individuals, families and nations.

DID YOU KNOW?

Around 80 000 children aged between 5 and 16 years suffer from severe depression.

KEY INFORMATION

Don't forget that ill health can lead to poor mental health.

KEY CONCEPT

Looking at risk factors

Learning objectives:

- recall the causes of some non-communicable diseases
- describe the impact of lifestyle on non-communicable diseases
- explain the impact of lifestyle on non-communicable diseases.

KEY WORDS

causal
 mechanism
risk factor

In developed countries, non-communicable disease is the largest cause of death.

Risk factors

Risk factors increase the chance of having a disease. Risk factors can be aspects of a person's lifestyle, and substances in the person's body or environment. They include diet, exercise, type of workplace, sexual habits, smoking, drinking and drug-taking.

Non-communicable diseases may be caused by the interaction of a number of factors:

- factors involved in cardiovascular disease may be diet/obesity, age, genetics and exercise.
- lung disease factors are smoking and cleanliness of the environment.
- alcohol, diet/obesity, genetics, drugs and viral infection may be involved in liver disease.
- genetics, diet/obesity and exercise may affect Type 2 diabetes.

Figure 4.4 Why is this person at risk of cardiovascular disease?

We cannot control some risk factors, such as genes and age, but we can control lifestyle factors, like drinking and smoking, to reduce the risk of contracting some diseases.

1. **What factors increase the risk of Type 2 diabetes?**
2. **How could the risk of liver disease be reduced?**

Causal mechanisms

A **causal mechanism** is one risk factor that may be partly responsible for a disease. Research has shown a causal mechanism for some risk factors.

Disease/condition	Risk factor with causal mechanism
cardiovascular disease	poor diet, smoking and lack of exercise
Type 2 diabetes	obesity
liver and brain damage	alcohol
lung disease, cancer	smoking
low birth weight and premature birth	smoking
abnormal foetal brain development	alcohol
cancer	carcinogens (including ionising radiation)

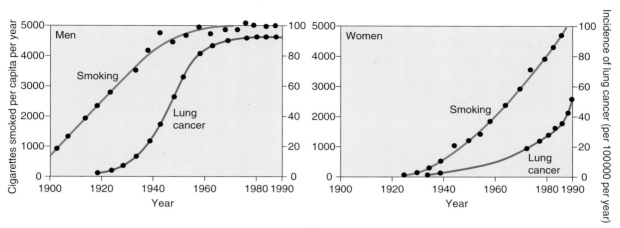

Figure 4.5 Smoking and lung cancer

3 Use the data in figure 4.5 to describe the smoking habits of men from 1900 to 1990. Describe what happened to the incidence of lung cancer during this time.

Time lag

There is a long time lag between when people start to smoke and the development of cancer. So, it took a long time to compile evidence to prove smoking was the causal mechanism for lung cancer.

Human and financial costs of smoking include:

- serious health problems, possible death, damage to unborn children, dangers of passive smoking
- cost of cigarettes, loss of family income and a depleted workforce
- financial burden of healthcare costs on local, national and global economies.

An isolated case of disease does not provide evidence for or against a risk factor. Large sample groups need to be monitored. Groups are randomly selected, or matched in terms of age and gender.

4 Explain the impact of lifestyle on non-communicable diseases.

5 Explain the human and financial impact of smoking.

Figure 4.6 Smoking does not cause lung cancer immediately

DID YOU KNOW?

Research shows that teenagers who eat fast food at least once a week are more likely to develop insulin resistance, a risk factor for Type 2 diabetes.

KEY INFORMATION

Remember: smoking is a causal mechanism for lung cancer but cancer is caused by the chemicals in cigarettes.

Exploring non-communicable diseases

Learning objectives:

- identify risk factors for cancer
- explain the differences between types of tumours
- explain the impact of non-communicable diseases.

KEY WORDS

benign
carcinogen
malignant
tumour

In the UK, over 33% of people will develop a form of cancer during their life.

Cancer facts

Sometimes cell division can become uncontrolled. This causes **tumours** to form. Tumours can be harmless and **benign**, or **malignant** and cancerous.

Doctors cannot explain why only some people develop cancer. Research shows that certain risk factors increase the chance of a person developing cancer. These include:

- smoking
- obesity
- some common viruses
- UV light
- age
- possible genetic causes.

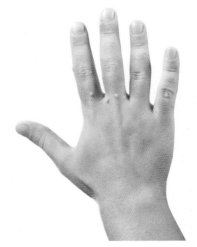

Figure 4.7 Warts are benign tumours

1 **How do tumours form?**

2 **Suggest which risk factor may increase the chance of:**

 a) **skin cancer**
 b) **lung cancer.**

Cancer in detail

Benign tumours divide slowly, do not spread and are harmless. Warts are benign tumours. They can be removed by simple surgery.

Malignant tumours divide very quickly. They can invade tissues around them and spread via the blood to different parts of the body, where they form secondary tumours. Malignant tumour cells are cancers.

Substances or viruses that increase the risk of cancer are called **carcinogens**; for example:

- tar in tobacco smoke
- asbestos
- human papilloma viruses (HPV).

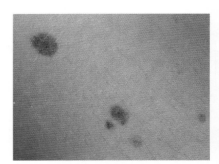
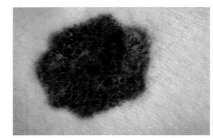

Figure 4.8 How are benign and malignant moles different?

Some risk factors that are associated with cancer cannot be changed, for example, aging and family history. Ways to reduce other risk factors include:

- not smoking
- staying out of the sun and using sun screen
- only drinking alcohol in moderation
- healthy diet and exercise.

3 Explain how benign and malignant tumours differ.

4 Describe some lifestyle changes to reduce the risk of cancer.

Smoking and cancer

Figure 4.9 Over 25% of UK cancer deaths are caused by smoking

Smoking is the biggest preventable cause of cancer worldwide. The link between smoking and cancer was known for many years before being widely accepted. Lung cancer symptoms include a cough, chest pains, shortness of breath, poor appetite and weight loss, and coughing up blood. Each person with lung cancer costs the NHS over £9000 every year. The total workforce is reduced and the economy is weakened.

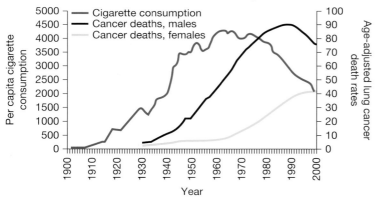

Figure 4.10 Is there evidence for a link between smoking and cancer?

5 Describe and explain the graph in Figure 4.10.

6 Suggest why there is a time lapse between the rise in cigarette consumption and the rise in lung cancer deaths.

7 'The data on the graph proves that smoking causes lung cancer.' Do you agree with this statement? Justify your answer.

DID YOU KNOW?

Breast, lung, prostate and bowel cancers are the most common in the UK. They account for 50% of diagnosed cancers.

KEY INFORMATION

Remember: malignant tumours cause cancer, whereas benign ones do not.

Analysing and evaluating data

Learning objectives:

- translate information between graphical and numerical forms
- use scatter diagrams to identify correlations
- evaluate the strength of evidence.

· ·

Analysing and evaluating data is a key skill for all scientists, but it is vitally important in medical research.

Looking for patterns

Scientists try to identify links between variables. Sometimes there is no link. When interpreting graphs:

- identify patterns or **trends**
- use axis labels and units when describing the graph; for example, 'as vaccination uptake increases, cases of measles fall'
- look for particular features of the graph, such as maximum and minimum values, the range, outliers and patterns that do not fit trends
- quote numbers to clarify descriptions.

Useful words to describe graphs are: increased, decreased, faster, slower, constant, plateau, maximum, minimum.

1 What do the data in Figure 4.11(a) show?

2 Describe the trends in the data shown in Figure 4.11(b). Are the two trends linked?

Exploring links in data

A **correlation** is an association between two sets of random data. A correlation does not prove that A affects B, or that B affects A. Both A and B could be caused by another variable. In medical research, scientists have to determine links between treatments and cures, or risk factors and disease. This is difficult if:

- the proposed cause only sometimes results in the disease
- the disease has many possible causes
- there is a long delay between proposed cause and effect.

If one variable increases as the other increases, this is a positive correlation. If one variable increases as the other decreases, it is a negative correlation. To prove a causal mechanism, an investigation needs to be carried out.

(a)

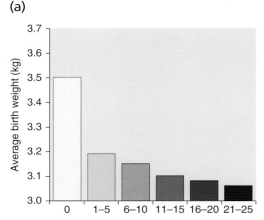

(b)

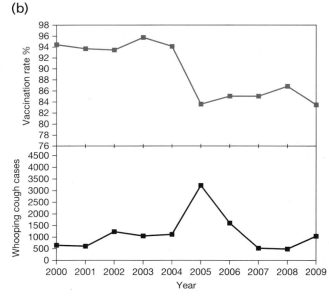

Figure 4.11 (a) Smoking in pregnancy and birthweight (b) Whooping cough incidence and vaccination uptake

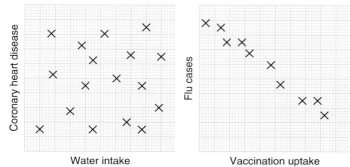

Figure 4.12 Do these graphs show correlation or not?

3 Explain the data in Figure 4.13.

4 Why is it often difficult to determine links between variables?

Evaluating data

When evaluating data think about these questions:

- Are the data reliable? Was there an appropriate control group?
- Are the data valid?
- Was the sample size sufficient?
- How could better data be collected? (Think about length of study, sample selection etc.)
- How could more accurate data be collected? Do the data answer the question?
- Are there any anomalies in the data? Can these be explained?

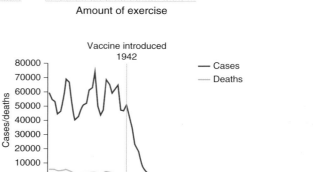

Figure 4.13 Diphtheria cases in the UK

- How confident are you that the evidence supports the conclusion? Has a causal relationship been proved? Could anyone use the same data to support a different conclusion?
- Are the data **biased**? Who completed the research? Do they have a personal interest in the data?

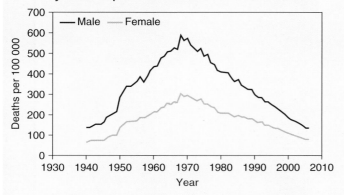

Figure 4.14 Deaths due to coronary heart disease in Australia

5 Describe the male trend in Figure 4.14. Compare both trends.

6 Evaluate the data in Figure 4.14.

COMMON MISCONCEPTION

Remember: a correlation is not the same as a causal mechanism.

DID YOU KNOW?

Researchers often need studies with large numbers of people carried out over long periods to detect small, subtle effects.

Studying pathogens

Learning objectives:

- recall the definition of a pathogen
- explain how communicable diseases can be controlled

Ebola haemorrhagic fever (EHF) first appeared in 1976 in two outbreaks. One was in Sudan, and the other was in a village near the Ebola River in the Democratic Republic of Congo.

What are pathogens?

Pathogens are microorganisms that cause infectious diseases. They depend on their hosts to provide the conditions and nutrients they need to grow and reproduce. Pathogens can be

- viruses, for example, measles
- bacteria, such as *Salmonella* which causes food poisoning
- **protists**, a protist causes malaria
- fungi, like rose black spot.

Pathogens can infect plants or animals. They are spread by direct contact, by water or through the air.

Figure 4.15 What type of pathogen causes rose black spot?

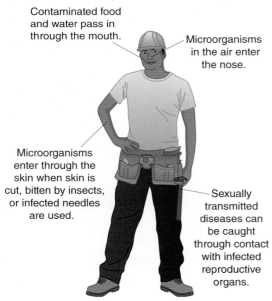

Contaminated food and water pass in through the mouth.

Microorganisms in the air enter the nose.

Microorganisms enter through the skin when skin is cut, bitten by insects, or infected needles are used.

Sexually transmitted diseases can be caught through contact with infected reproductive organs.

Figure 4.16 How are pathogens spread?

1. Give four examples of pathogens.
2. Describe how pathogens spread.

Pathogens in detail

Bacteria and viruses can reproduce rapidly in the body.

- Bacteria produce toxins that damage tissues and make us feel ill.
- Viruses live and reproduce inside cells, damaging them.

The toxins released by pathogens cause the symptoms of infection. These include high temperatures, nausea, headaches and rashes. People get a communicable disease because they receive the pathogen from someone who has it. Measures to prevent or reduce the spread of diseases include:

- simple hygiene, such as covering your mouth when coughing, using a handkerchief when sneezing and washing your hands after using the toilet
- isolation of infected individuals
- destroying vectors, for example mosquitoes are the malaria vector
- vaccination.

Figure 4.17 Why do doctors and nurses wear protective clothing?

3 Why do pathogens make us feel ill?

4 Explain how the spread of diseases can be reduced or prevented.

A case study

Ebola haemorrhagic fever (EHF) is a deadly virus: 50–90% of people infected with EHF die. The last outbreak of EHF began in December 2013, in Guinea, and spread to Nigeria, Sierra Leone and Liberia.

Fruit bats carry EHF and may infect animals and humans. The disease is caught by direct contact with body fluids (blood and saliva) of an infected individual. Symptoms include fever, headache, diarrhoea, nausea and rashes. Currently, there is no cure for EHF but two vaccines are under development.

Prevention measures include:

- reducing the risk of contact with infected animals
- wearing protective clothing
- washing hands frequently
- isolation of infected people and safe burials of the dead
- travel restrictions.

5 Evaluate the control measures for EHF.

Figure 4.18 Fruit bat meat is a delicacy in West Africa

DID YOU KNOW?

In 1918, Spanish influenza killed 40–50 million people.

ADVICE

Remember that pathogens come in many different forms, and that each one causes a different disease.

Learning about viral diseases

Learning objectives:

- describe the symptoms of some viral diseases
- describe the transmission and control of some viral diseases
- explain how some viral diseases are spread.

Communicable diseases are caused by pathogens that pass from one person to another.

Viral diseases

Viruses are very small pathogens. They are not living cells. They have a strand of genetic material inside a protein coat. The genetic material replicates inside host cells to make new viruses, which are then released.

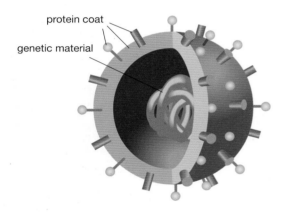

protein coat

genetic material

Figure 4.19 How are viruses different from living cells?

Examples of viral diseases are:

- measles, a serious illness. Symptoms include fever and a red rash over the skin. It can be fatal if there are complications.
- HIV, which initially causes a flu-like illness.
- tobacco mosaic virus (TMV), which affects over 150 plant species including tomatoes. Leaves have discoloured 'mosaic' patterns. This reduces photosynthesis and affects plant growth, and the quality and number of fruits are affected.

1 **How do viruses reproduce?**

2 **Describe the symptoms of two viral diseases.**

Figure 4.20 Suggest how TMV got its name

Spread and control

Viruses spread and are controlled in various ways.

Virus	Mechanism for spread	Control
Measles	**Droplet infection** Talking, coughing and sneezing all cause the expulsion of tiny droplets into the air. Inhaling droplets that carry viruses causes measles to spread.	Most young children have **vaccinations** to protect them against this possibly fatal disease.
HIV	**Direct contact** HIV is spread by sexual contact or exchange of body fluids such as blood. This can occur if drug users share needles.	**Antiretroviral drugs** are prescribed to stop the virus entering the lymph nodes.
TMV	**Enters plants via wounds** The virus gains access through a break in the skin or plant epidermis.	It is controlled by: removal of infected plant material, controlling pests, sterilising tools (using heat), washing hands after handling infected plants.

Figure 4.21 Suggest why these commuters are wearing masks

3 Describe how HIV is spread and suggest control measures.

4 Describe how TMV is spread and controlled.

Counting the cost

Viral diseases are harder to treat than bacterial diseases. Bacteria are living cells that survive outside body cells. They can be treated with prescribed drugs. Viruses are only found inside host cells, where they are protected from drugs. This is why we are vaccinated against some viral diseases.

HIV is very difficult to control. If it is not successfully controlled by antiretroviral drugs, the virus enters the lymph nodes and attacks the body's immune cells. Late-stage HIV, or AIDS, occurs when the body's immune system is no longer able to deal with other infections or cancers.

5 Explain how HIV leads to AIDS.

6 Why is it harder to treat viral diseases than bacterial diseases?

7 Compare the mechanism for the spread of measles with that of HIV. Explain which disease is most infectious.

DID YOU KNOW?

Each type of virus attacks a specific cell. For example, the measles virus attacks skin and sensory nerve cells.

KEY INFORMATION

Remember: viruses are not living things. They have many different shapes (just one of which is shown in Figure 4.19).

Studying bacterial diseases

Learning objectives:

- describe the symptoms of some bacterial diseases
- explain how some bacterial diseases can be controlled
- compare and contrast bacterial and viral diseases.

KEY WORDS

diarrhoea
gonorrhoea
Salmonella

Some pathogens are bacterial. Bacteria can reproduce very quickly in our bodies.

Bacterial diseases

Many bacteria are not harmful and some are actually very useful. We use them to make cheese and yoghurt, to break down our waste and make medicines. Some bacteria, however, are pathogens and cause diseases, infecting both plants and animals.

Examples of bacterial diseases are:

- *Salmonella*, which causes food poisoning. A build-up of toxic bacterial waste products causes symptoms that include:
  - › fever
  - › abdominal cramps
  - › vomiting
  - › **diarrhoea**.
- **Gonorrhoea**, a sexually transmitted disease. The toxic bacterial products cause symptoms that include:
  - › a thick yellow or green discharge from the vagina or penis
  - › pain when urinating.

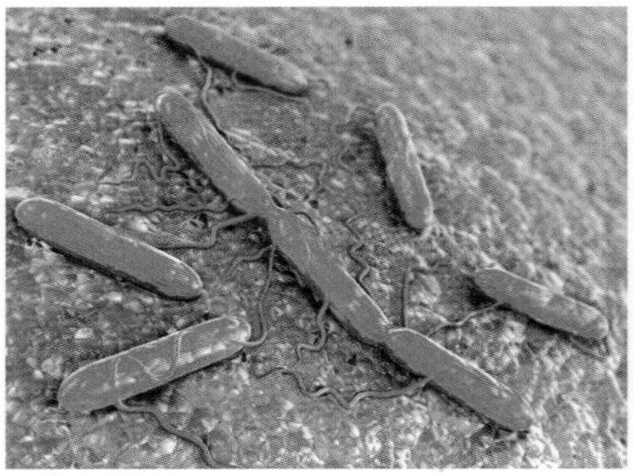

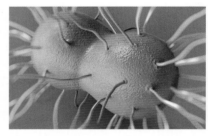

Figure 4.22 *Salmonella* bacteria cause food poisoning

1 **Name some bacterial diseases.**

2 **Describe the symptoms of** *Salmonella* **and gonorrhoea.**

Spread and control

Bacteria are spread and controlled in various ways.

Salmonella is spread by:

- ingesting (eating) food that is contaminated with *Salmonella* bacteria
- preparing food in unhygienic conditions; for example, using contaminated knives or chopping boards.

In the UK, the spread of *Salmonella* is controlled by vaccinating poultry against the bacterium. Food should be prepared in hygienic conditions and cooked thoroughly. Washing hands before and after food preparation and after using the bathroom also help.

Figure 4.24 Why must food be cooked thoroughly?

Figure 4.23 How are these gonorrhoea bacteria different from *Salmonella* bacteria?

Gonorrhoea is spread by sexual contact. The spread can be controlled by:

- treatment with antibiotics. In the past, the disease was easily treated with the antibiotic penicillin, but in recent years many resistant strains of the bacteria have appeared
- use of a barrier method of contraception, such as a condom, to prevent contact.

3 **Describe how** *Salmonella* **is spread.**

4 **Explain how gonorrhoea can be controlled.**

HIGHER TIER ONLY

Symptom delay

People with communicable infections do not develop symptoms as soon as they are infected with the pathogen. There are distinct stages of infection:

- The pathogen enters an organism.
- The pathogen reproduces rapidly in ideal conditions to increase numbers. This is the incubation period.
- Pathogens make harmful toxins, which build up. The more bacteria that are present, the quicker the toxins build up.
- Symptoms develop, for example, fever and a headache.

The graph shows *Salmonella* cases over a period of 12 months.

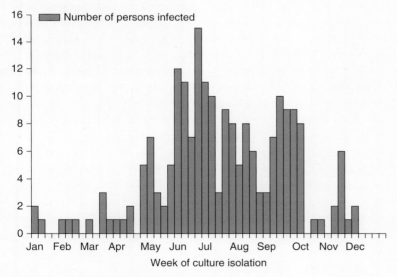

Figure 4.25 *Salmonella* cases over 1 year

5 **Describe the pattern of** *Salmonella* **cases shown in the bar chart and suggest explanations for it.**

6 **Explain how knowledge of the spread of bacterial diseases can lead to their control.**

7 **Compare and contrast bacterial and viral diseases.**

DID YOU KNOW?

In 2015 there was an outbreak of a highly drug-resistant strain of gonorrhoea bacteria in the UK.

KEY INFORMATION

Remember: bacteria are living things, unlike viruses. Bacterial diseases can be treated with antibiotics, unlike viral diseases.

Looking at fungal diseases

Learning objectives:

- recall the name and symptoms of a fungal disease
- describe the transmission and treatment of rose black spot
- explain how rose black spot affects the growth of the plant.

Rose black spot is the most serious disease of wild and cultivated roses. It is caused by a fungus.

What is rose black spot?

Diplocarpon rosae is a parasitic fungus that causes black spot on roses. It needs warm, wet conditions to grow and spread; British summers are ideal.

The fungus produces black or purple spots on the upper surface of leaves. The spots spread rapidly and sometimes the string-like fungus can be seen. The leaves often turn yellow around the spots and drop early, even though other plant parts are not affected. Sometimes small, black scab-like spots develop on young stems.

Figure 4.26 Newly sprouted leaves are susceptible to black spot

Figure 4.27 Black spot can cause black lesions on stems

In recent years, rose black spot has become more common in urban gardens. Untreated, black spot can quickly affect all the roses in a garden, resulting in plants with bare stems.

1. What are the symptoms of rose black spot?
2. What conditions does the fungus need to grow?

Transmission and treatment

The fungus produces **spores** (similar to seeds). The spores are released in wet, humid conditions, for example, when it rains or the plants are watered. Wind also helps spores to disperse. The optimum temperature for fungal growth is about 24°C. After about 7 hours of wet, hot conditions the spores germinate. Symptoms start to appear on leaves 3–10 days after infection. Spores are then produced every 3 weeks throughout the growing season. They overwinter on shoots, but can survive on dropped leaves and in soil.

Treatments include:

- immediately removing infected dropped leaves from the soil and burning them
- **pruning** shoots in the spring and burning all cut stems
- not composting infected leaves and stems
- treating infections with fungicides to kill the pathogen
- putting manure or **mulch** around the plants in spring to prevent fungal spores reaching the stems.

Figure 4.28 How does pruning help prevent the spread of the fungus?

3 Describe how black spot is transmitted.

4 Explain three treatments for black spot.

How does the pathogen affect growth?

Rose growth is greatly reduced because affected leaves are unable to photosynthesise efficiently. The fungus grows in the mesophyll, penetrating cells and intercellular spaces. This causes irreversible structural changes in affected cells, for example, by damaging cell membranes which causes leakage of essential nutrients into the intercellular spaces. The cells can then no longer photosynthesise to make the food the plant needs to survive.

5 Explain how the black spot fungus affects the growth of roses.

DID YOU KNOW?

The conidia (asexual spores) of black spot germinate on the leaf. Hyphae (root-like structures) grow out and rupture the cuticle, penetrating mesophyll cell membranes to get nutrients.

KEY INFORMATION

Remember: when a pathogen damages leaves, the plant cannot photosynthesise as much, so growth is reduced.

Learning about malaria

KEY WORDS

protist
vector

Learning objectives:

- recall that malaria is a protist disease
- describe the lifecycle of the malarial vector

Malaria is the most prolific parasitic disease of humans in the world. More children die from malaria than any other communicable disease.

What is malaria?

The pathogens that cause malaria are called **protists**. Protists are single-celled organisms. The protist that causes malaria is called *Plasmodium*. For part of its lifecycle, it lives in human blood.

The protist is spread by female mosquitoes, which feed on blood. They suck infected blood from someone with malaria and then pass the protists on when they feed on a new person.

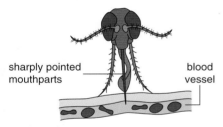

sharply pointed mouthparts

blood vessel

Figure 4.29 How do you catch malaria?

Organisms that spread disease, rather than causing it themselves, are called **vectors**. The mosquito is the vector for malaria.

Symptoms of malaria appear from 7–18 days to over a year after infection and include:

- recurrent episodes of fever
- sweats and chills
- muscle pains
- headaches
- diarrhoea
- cough.

Malaria can be fatal if not treated.

1 **What causes malaria?**

2 **What are the symptoms of malaria?**

3 **What is a vector?**

Spreading malaria

The mosquito has a complicated lifecycle. Mosquitoes breed and lay eggs in still water. The eggs hatch and each larva develops into a pupa in the water. When an adult mosquito hatches, it rests on the water surface to let its body dry and harden so it can fly.

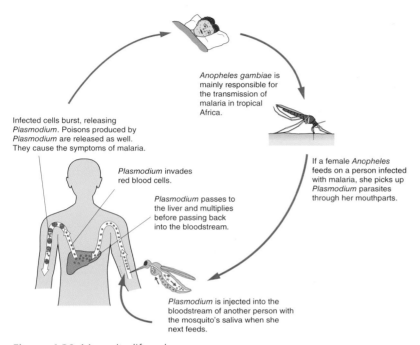

Anopheles gambiae is mainly responsible for the transmission of malaria in tropical Africa.

Infected cells burst, releasing *Plasmodium*. Poisons produced by *Plasmodium* are released as well. They cause the symptoms of malaria.

If a female *Anopheles* feeds on a person infected with malaria, she picks up *Plasmodium* parasites through her mouthparts.

Plasmodium invades red blood cells.

Plasmodium passes to the liver and multiplies before passing back into the bloodstream.

Plasmodium is injected into the bloodstream of another person with the mosquito's saliva when she next feeds.

Figure 4.30 Mosquito lifecycle

Malaria is controlled by breaking the lifecycle of the vector or by avoiding contact with it. Control measures include:

- spraying pools of water with insecticide to kill the mosquitoes
- draining stagnant water pools
- spraying pools with oil to prevent the larvae from breathing
- using mosquito nets to avoid bites
- taking drugs to kill the protist in the blood.

Malaria can also be spread by using dirty needles for injections or can be passed from a pregnant woman to her baby.

4 Explain how malaria is spread.

5 Describe the lifecycle of the malarial vector.

DID YOU KNOW?

Malaria is not found in the UK, but about 1400 people were diagnosed with it after returning from holiday in 2012. Two people died.

Protecting the body

Learning objectives:

- describe how the body protects itself from pathogens
- explain how the body protects itself from pathogens
- explain how communicable diseases can be spread.

The body has defence mechanisms for each way that pathogens are transmitted.

How does the body defend itself?

Millions of pathogens are around us each day, but your body protects you from being infected:

- Your skin acts as a barrier and produces antimicrobial secretions via glands in the skin.

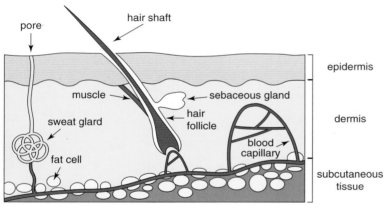

Figure 4.31 Structure of the skin

- The nose traps particles that may contain pathogens.
- Your trachea and bronchi secrete mucus, which traps pathogens.
- The stomach produces acid, which kills the majority of pathogens that enter via the mouth.
- Platelets (cell fragments in your blood) start the clotting process at wound sites. Clots dry to form scabs which seal the wound.

1 How does the body protect itself from pathogens?

2 Which of these defence mechanisms will protect you against pathogens in water and in food?

Hygiene and wounds

When you cough or sneeze, thousands of tiny drops of liquid are sprayed into the air. If you have a disease, the droplets may contain pathogens. This is why you should use a paper tissue and then put it in the bin. Coughing up phlegm and then spitting it out can also spread infections to those around you. Children playing may fall on it and any pathogens can enter the body through wounds in the skin.

When the skin is cut, platelets in the blood are exposed to the air at the wound site. They make protein fibres (fibrin) that form a mesh over the wound. The platelets and red blood cells get caught in the fibres to form a clot.

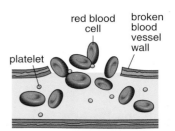

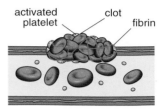

3 Explain how coughing and sneezing spread pathogens.

4 How do clots form over wounds? Explain how the formation of a scab helps to protect the body.

Figure 4.32 How are wounds sealed?

HIGHER TIER ONLY

Defence mechanisms in detail

Skin protects the body from physical damage, infection and dehydration. The outer layer of skin cells is dry and dead. Pathogens cannot easily penetrate these dead cells. Sebaceous glands in the skin produce antimicrobial oils.

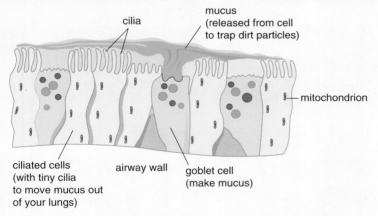

Figure 4.33 How does the respiratory system protect the body?

Every time you breathe, you take in many microbes. Hairs in the nose trap larger microbes and dust particles. The trachea and bronchi have a ciliated epithelium. **Cilia** are tiny hair-like structures. **Goblet cells** in the epithelium produce mucus; mucus traps smaller dust particles and microbes. Cilia beat together to waft mucus to the back of the throat, where it is swallowed. Ciliated cells have many mitochondria to supply the energy needed to do this.

5 How does skin act as a defence against pathogens?

6 Explain the adaptations of the respiratory system to protect us against pathogens.

7 Smoking can damage and paralyse the cilia. Explain why smokers are more susceptible to respiratory infections.

DID YOU KNOW?

It has been estimated that a sneeze expels mucus at 160 km/h.

REMEMBER!

Cilia are like hairs; they are not cells and cannot be killed.

Exploring white blood cells

Learning objectives:

- describe phagocytosis
- explain how antibody production can lead to immunity
- explain the specificity of immune system responses.

KEY WORDS

antibody
antitoxin
immunity
lymphocyte
phagocyte

Although the body's defence mechanisms prevent many microbes from entering the body, some will succeed.

White blood cells

The immune system recognises and destroys pathogens that enter the body. White blood cells are an important part of the immune system.

They attack invading pathogens. If a pathogen enters the body, white blood cells defend it by:

- ingesting pathogens (phagocytosis)
- producing antibodies
- producing antitoxins.

There are two main groups of white blood cell: **phagocytes** and **lymphocytes**. Phagocytes can leave the blood by squeezing through capillaries to enter tissues that are being attacked. They move towards pathogens or toxins and ingest them. This is called phagocytosis.

Figure 4.34 White blood cells are the largest blood cells

1 A phagocyte moves towards a bacterium.

2 The phagocyte pushes a sleeve of cytoplasm outwards to surround the bacterium.

3 The bacterium is now enclosed in a vacuole inside the cell. It is then killed and digested by enzymes.

Figure 4.35 Phagocytes ingest pathogens during phagocytosis

1 How do white blood cells protect the body?

2 What happens to a pathogen during phagocytosis?

Antitoxins and antibodies

Pathogens make us feel ill because they release toxins into our body. White blood cells, called lymphocytes, produce **antitoxins** to neutralise toxins made by the pathogen. Antitoxins combine with the toxin to make a safe chemical. Antitoxins are specific to a particular toxin.

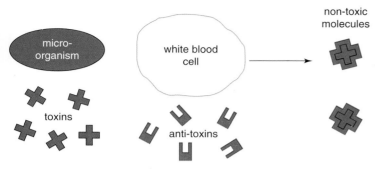

Figure 4.36 Antitoxins neutralise toxins made by pathogens

Lymphocytes also produce chemicals called **antibodies** that destroy pathogens. Lymphocytes recognise when pathogens are present. They quickly reproduce to make lots of antibodies that:

- cause cell lysis (the pathogens burst)
- bind to the pathogens and destroy them
- cover the pathogens, sticking them together. Phagocytes then ingest them.

If the same type of pathogen enters the body again, lymphocyte cells recognise it and immediately make lots of antibodies. This is **immunity**: the person is immune to that disease.

3 How do antitoxins work?

4 Explain how antibody production leads to immunity.

5 Explain how phagocytosis and antitoxin production protect the body.

Producing antibodies

Each lymphocyte has a specific antibody to attack a specific pathogen. Pathogens carry chemicals called antigens on their surface. The appropriate antibodies lock onto the matching antigens, sticking the pathogens together to destroy them.

HIV damages white blood cells, meaning that they cannot make antibodies or kill other infected cells. When HIV has destroyed sufficient white blood cells, the body cannot make appropriate immune responses. Then the person has AIDS. They have fewer or no lymphocytes to recognise simple infections and release antibodies. This means that they can die from simple infections that healthy people can overcome.

6 Explain the specificity of antibodies.

7 Why can a person with AIDS die from a simple infection?

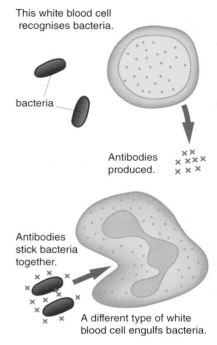

Figure 4.37 How do these white blood cells work?

DID YOU KNOW?

Only 1% of all the cells in blood are white blood cells. There are 5000–10000 of them in a microlitre of blood.

COMMON MISCONCEPTION

Remember: white blood cells do not *eat* pathogens; they *ingest* them.

Using antibiotics and painkillers

Learning objectives:

- describe the uses of antibiotics and painkillers
- explain how antibiotics and painkillers can be used to treat diseases
- explain the limitations of antibiotics.

KEY WORDS

antiviral
aspirin
opiates
penicillin

You can take medicines to make you feel better when you have an infectious disease.

Useful drugs

A drug is any chemical that alters how the body works. Medicines contain useful drugs. Many medicines do not affect the pathogen that makes you ill. They just relieve the symptoms caused by the infection, for example, painkillers and cough medicines. Other medicines work inside the body to kill bacterial pathogens: these are antibiotics, for example, **penicillin**.

Antibiotics work by interfering with the pathogen's metabolism; for example, with processes that make bacterial cell walls. Antibiotics do not affect human cells and they do not kill viral, protist or fungal pathogens. They only kill bacterial pathogens.

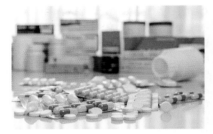

Figure 4.38 Antibiotics and painkillers are medicinal drugs

1. **What is a medicine?**
2. **How are painkillers and antibiotics similar and different?**

How do medicines work?

There are many different antibiotics. Some antibiotics work against one type of bacterial infection. Others are used to treat many different bacterial infections. It is important to use the correct antibiotic for a specific bacterium.

If a bacterium cannot be killed by an antibiotic, it is resistant to that antibiotic. A prescribed course of antibiotics has the correct amount to kill the bacterial pathogen completely. This is why you must always take the complete antibiotic course.

The use of antibiotics has greatly reduced deaths from infectious bacterial diseases. But, misuse of antibiotics has led to the emergence of strains of bacteria that are antibiotic resistant (for more on this, see Chapter 7, Variation and evolution).

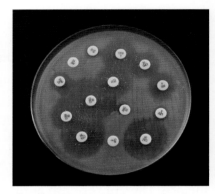

Figure 4.39 Antibiotic test: the size of the clear area around the antibiotic disc indicates the effectiveness of the antibiotic against the bacteria growing on the agar plate

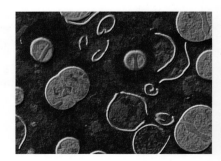

Figure 4.40 How has the antibiotic affected these bacteria?

When you are injured, sensory nerve endings send pain messages to your brain. Painkillers stop these nerve impulses, so you feel little or no pain. Many painkillers are based on two natural drugs:

- **aspirin**, from willow bark
- **opiates**, from poppies.

3 **Explain how painkillers work.**

4 **Compare the use of painkillers and antibiotics.**

Antiviral drugs

Antibiotics cannot be used to kill viral pathogens. This is because viruses live and reproduce inside cells. Antibiotics do not harm body cells and the viruses inside are protected. **Antiviral** drugs are used to treat viral infections. Antivirals are specific to a particular virus. While antibiotics can kill bacteria, antiviral drugs only slow down viral development. It is difficult to develop drugs which kill viruses without also damaging the body's tissues.

5 **Explain the limitations of antibiotics.**

6 **The graph shows the bacteria present during an infection.**

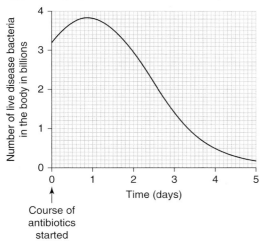

Figure 4.41 Graph showing the effect of an antibiotic on bacterial numbers

Explain

a why the numbers increase after antibiotics were taken.
b why the full course of antibiotics needs to be taken.

DID YOU KNOW?

Alexander Fleming discovered penicillin, the first antibiotic, in 1928.

KEY INFORMATION

Remember: painkillers relieve infection symptoms. Antibiotics kill bacteria.

Building immunity

Learning objectives:

- recall how vaccinations prevent infection
- explain how mass vaccination programmes reduce the spread of a disease
- evaluate the global use of vaccination

One in three people who caught smallpox in the 1600s died from it. Smallpox is now almost eradicated globally.

What is vaccination?

A **vaccination** introduces a small quantity of an inactive or dead form of a pathogen into the body to protect us from disease. Lymphocytes produce antibodies to fight the 'infection' but we don't actually become ill. When live pathogens of the same type infect you, your immune system starts to protect you immediately.

Different **vaccines** are needed for specific pathogens. For example, polio, whooping cough, flu and HPV (human papilloma virus) all have a different vaccine. MMR vaccinations contain three vaccines (for measles, mumps and rubella).

Vaccinations are usually given to children and people who are going to travel to countries where there is a risk of serious disease. Vaccines are given by injection, orally or nasal sprays.

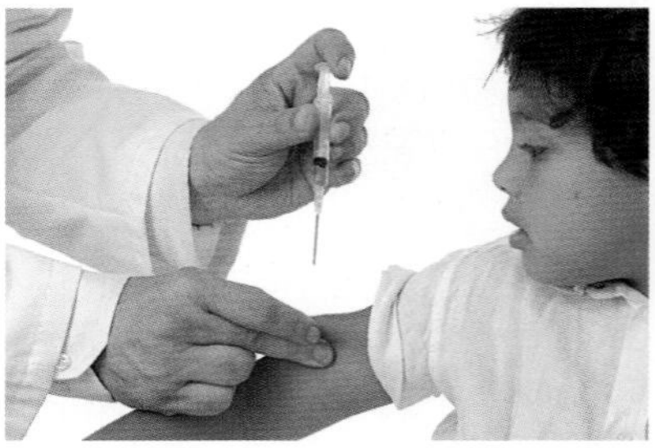

1 What is a vaccination?

2 What do vaccinations do?

Figure 4.42 Suggest why vaccinations are given to children

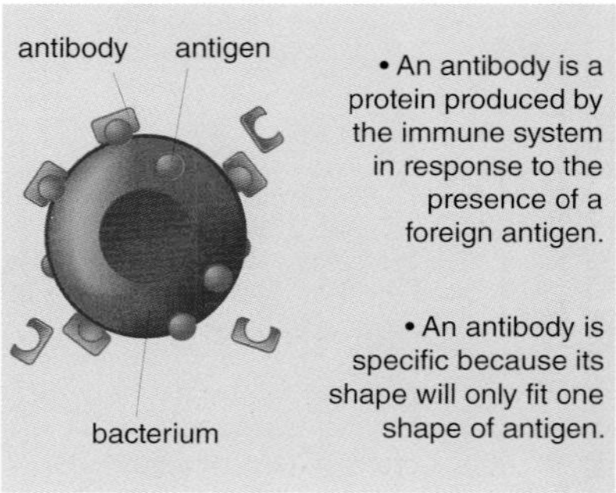

antibody   antigen

- An antibody is a protein produced by the immune system in response to the presence of a foreign antigen.

- An antibody is specific because its shape will only fit one shape of antigen.

bacterium

Figure 4.43 Antibodies are specific to a particular antigen on a pathogen

How do vaccines work?

After the vaccination:

- lymphocytes detect antigens on the dead or inactive pathogen; they produce a specific antibody for these antigens
- antibodies lock onto the antigens
- lymphocytes 'remember' the shape of the antigens
- when there is a real infection due to a live pathogen entering the body, lymphocytes instantly recognise the pathogen because it has the same antigens as the vaccine
- lymphocytes quickly make many specific antibodies
- antibodies lock onto the pathogens and kill them before they have a chance to make you feel ill. This is **immunity**.

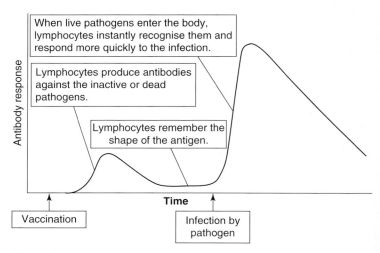

When live pathogens enter the body, lymphocytes instantly recognise them and respond more quickly to the infection.

Lymphocytes produce antibodies against the inactive or dead pathogens.

Lymphocytes remember the shape of the antigen.

Antibody response

Time

Vaccination

Infection by pathogen

Figure 4.44 The immune response

Mass vaccination programmes increase the number of people who are immune to a pathogen, making it difficult for the pathogen to pass to people who are not immunised. If a large proportion of the population is immune to a pathogen, its spread is very much reduced. Global vaccination programmes have eradicated polio, except in Pakistan and Afghanistan. Nigeria did not have any new polio infections in 2015.

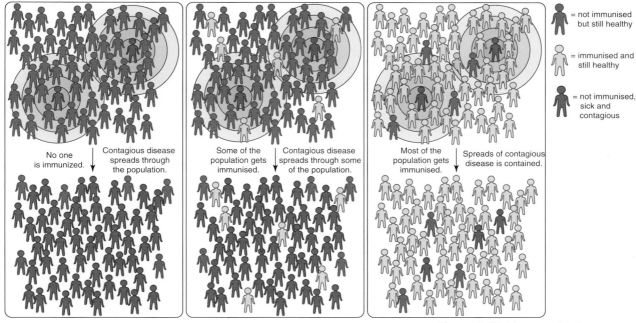

Figure 4.45 Mass vaccination programmes reduce the spread of disease

3 **Look at Figure 4.44.**
   a **Why were antibodies produced more quickly after the second infection?**
   b **Will the person be ill after the second infection?**

4 **How do mass vaccination programmes reduce the spread of disease?**

Vaccination and populations

Some viruses frequently mutate into new strains. This means that lymphocytes do not recognise the new strains. As a result, new vaccines are made regularly for some diseases, such as flu.

Some antibodies give lifelong protection, such as those that are active against measles. Sometimes antibodies fall below the critical level needed for immunity, as happens in the case of tetanus. Booster doses are needed to increase the antibodies again.

5 **Explain why new flu vaccines are made each year.**

6 **People now travel widely across the world. Evaluate the global use of vaccination.**

DID YOU KNOW?

Flu vaccine is stored in eggs because it needs to be in a living organism to survive.

ADVICE

Remember: lymphocytes make *antibodies*. Pathogens have antigens.

Making new drugs

KEY WORDS

dose
efficacy
placebo

Learning objectives:

- recall some traditional drugs and their origins
- describe how new drugs are developed
- explain why 'double-blind' trials are conducted.

Doctors, politicians, drug companies and scientists all have roles in deciding the development of new drugs.

Making new drugs from old

Researchers sometimes use traditional medicines to start developing new drugs. Traditionally, drugs were extracted from plants and microorganisms, for example:

- the heart drug digitalis comes from foxgloves
- the painkiller aspirin comes from willow trees
- penicillin was discovered by Alexander Fleming in *Penicillium* mould.

Chemists working for pharmaceutical companies formulate most new drugs. The starting point is often chemicals that are extracted from plants.

New drugs are tested and trialled before being prescribed to ensure they are:

- effective – able to prevent or cure a disease, or make you feel better; this is the drug's **efficacy**
- safe –not too toxic or with any undesirable side effects
- stable – it must be possible to store any new drug for a period of time.

1. Name three traditional drugs and state the source of each.

2. Why do new medicinal drugs need to tested and trialled?

Developing new drugs

Stages in drug development are:

- preclinical testing in laboratories (using cells, tissues and live animals) to find out side effects and efficacy
- clinical trials, which use healthy volunteers and other patients. Low **doses** of the drug are given at the start. Then, if the drug is deemed safe, further trials are performed to find the optimum dose. Clinical trials are split into phases, as shown in Figure 4.47.

Figure 4.46 Suggest why research may start with traditional medicines

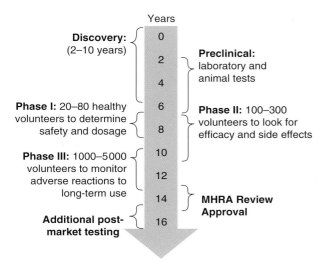

Figure 4.47 Developing new drugs

Trial results are peer reviewed by medical professionals and, only then, are they published.

Clinical trials involve some risk because unexpected side effects can occur.

3 Describe the stages in the development of a drug.

4 Why are potential drugs tested on human cells, animals and healthy human volunteers before being tested on patients?

Double-blind trials

Double-blind trials are when patients are allocated randomly to groups, so that doctors and patients do not know, until the trial is complete, if they are taking:

- the new drug (this is the test group)
- a **placebo** (a treatment that does not contain the drug). This is the control group.

Clinical trials must not be influenced by the people involved in them, whether patients, doctors or employees of pharmaceutical companies. Sometimes, people feel better just because they think they will if they take a medicine. This is the 'placebo effect'.

5 Why are double-blind trials conducted?

6 What is the placebo effect?

7 In a blind trial, patients don't know which drug they are receiving but the doctors do. Results from such trials are not always reliable. Suggest why this might be the case.

DID YOU KNOW?

Before drug development was regulated by law, a drug called thalidomide was used to treat morning sickness in pregnant women. It caused severe limb deformities in foetuses.

ADVICE

Remember: a placebo is a treatment that does not contain the drug.

Investigating monoclonal antibodies

KEY WORDS

hybridoma
monoclonal
antibodies

Learning objectives:

- describe uses of monoclonal antibodies
- explain how monoclonal antibodies are produced
- evaluate the use of monoclonal antibodies.

Antibodies have many important uses apart from fighting infections in our bodies.

HIGHER TIER ONLY

Monoclonal antibodies

Monoclonal antibodies (mABs) come from cells that are cloned from one cell. They can be produced and used over a long period of time. They are used:

- For diagnosis – in pregnancy tests. mABs are used to measure the levels of hormones and other chemicals in blood, or to detect pathogens.
- In research – they can be used to locate or identify specific molecules in a cell or tissue by binding to them with a fluorescent dye.
- To treat some diseases – for cancer treatment, mABs can be bound to a radioactive substance, a toxic drug or a chemical which stops cells growing and dividing. It delivers the substance to the cancer cells without harming other cells in the body.

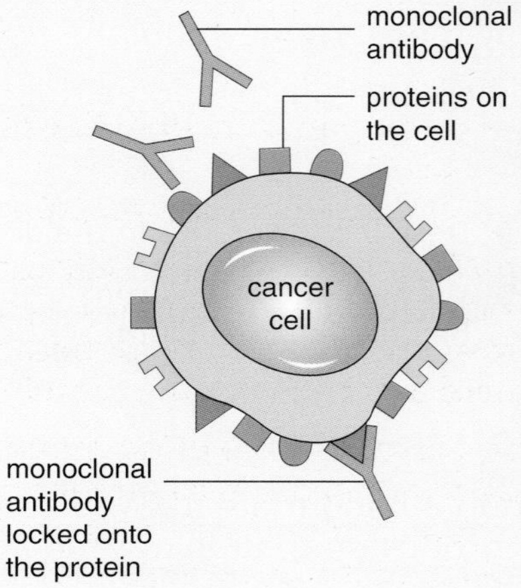

Figure 4.48 mABs are used to treat cancer

1. What are monoclonal antibodies?

2. Describe two uses of monoclonal antibodies.

Making monoclonal antibodies

mABs are specific to one binding site on one protein antigen. As a result, they are able to target a specific chemical or a specific cell type.

Once a lymphocyte starts to make antibodies, it cannot divide anymore. mABs are produced by this process:

- A mouse is injected with a specific antigen to stimulate lymphocytes to make a particular antibody.
- Extracted lymphocytes are then combined with a particular kind of tumour cell. Detergents are used to break down the cell membranes of both cells to help them fuse.

- This makes a **hybridoma** cell, which can divide and make antibodies.
- Single hybridoma cells are cloned to produce many identical cells that all produce the same antibody. These are mABs.
- A large amount of mABs are collected and purified.

3 Why are mice injected with antigens?

4 Why is detergent added to the lymphocyte and tumour cells?

5 Explain how hybridoma cells make monoclonal antibodies.

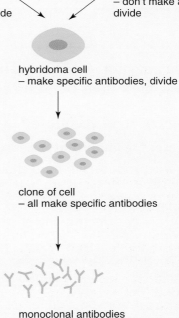

Future use of monoclonal antibodies

Using mABs in humans creates more side effects than expected including:

- chills or fever
- itchy rashes
- nausea
- breathlessness or, wheezing
- headaches
- changes in blood pressure.

mABs are not yet as widely used as everyone hoped they would be when they were first developed. Immune responses to them occur because they are produced in mice and have foreign proteins.

An advantage of using monoclonal antibodies in the treatment of cancer is that healthy body cells are unaffected, because mABs only combine with the specific cancer cells being treated, unlike drug and radiotherapy treatments which are less targeted.

Figure 4.49 How are monoclonal antibodies made?

6 Evaluate the use of monoclonal antibodies as a form of cancer treatment over the use of radiotherapy or chemotherapy.

DID YOU KNOW?

All drugs containing monoclonal antibodies have names that include 'mab' at the end of their generic name, for example trastuzumab (Herceptin).

REMEMBER!

mABs are used for diagnosis, research, treating cancer and in laboratories for blood analysis.

Looking at plant diseases

Learning objectives:

- recall the causes of plant diseases
- describe the symptoms and identification methods of some plant diseases
- explain the use of monoclonal antibodies in identifying plant pathogens.

Plants are affected by viral, bacterial and fungal pathogens, as well as by pests such as insects.

Plant pests and diseases

Plant diseases are important because:

- plants are producers in food chains
- they destroy crops and are hard to control.

The first plant virus to be identified was tobacco mosaic virus (TMV). TMV affects many plants including tomatoes. Symptoms are:

- mottling or discoloured leaves
- curled leaves
- stunted growth
- yellow streaks or spots on leaves.

Aphids (greenfly) are destructive small insects that reproduce quickly. Symptoms of aphid infestation include decreased growth rate, mottled or yellow leaves, wilting, low yields and death.

Black spot fungus causes black or purple spots on the upper surface of leaves (see topic 4.8 'Looking at fungal diseases' for more on black spot).

Just like us, plants need mineral ions to be healthy. Plants with mineral deficiencies have poor growth.

- Nitrate is needed to make amino acids, which are used in protein synthesis for growth. Plants that are deficient in nitrate have stunted growth.
- Magnesium is needed to make chlorophyll. Lack of magnesium results in **chlorosis** (yellow leaves).

1 **What are the causes of plant disease?**

2 **Why are magnesium and nitrate ions needed by plants?**

Figure 4.50 Plant diseases reduce biomass in food chains

Action and control

TMV destroys chloroplasts and slows down photosynthesis. Control measures include:

- removal of weeds (they may have TMV)
- removal of infected plants
- disinfection of all tools.

Identification of TMV is difficult. Symptoms are similar to other plant diseases.

Aphids damage plants and act as disease vectors, carrying pathogens from plant to plant. Aphids pierce phloem tubes using specially adapted mouthparts. They feed on sap, lowering turgor pressure and causing wilting. Viruses are transferred to the phloem by saliva during feeding.

Aphid infestations are easily spotted. They are controlled by natural predators (such as ladybirds) and insecticides.

Figure 4.51 Ladybird larva feeding on aphids

Treating mineral deficiency

Mineral deficiencies can be identified using chemical tests and the deficient mineral can be added to the soil. Sometimes, it is more convenient to add a general-purpose fertiliser.

3 Describe the symptoms and identification of TMV.

4 Explain why magnesium ions are needed for healthy growth.

HIGHER TIER ONLY

Detecting and identifying plant diseases

Plant diseases can be indicated by:

- stunted growth
- spots on leaves
- areas of decay (rot)
- growths
- malformed stems or leaves
- discolouration
- the presence of pests.

Figure 4.52 Galls (growths) are formed by insects or mites feeding or laying eggs

Identification of diseases can be made by:

- reference to a gardening manual or website
- taking infected plants to a laboratory to identify the pathogen
- using testing kits that contain monoclonal antibodies. Plants do not produce antibodies. The antibodies are obtained from animal plasma, after a rabbit, for example, has been injected with the plant virus or an antigen of the virus.

5 Explain the production and use of monoclonal antibodies for identifying plant viruses.

DID YOU KNOW?

Adult female aphids produce daughters without mating, so the population grows quickly.

ADVICE

Remember: plant diseases are caused by fungi, viruses, insects and mineral deficiencies.

Learning about plant defences

Learning objectives:

- recall some physical plant defence responses
- explain how plant defence systems help them survive.

KEY WORDS

antibacterial
  chemicals
mimicry
rust

Plants are attacked by pathogens and herbivores at all stages of their lifecycle.

Physical plant defence responses

Plant pathogens and pests often destroy leaves, reducing photosynthesis. Plant defences minimise this damage.

Physical defences are barriers preventing microbial pathogens entering undamaged plants. They include:

- layers of dead cells around stems (bark on trees) that prevent pests from entering the living cells underneath
- tough waxy leaf cuticles that prevent pathogens from entering the epidermis
- cellulose cell walls which prevent pathogens from entering cells.

Scientists are developing disease-resistant crops. Fungi cause plant diseases called **rusts** in apples, pears, oats and coffee. Rust-resistant plants help maintain crop levels.

1 **Describe how the waxy cuticles on leaves are a plant defence mechanism.**

2 **How do dead cells on the stem and bark protect plants?**

Figure 4.53 Aloe vera has evolved a very thick cuticle

Mechanical defences

Mechanical defence adaptations in plants include:

- Thorns and hairs – thorns impale insects and prevent egg laying. Hairs prevent larvae reaching the epidermis to feed. Thorns also deter grazing herbivores and wear down their teeth.
- Some leaves droop or curl when touched – mimosa stems release chemicals when touched. Water then exits the mimosa's cell vacuoles, causing cells to collapse and leaves to curl.
- **Mimicry** – features trick animals into not feeding or not laying eggs. Some grasses have anthers that look like aphids and hollyhocks have stem markings that look like aphids; both help prevent aphid attacks.

3 **Explain two plant physical defence responses.**

4 **Explain two plant mechanical defence responses.**

Figure 4.54 Passion flower leaves grow false 'butterfly eggs' to stop butterflies laying eggs

Chemical defence responses

Chemical defences can make plants taste unpleasant to grazers or may cause death by interfering with a pathogen's metabolic pathways. Chemical defences include:

- Production of **antibacterial chemicals**, amounts of which are increased when the plant is attacked by herbivores or pathogens.
- Production of poisons which taste bad to deter herbivores. Some poisons, however, are useful to humans.

Some disease-resistant crops that are used in agriculture are clones that can be produced quickly and easily from meristems.

5 **Explain two plant chemical defence responses.**

6 **Evaluate physical and chemical defence responses in plants.**

Figure 4.55 Deadly nightshade produces a poison called atropine which is used to treat cardiac arrest

DID YOU KNOW?

Quinine, produced by cinchona trees, is used to treat malaria. Foxgloves produce digitalis, which is used in heart medication.

REMEMBER!

Recall the differences between physical and mechanical defences.

DID YOU KNOW?

Mint produces menthol which is an insect toxin. Witch hazel also produces the insect toxin tannin. Both plants are also thought to produce antibacterial chemicals.

MATHS SKILLS

Sampling and scientific data

Learning objectives:

- understand why sampling is used in science
- be able to explain different sampling techniques
- be able to extract and interpret information from graphs.

KEY WORDS

bias
systematic

When undertaking clinical investigations, scientists are unable to test all people, so sampling is used.

Populations and samples

We cannot study every individual in a population, so we use a sample of the population.

Choosing the right size of sample is important. Carrying out studies costs money so scientists don't want to use sample groups that are large but the risk with small groups is that they aren't representative enough of the whole population. This is known as a 'sampling error' and it means that the results of the study don't indicate well what the population as a whole is like.

Scientists were investigating the effectiveness of a new antibiotic. For an antibiotic to be effective it should cure 95–100% of the cases of an illness.

Imagine that the population taking the antibiotic is represented by a container of coloured balls. Each ball is a patient. Red balls represent patients who are not cured by the antibiotic. The blue balls represent the patients who are cured.

The scientists take a small sample, of 10 balls. There is 1 red ball and 9 blue balls:

$$\frac{\text{Patients not cured (red balls)}}{\text{All patients (red and blue balls)}} \times 100 = \% \text{ chance of not being cured}$$

So

$$\frac{1}{10} \times 100 = 0.1 \times 100 = 10\% \text{ chance of not being cured.}$$

This suggests that the antibiotic is only effective for 90% of patients treated.

However, in another sample, this time of 100 balls (patients), there are only two red balls:

$$\frac{2}{100} \times 100 = 0.02 \times 100 = 2\% \text{ chance of not being cured.}$$

This larger sample suggests that the antibiotic is 98% effective.

DID YOU KNOW?

Thalidomide relieved morning sickness in pregnant women. Pregnant women were not in the sample used to test thalidomide. Many babies born to mothers who took thalidomide had limb abnormalities.

ADVICE

Remember: a small error in a small sample can mean a large error.

When using a small sample size, a sampling error can easily be made, which makes the outcome unreliable. To avoid this sampling error, large-scale clinical trials that test new drugs in humans are likely to involve thousands of patients.

1 Why do scientists use samples for investigations?

2 Figure 4.57 shows more samples in the antibiotic investigation.
   a Calculate the proportion of people who were cured in each sample.
   b How does sample size effect the results of the test?

Sampling techniques

Research aims to find consistent patterns in repeat samples, which show relationships. Scientists select samples for clinical trials by:

- using random samples
- grouping individuals by characteristics, then randomly selecting people from each group
- focusing on a particular subset of a population.

The sampling technique used can affect the outcomes.

a) b)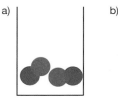

c) 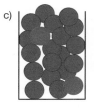 d)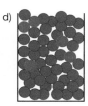

Figure 4.56 How does sample size affect results?

3 Why must care be taken when selecting the sampling method to be used?

4 Scientists studied 31 000 heart disease patients. They compared patients with high heart rates (more than 80 beats per minute) and low heart rates (less than 58 beats per minute). How could the sampling method be improved?

Looking at bias

Sample size needs to be considered as part of the research process. For a small population, 5–10% is a large enough sample size. For a larger population of size 'n', sample size = $\sqrt{n}$.

The symbol '$\sqrt{}$' means square root, so

$\sqrt{25}$ means the 'square root of 25'

The square root is the inverse of the square of a number.

For a population of 900, the sample size = $\sqrt{900}$ = 30.

There is a square root key on a scientific calculator.

Bias refers to **systematic** errors, which can give false outcomes. These must be eliminated wherever possible. In clinical trials, control groups are used (these are people with similar diet, weight etc.), along with an approach called blinding (see 'Making new drugs', section 4.14). Neither the patients nor the doctors can tell which are the test or the placebo drugs, so they don't know which patients get the test drug.

5 Explain fully how bias in clinical trials can be minimised.

6 What sample size would you suggest for trials involving these total populations?
   a 36 patients
   b 169 patients
   c 729 patients
   d 136 161 patients

Check your progress

You should be able to:

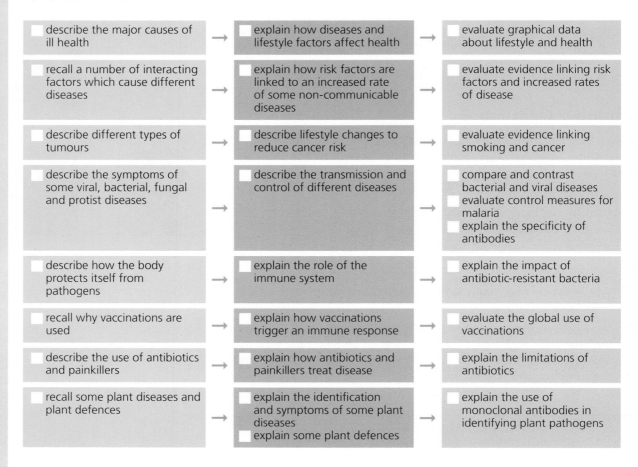

describe the major causes of ill health → explain how diseases and lifestyle factors affect health → evaluate graphical data about lifestyle and health

recall a number of interacting factors which cause different diseases → explain how risk factors are linked to an increased rate of some non-communicable diseases → evaluate evidence linking risk factors and increased rates of disease

describe different types of tumours → describe lifestyle changes to reduce cancer risk → evaluate evidence linking smoking and cancer

describe the symptoms of some viral, bacterial, fungal and protist diseases → describe the transmission and control of different diseases → compare and contrast bacterial and viral diseases
evaluate control measures for malaria
explain the specificity of antibodies

describe how the body protects itself from pathogens → explain the role of the immune system → explain the impact of antibiotic-resistant bacteria

recall why vaccinations are used → explain how vaccinations trigger an immune response → evaluate the global use of vaccinations

describe the use of antibiotics and painkillers → explain how antibiotics and painkillers treat disease → explain the limitations of antibiotics

recall some plant diseases and plant defences → explain the identification and symptoms of some plant diseases
explain some plant defences → explain the use of monoclonal antibodies in identifying plant pathogens

Worked example

Viruses are pathogens that cause many diseases. Babies and children are vaccinated to protect them from many viruses.

1 a Why are antibiotics not used to kill viruses in the body?

They cannot kill them.

> The correct answer is that viruses live inside cells, where they are protected from the action of antibiotics.

b What is in a vaccine?

A weak form of the virus.

> Correct answer identified, although 'weak form' could be improved by replacing it with 'inactive or dead form'.

c What does the vaccine do in the child's body?

It makes antibodies.

> This is inaccurate. White blood cells/lymphocytes make antibodies.

d Why do pathogens make us feel ill?

They produce poisons.

> This is partially correct, but the answer should also include the fact that pathogens damage cells.

2 a One type of tumour is known as benign. Name the other type of tumour and explain how it is different from a benign tumour.

Cancer tumour spreads through the body.

> Incorrect tumour type identified. Correct answer is malignant tumour, but the difference is correct.

b Name a risk factor for lung cancer.

Smoking

> Correct answer identified

3 a Describe how malaria is spread by mosquitoes.

Mosquitoes suck the blood of an infected person.

> Answer should also include how the mosquito then passes the protist pathogen/*Plasmodium* to an uninfected person.

b Explain how malaria is controlled by insecticide.

Insecticide kills eggs.

> Answer should mention mosquito eggs, and also include that mosquitoes lay their eggs in water.

End of chapter questions

Getting started

1. Define the term 'vector'. `1 Mark`

2. Describe how the stomach helps defend the body against pathogens. `1 Mark`

3. Name two risk factors for developing cardiovascular disease. `2 Marks`

4. What type of pathogen causes rose black spot? `1 Mark`

5. What type of blood cells destroy bacterial pathogens? `1 Mark`

6. A blood test reveals that Victoria has a white cell count of 4 000 000 000. `1 Mark`

   Write this number in standard form.

7. Smoking damages cilia in the bronchi. How does this affect the smoker? `1 Mark`

8. Hebe plants have tough, waxy cuticles. How does this protect them from aphids? `2 Marks`

Going further

9. Name a disease that is caused by a protist pathogen. `1 Mark`

10. What do vaccines contain? `1 Mark`

11. Give two examples of physical defences that help prevent pathogens entering the plant. `2 Marks`

12. The first plant virus to be identified was tobacco mosaic virus (TMV). The virus causes 'mosaic'-like mottling and yellow spots on the leaves. The virus damages chloroplasts.

    Suggest why plants infected with TMV have stunted growth and discoloured leaves. `4 Marks`

13. The graph shows the effect of two risk factors on mortality. Describe the patterns in the data. `2 Marks`

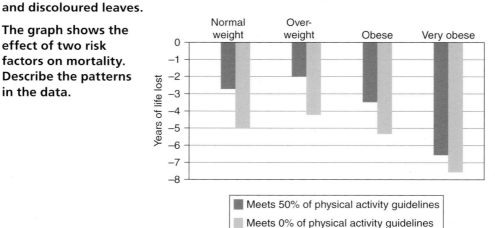

More challenging

14. Why is an antibody specific? `1 Mark`

15. Why is it difficult to develop drugs that kill viruses? `1 Mark`

16. Explain the use of antibiotics against bacteria and viruses. `2 Marks`

17 A group of scientists are investigating the effectiveness of a new antibiotic used to treat *streptococcus pyogenes* bacteria. They test the antibiotic by spreading *Streptococcus pyogenes* bacteria on an agar jelly plate and place a small disc of filter paper containing the antibiotic in the centre of the dish. They then measured the radius of the 'zone of inhibition' shown in the diagram.

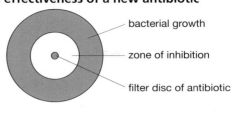

bacterial growth

zone of inhibition

filter disc of antibiotic

a The radius of the zone of inhibition was 12 mm. Calculate the area using the equation: Area = πr^2

Use the value $\pi = 3.14$

`1 Mark`

For an antibiotic to be considered effective it should cure 95–100% of the cases of an illness. Out of a sample of 568 patients given the new antibiotic, 483 were cured.

b The scientists used aseptic techniques during the investigations in order to avoid contamination. Explain why is this important.

`2 Marks`

c Calculate the chance of not being cured.

`1 Mark`

d The patients sampled were all university students aged between 18 and 25. Suggest how this may have affected the results of the investigation.

`2 Marks`

Most demanding

18 Explain how monoclonal antibodies are produced from a hybridoma cell.

`2 Marks`

19 Cholera is a potentially fatal bacterial infection caused by consuming contaminated food or water. In 1997 there were 150 000 cases of cholera and in 1998 there were 290 000.

Calculate the percentage increase in cases between 1997 and 1998. Give your answer to 1 decimal place.

`2 Marks`

20 The graph below shows smoking and lung cancer rates in the UK.

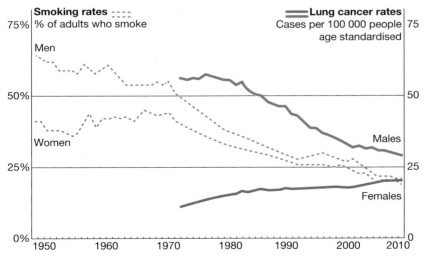

Smoking rates ----
% of adults who smoke

Lung cancer rates
Cases per 100 000 people
age standardised

Men

Women

Males

Females

`6 Marks`

Describe the trends in the data and evaluate the evidence for a link between smoking and lung cancer.

`Total: 40 Marks`

COORDINATION AND CONTROL

ORGANS WORK TOGETHER AS SYSTEMS.

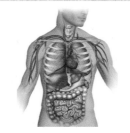

- Organs are aggregations of tissues.
- Organ systems work together to form organisms.
- Glands in the digestive system produce digestive enzymes.

METABOLISM INVOLVES CHEMICAL SYNTHESIS AND BREAKDOWN OF SUBSTANCES.

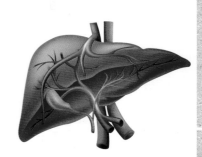

- Products of digestion are used to build carbohydrates, lipids and proteins.
- In animals, carbohydrate is stored as glycogen.
- Glucose is required by all cells for respiration; carbon dioxide and water are waste products.
- Carbon dioxide is excreted by the lungs.
- Excess protein is converted to urea.
- Other waste products from cells are transported in the blood and excreted by the kidneys.

HOW HUMANS REPRODUCE.

- The male and female reproductive systems are structured differently and adapted to their specialised functions.
- Male and female gametes join together in the process known as fertilisation.
- The fertilised cell develops to form an embryo during pregnancy.
- Some diseases are transmitted sexually and this can be prevented by some forms of contraception.

LIFE ON EARTH IS DEPENDENT ON THE LIFE PROCESSES OF PLANTS.

- Leaves are adapted to absorb the light energy needed for photosynthesis.
- Higher plants use flowers to reproduce.
- Fertilisation is followed by the production of seeds and sometimes fruit.

IN THIS CHAPTER YOU WILL FIND OUT ABOUT:

CONDITIONS IN THE BODY, PROCESSES AND ORGAN SYSTEMS ARE COORDINATED AND CONTROLLED.

- Regulation of the internal conditions in the body is called homeostasis.
- The nervous and endocrine systems are involved in this coordination and control.
- The nervous system works using electrical impulses, transmitted using nerves; the endocrine system uses chemicals called hormones, which are secreted by endocrine glands.
- In the nervous system, receptors can be grouped into sense organs, such as the eye.

- Different regions of the brain coordinate our responses, though spinal reflexes can by-pass it.
- The brain is a delicate and complicated structure, and this has consequences in its mapping and treatment of nervous system disorders.
- Temperature regulation is controlled by the brain and involves both the endocrine and nervous systems.

CONTROL OF METABOLISM AND LEVELS OF CHEMICALS IN THE BODY.

- The concentrations of glucose, water and salts must be kept within strict limits.
- Glucose concentrations and water balance are controlled by hormones.
- Lack of insulin, or a loss of sensitivity to it, causes a condition called diabetes, which must be controlled.

- The kidneys filter substances from the blood, excrete the waste products and reabsorb substances that are useful.
- Kidney failure requires dialysis or a transplant, and there are advantages and disadvantages to both.
- The control of hormone secretion by many glands is by negative feedback.

CONTROL OF SEXUAL DEVELOPMENT AND HUMAN REPRODUCTION.

- Reproductive hormones cause secondary sexual characteristics to develop.
- Pituitary gland hormones regulate egg development and release, and along with reproductive hormones, prepare the body for a possible pregnancy.

- Different methods of contraception help to prevent unwanted births.
- Fertility drugs and *in-vitro* fertilisation are possible solutions to infertility.

PLANTS RESPOND TO STIMULI TO CONTROL IMPORTANT PROCESSES.

- Growth movements are called tropisms and these are brought about by hormones called auxins.
- Other hormones include gibberellins and ethene, and these control seed

germination, cell division and fruit ripening.
- Plant hormones have many applications in horticulture and agriculture.

Homeostasis

Learning objectives:

- explain the importance of homeostasis in regulating internal conditions in the body
- recall that these control systems involve nervous or chemical responses
- describe how control systems involve receptors, coordination centres and effectors.

KEY WORDS

endocrine system
homeostasis
hormone
nervous system
target organ

Eve is running a marathon. While she is running, mechanisms in her body will try to keep internal conditions as constant as possible.

What changes in Eve's body?

As Eve is running, the rate at which her body cells respire increases. Her heart rate, breathing rate and breath volume increase during exercise to supply her muscles with more oxygenated blood. Her muscle cells break down glucose to release the energy she needs. Thermal energy is also released.

The thermal energy is carried around Eve's body by her blood. Normal body temperature is 37°C, the optimum temperature for enzyme action and other cell functions. The human brain, for instance, is very sensitive to changes in temperature.

Eve must control her body temperature within strict limits.

One way she does this, as she runs, is to sweat. But sweating will also affect the water level in her body. She takes in extra water from sports drinks during the race. The drinks also contain some glucose. Her stores of glycogen will decrease during the race and, as she suffers from fatigue at the end of the race, her blood glucose concentration falls.

Figure 5.1 Conditions in Eve's body will change during the marathon

1 **Why does body temperature need to be kept constant?**

2 **Name two other things that have to be controlled by the body.**

Homeostasis

Homeostasis is the regulation of internal conditions in the body. It is necessary to maintain the optimum conditions for body function. These internal conditions can change as a result of processes within the body and as external conditions change.

The control systems involved in homeostasis are 'automatic' – they happen all the time or if conditions change – we don't have to think about taking any action. The body systems that are responsible for homeostasis are:

- the **nervous system** – uses electrical impulses to communicate
- the **endocrine system** – uses chemical molecules to communicate.

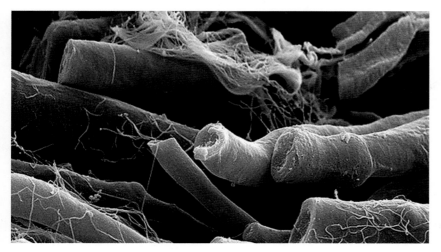

Figure 5.2 The long regions of nerve cells are bundled together as nerves

③ Write down a definition of homeostasis.

④ Which two body systems are responsible for homeostasis?

Control systems

All control systems in the body, whether nervous or endocrine, have the same pattern.

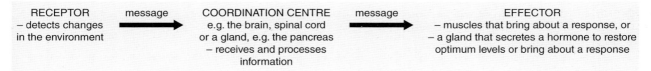

RECEPTOR — detects changes in the environment   → message →   COORDINATION CENTRE e.g. the brain, spinal cord or a gland, e.g. the pancreas — receives and processes information   → message →   EFFECTOR — muscles that bring about a response, or — a gland that secretes a hormone to restore optimum levels or bring about a response

Figure 5.3 The components of the body systems that are responsible for homeostasis

The nervous system and endocrine system are different in nature but, in practice, the two systems interact with and regulate each other.

	Nervous system	Endocrine system
response	rapid and short duration	slower but acts for longer
nature of message	nerve impulse – electrical	a **hormone** – chemical
action	carried in nerves to specific location, e.g. muscle	carried in blood to all organs, but affects the **target organ** only

The nervous and endocrine systems compared

⑤ How are changes detected by the body?

⑥ Compare and contrast the nervous system and the endocrine system.

DID YOU KNOW?

The squid has giant nerve cells, up to 1 mm in diameter. Study of these has been key to our understanding of how nerve impulses are transmitted.

ADVICE

Remember the sequence of receptor → coordination centre → effector.

The nervous system

Learning objectives:

- explain how the nervous system is adapted to its functions
- describe the structure of the central nervous system and nerves.

KEY WORDS

central nervous
  system
myelin sheath
neurone
receptor

Nerve cells communicate with each other and with muscles, glands and other structures. There are critical periods during our lives when nerve cells are able to make these connections with other nerve cells. That is why we are able to learn different skills at different times – usually early on – in our lives.

The structure of the nervous system

Our nervous system enables us to detect our surroundings, and coordinate our behaviour.

The structure of the nervous system is well adapted to these functions.

1. What is the function of the nervous system?

2. What are the two parts of the nervous system?

Nerves

The nervous system consists of nerve cells or **neurones**. Neurones are specialised for transmitting messages in the form of an electrical impulse.

The part of the cell containing the nucleus is called the cell body. The cell body of all neurones is found in the CNS. Neurones, however, have an extended shape so that they can carry nerve impulses from one part of the body to another. Neurones also have fine branches at their tips to communicate with other neurones.

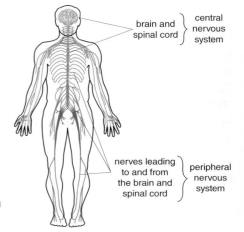

Figure 5.4 The nervous system is made up of the **central nervous system** (CNS) and the peripheral nervous system (PNS)

Figure 5.5 The projections that extend from a nerve cell communicate with other nerve cells

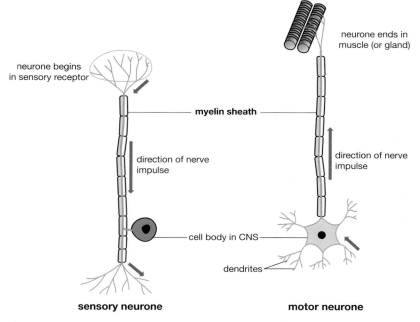

neurone ends in muscle (or gland)

neurone begins in sensory receptor

myelin sheath

direction of nerve impulse

direction of nerve impulse

cell body in CNS

dendrites

sensory neurone

motor neurone

Figure 5.6 Two types of neurone. The long regions of neurones are bundled together as 'nerves'.

Receptors are cells that detect any changes in the environment. Receptors are sometimes grouped to form sense organs. Sensory neurones relay nerve impulses from these receptors to the CNS.

The CNS processes the information and coordinates how the body should respond.

Motor neurones relay impulses from the CNS to the effector – for instance, a muscle or gland.

The sequence of events is:

stimulus → receptor → coordinator → effector → response

3 What is the scientific name for a nerve cell?

4 Describe the pathway of a nerve impulse, beginning with the stimulus and ending with the response.

MAKING CONNECTIONS

Link the information given here back to topic 1.7 and how nerve cells are adapted to their functions.

DID YOU KNOW?

If the myelin sheath does not develop properly, or becomes or inflamed or damaged – in conditions such as multiple sclerosis, and diseases such as leprosy – the transmission of nerve impulses can be seriously affected.

Reflex actions

Learning objectives:

- explain the importance of reflex actions
- describe the path of a reflex arc
- explain how the structures in the reflex arc relate to their function.

KEY WORDS

reflex action
reflex arc
relay neurone
synapse

Nasim's doctor tests his reflexes by tapping the tendon just below the knee cap. Nasim's leg kicks up.

Reflexes are related to survival

Reflex actions are rapid, automatic responses to a stimulus. We do not have to think about them. Reflex actions form the basis of behaviour in simpler organisms. In humans, they prevent us from getting hurt. In other animals, our human ancestors and babies they are also related to survival.

Some reflex actions include:

- removing our hand from a hot or sharp object
- the grasping reflex, in which a baby grips a finger
- blinking our eyes if an object approaches rapidly.
- the pupil reflex, whereby the pupil gets wider in dim light and narrower in bright light.

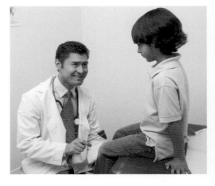

Figure 5.7 The knee jerk reflex. A normal reaction time is around 50 ms

1. What is a reflex action?

2. Why are reflex actions important?

The reflex arc

Our spinal reflexes do not involve the brain. Or, at least, not to begin with. In a reflex action, the nerve impulse follows a pathway called the **reflex arc**.

Figure 5.8 shows the pathway taken by a nerve impulse when a person puts a hand on a hot object.

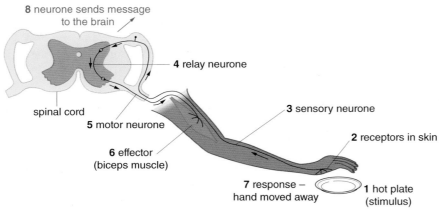

8 neurone sends message to the brain

4 relay neurone

spinal cord

5 motor neurone

3 sensory neurone

2 receptors in skin

6 effector (biceps muscle)

7 response – hand moved away

1 hot plate (stimulus)

Figure 5.8 The reflex arc. Follow the numbers 1–8. The pathway is through the spinal cord

The pathway includes:

- a sensory neurone – transmits nerve impulses from the receptor to the CNS
- a **relay neurone**, in the spinal cord – transmits the impulses from the sensory to the motor neurone
- a motor neurone – sends impulses from the CNS to the effector.

In this case, the effector is the biceps muscle, which moves the arm. The hand is moved away from the hot object.

Other neurones in the spinal cord link via synapses with those of the reflex arc, a message is sent to the brain *after* the hand has been removed (number 8 in the figure). It tells us that the plate was hot.

3 Name the nervous pathway that a nerve impulse takes during a reflex action.

4 Explain how the parts of this pathway relate to their function.

Linking nerves

The three neurones in the reflex arc do not link *physically*. There is a gap – called a **synapse** – between each pair. This means that *many* neurones can connect with each other. In the brain, neurones can link up with up to 10 000 others.

Nerve impulses pass across a synapse with the help of chemical transmitter molecules.

5.3

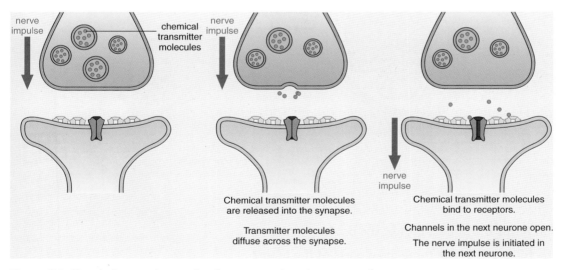

Figure 5.9 Chemical transmitter molecules cause an impulse to move from one neurone to the next

5 What is the gap between neurones called?

6 How does a nerve impulse travel from one nerve to the next?

7 Compose a flow diagram to describe and explain the sequence of events in a reflex arc.

The brain

Learning objectives:

- recall that the brain controls complex behaviour using billions of interconnected neurones
- identify the three main regions of the brain and describe their functions
- describe how the regions of the brain are mapped.

KEY WORDS

cerebellum
cerebral cortex
medulla
magnetic resonance imaging
non-invasive

The brain coordinates and controls all our activities and behaviour – from solving maths problems, writing and playing music, and sporting activities, to behaving aggressively and falling in love.

The regions of the brain

The brain consists of three main regions:

- the **cerebral cortex**
- the **cerebellum**
- the **medulla**.

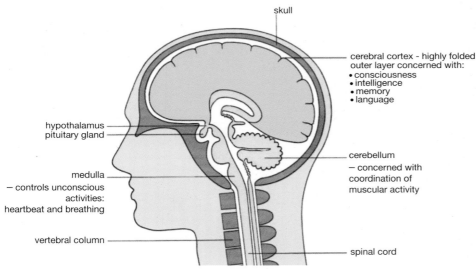

Figure 5.10 The location and functions of the main parts of the human brain

skull

cerebral cortex - highly folded outer layer concerned with:
• consciousness
• intelligence
• memory
• language

hypothalamus
pituitary gland

cerebellum
— concerned with coordination of muscular activity

medulla
— controls unconscious activities:
heartbeat and breathing

vertebral column

spinal cord

1 Which region forms the largest part of the brain?

2 What are the functions of the three main parts of the brain?

HIGHER TIER ONLY

Brain mapping

Owing to its complexity and delicate nature – it has the texture of blancmange – investigating the brain is difficult and must be done with caution.

Electroencephalograms – EEGs – are used to monior abnormal electrical activity in the brain, but can also be used in brain mapping.

In transcranial magnetic stimulation (TMS), a magnetic field changes the electrical activity in parts of the brain targeted.

Changes in the patient's behaviour as different areas are stimulated are used to map the brain.

Figure 5.11 Non-invasive mapping: electroencephalography and transcranial magnetic stimulation

Early evidence of brain function came from observing the effects of brain damage. Later, people were studied by electrically stimulating regions of an exposed brain. It's a technique still used today.

But brain maps can now be built up using **non-invasive** techniques that are *external* to the body.

Imaging techniques are also essential in diagnosing disorders of the nervous system. **Magnetic resonance imaging (MRI)** is one of the most important techniques. MRI scanning uses strong magnetic fields and radiowaves to produce very detailed images of the nervous system (or any other region of the human body).

MRI scanning is very safe. It's non-invasive and doesn't use ionising radiation. It's therefore safer than the alternatives – a CT scan or a PET scan.

3 What invasive techniques have been used to map the brain?

4 Why are MRI scanners used?

Treating nervous system disorders

In many cases where the nervous system has been damaged, repair is simply not possible.

Surgery is often needed to save a life, such as by removing a tumour, draining excess fluid (from a bleed or infection) or adding a brain implant. But it does carry risks, such as infection and the possibility of a stroke. Surgery has to be carried out with minimal damage to the surrounding tissue. But in the case of a tumour, cancer cells must be removed as completely as possible. For this reason, surgery is followed up (or replaced) by radiotherapy and chemotherapy to kill remaining cancer cells. Both procedures affect normal cells as well and have serious side effects.

In future, there may be safer and more effective alternatives. Monoclonal antibodies and gene therapy may be used to treat brain cancer, while stem cells offer hope for repairing damaged nervous tissue.

5 Describe two situations in which surgery is used on the nervous system.

6 Discuss benefits and risks of procedures carried out on the nervous system.

7 Investigating and treating brain disorders is very difficult compared with other parts of the body. Why do you think this is?

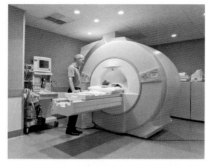

Figure 5.12 The patient is moved into a cylindrical MRI scanner. MRI scans show up damage, inflammation and tumours, and give information on blood flow

DID YOU KNOW?

An MRI scan works by giving precise locations of the protons in the body's hydrogen atoms.

KEY SKILL

You should be able to evaluate benefits and risks of procedures that are used to investigate and treat the nervous system.

REQUIRED PRACTICAL

Investigating reaction time

Learning objectives:

- select appropriate apparatus and techniques for the measurement of biological processes
- carry out physiological experiments safely
- use appropriate techniques in problem-solving contexts.

Our reflex actions protect us from harm. Quick reactions are also important in sport.

Measuring reaction time

A group of students worked in pairs to find their **reaction times** using the ruler drop test.

One student dropped a 30 cm ruler while another student caught it between their outstretched thumb and index finger of their **dominant hand**.

In the test, the release of the ruler is detected by our eyes and a message is sent to the sensory region of our brain. A message is then sent to another part, the motor region, which instructs muscles in our hand to contract.

After the student has carried out the test ten times, they rest their hand and then drink a cup of coffee. The student takes the test again to investigate the effect of coffee on their performance.

These pages are designed ❗ to help you think about aspects of the investigation rather than to guide you through it step by step.

Test number	Experiment 1: Normal student	Experiment 2: Student after drinking coffee
	Distance the ruler dropped (mm)	
1	119	98
2	116	98
3	117	92
4	113	91
5	150	92
6	113	93
7	108	92
8	109	92
9	108	91
10	107	91

The student's results for the ruler drop test

Figure 5.13 In cricket, fielders need fast reaction times

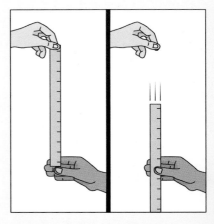

Figure 5.14 The ruler drop test

1 Write a risk assessment for this experiment.

2 Identify any anomalous results.

3 Calculate the average distance fallen by the ruler before and after drinking coffee.

Calculating reaction time

The distance travelled by the ruler before it is caught gives an indication of a student's reaction time, but not the reaction time itself.

Another group of students find a formula on the Internet that is used to calculate reaction time:

where t = time in seconds

d = distance in metres,

a = acceleration as a result of gravity = 9.81 m/s²

They use this formula to calculate their reaction times.

$$t = \sqrt{\frac{2d}{a}}$$

4 Calculate the mean reaction time for the student before and after coffee.

5 The ingredient in coffee that affects our nervous system is caffeine. When testing its effect on other students, explain why the experiment must be carefully controlled to produce **valid** results.

Pooling class results

All the students in the year group measured their minimum reaction time. They used an alternative test on the computer. They had to click on the mouse when the screen changed colour. The reaction time was measured by the computer timer.

The students produced a tally of those falling into different ranges of reaction time.

Mean reaction time (ms)	101–200	201–250	251–275	276–300	301–325	326–350	351–375	376–400	401–500
Number of students within range	1	3	16	24	33	14	6	2	1

The tally of student reaction times across the year group

6 What type of chart would be best suited to displaying the data shown above? Draw a chart of the results.

7 Determine the median and modal reaction time categories.

8 Suggest why the computer method for measuring reaction time may be better than the ruler drop method.

The eye

Learning objectives:

- relate the structures of the eye to their functions
- understand how the eye is adapted to seeing in colour and in dim light.

Humans and related primates can see the colours blue, green and red. But most mammals can distinguish combinations of just two colours.

The human eye

The structures of the human eye are all concerned with focusing an image of what we see onto light-sensitive receptor cells in the **retina**.

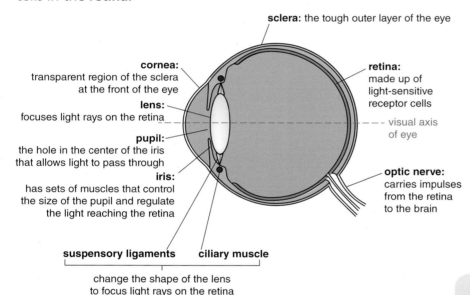

sclera: the tough outer layer of the eye

cornea: transparent region of the sclera at the front of the eye

lens: focuses light rays on the retina

pupil: the hole in the center of the iris that allows light to pass through

iris: has sets of muscles that control the size of the pupil and regulate the light reaching the retina

retina: made up of light-sensitive receptor cells

visual axis of eye

optic nerve: carries impulses from the retina to the brain

suspensory ligaments   ciliary muscle
change the shape of the lens to focus light rays on the retina

Figure 5.15 The structure of the eye

1. Which part of the eye has receptor cells that are sensitive to light?

2. Give one function of:
   - the **iris**
   - the **ciliary muscles and suspensory ligaments**
   - the **optic nerve**.

DID YOU KNOW?

Cone cells connect with *individual* neurones that send messages to the brain. This means that our colour vision is acute – we can see in fine detail. As we move away from the eye's central vision, there are fewer cones, mostly rods. Our peripheral vision is in black and white and is much less acute.

Seeing in colour

The receptor cells in the eye convert light into the electrical energy of a nerve impulse.

Receptor cells in the retina that perceive colour are called **cones**.

3 Name the receptor cells that are sensitive to colour.

4 Why is our colour vision acute?

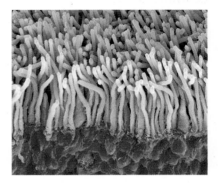

Figure 5.16 A scanning electron micrograph of the retina. Cones are shown in green

Seeing in dim light

It's also an advantage for humans to be able to see at low light intensities.

In dim – or bright – light, two sets of muscles work to regulate the amount of light falling on the retina.

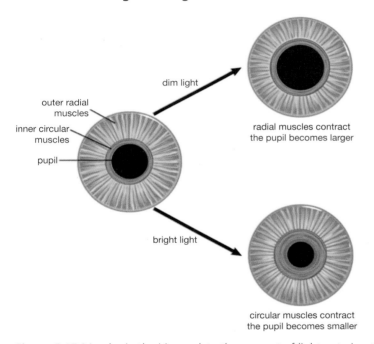

dim light

radial muscles contract
the pupil becomes larger

outer radial muscles

inner circular muscles

pupil

bright light

circular muscles contract
the pupil becomes smaller

Figure 5.17 Muscles in the iris regulate the amount of light entering the eye

A different type of receptor cell is also responsible for our vision at low light intensities – **rods**. Rods are around 1000 times more sensitive to light than cones. It's also much less acute.

5 How is the iris adapted to low-intensity vision?

6 Explain why rods enable us to see in low-intensity light.

ADVICE

Think about how the structures of the eye are adapted to their function and how the eye adapts to dim light.

Seeing in focus

Learning objectives:

- relate the structures of the eye to their functions
- understand how the eye is able to focus on near or distant objects.

As we get older, we find it more difficult to focus on close objects. One reason is because the lens hardens. Scientists are researching the possibility of refilling lenses with a soft polymer to restore normal vision.

Focusing an image on the retina

When light rays enter the eye, they are refracted. **Refraction** is the bending of light rays as they travel from one medium to another.

Structures of the human eye refract light rays so as to focus an image of what we see onto light-sensitive receptor cells in the retina.

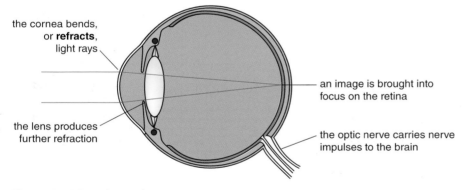

the cornea bends, or **refracts**, light rays

the lens produces further refraction

an image is brought into focus on the retina

the optic nerve carries nerve impulses to the brain

Figure 5.18 Focusing an image

First the cornea refracts light rays towards the lens. The lens refracts the light some more. Figure 5.18 shows two incoming light rays being focused to a point on the retina in normal vision.

1. What is refraction?
2. What structures in the eye take part in the refraction of light?

Near and far objects

Incoming light rays from a distant object are almost parallel. As they reach the eye, the lens has to do little to bring them into focus.

Light rays from a near object diverge. To focus them, the lens changes its shape. It becomes thicker.

DID YOU KNOW?

As we age, the lens in our eye changes from colourless to yellow. The intense light of the Sun damages the lens. When a clouding of the lens occurs, this is called a cataract.

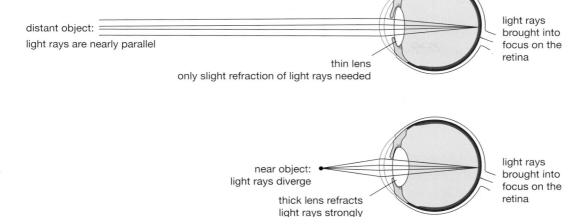

distant object:
light rays are nearly parallel

light rays brought into focus on the retina

thin lens
only slight refraction of light rays needed

near object:
light rays diverge

thick lens refracts
light rays strongly

light rays brought into focus on the retina

Figure 5.19 The lens changes shape in order to focus an image of an object

3 **Describe how light rays approach the eye from distant and near objects.**

4 **How does the shape of the lens change when focusing light rays from a near object?**

Accommodation

The cornea is fixed in its shape. The lens, however, can change its shape to focus on objects that are located at different distances. This is **accommodation**.

The **ciliary muscles** form a circular ring of muscle around the lens. This can change the shape of the lens. Attached to the ciliary muscle, and holding the lens in place in its elastic capsule, are the **suspensory ligaments**. These can be stretched or loose.

At rest, the ciliary muscles are relaxed and the taut suspensory ligaments stretch the lens into a thin shape.

When we focus on a close object, the ring of ciliary muscles contracts. The ring decreases in diameter. This reduction in diameter releases the tension on the suspensory ligaments. This allows the lens to bulge and become thicker. The light rays are refracted more.

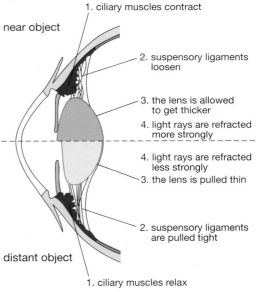

near object

1. ciliary muscles contract

2. suspensory ligaments loosen

3. the lens is allowed to get thicker

4. light rays are refracted more strongly

4. light rays are refracted less strongly

3. the lens is pulled thin

2. suspensory ligaments are pulled tight

distant object

1. ciliary muscles relax

Figure 5.20 The process of accommodation changes the shape of the lens

5 **What is the name of the process that changes the shape of the lens?**

6 **Create a table to compare the processes of focusing on a nearby object and a distant object.**

COMMON MISCONCEPTION

Many students think that the lens alone is responsible for refracting light rays. In fact, the cornea makes up *70–80%* of the refracting power of the eye.

Eye defects

Learning objectives:

- understand that, in myopia and hyperopia, the eye cannot focus light rays on the retina
- demonstrate how techniques are used to correct eye defects.

KEY WORDS

gel
hyperopia
myopia
laser surgery

It's estimated that 40 people every day in the UK lose their sight. And yet, only 2% of medical funding is currently available for eye disease and sight loss.

Short-sightedness

Short-sightedness, or **myopia**, is when people can see objects at short distances away, but struggle to see objects at a distance. It occurs when:

- the eyeball is too long for the strength of the lens

  or

- the cornea is too sharply curved.

When the eye attempts to produce an image, it falls short of the retina.

This can be corrected by a concave lens.

1. List the causes of short-sightedness.

2. Explain how a concave lens can be used to correct short-sightedness.

Long-sightedness

Long-sightedness, or **hyperopia**, is when people can see objects at long distances away, but struggle to see near objects. It occurs when:

- the lens is too weak – it is not thick enough

  or

- the eyeball is too short

  or

- the cornea is not curved enough.

When the eye attempts to produce an image, it falls *behind* the retina. It can be corrected by a convex lens.

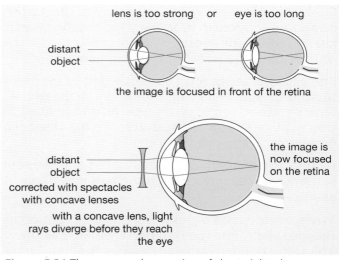

lens is too strong   or   eye is too long

distant object

the image is focused in front of the retina

distant object

corrected with spectacles with concave lenses

with a concave lens, light rays diverge before they reach the eye

the image is now focused on the retina

Figure 5.21 The cause and correction of short-sightedness

DID YOU KNOW?

Up to one in three people in the UK is short-sighted. The cause isn't known, but it's clear that there's a genetic link.

3 List the causes of long-sightedness.

4 Explain how a convex lens can be used to correct long-sightedness.

Figure 5.22 A flap is cut in the cornea and folded back. A laser is used to change the shape of the cornea, and the flap is repositioned

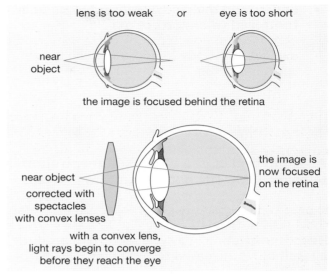

Figure 5.23 The cause and correction of long-sightedness

New technologies

Some new technologies are available to correct eye defects.

Contact lenses work in the same way as spectacles – by correcting refractive error. They are simply a thin lens placed on the surface of the eye.

Today, most contact lenses used are described as *soft*. Most are made from a **gel** based on silicone. These are freely permeable to oxygen, allowing oxygen to be transmitted to the cornea. They're considered more comfortable than *hard* contact lenses, which are more rigid and durable. Today's hard lenses are also gas permeable.

Modern contact lenses can be worn for varying periods of time. Some are disposable after a day's use; others require cleaning and disinfection.

Some surgical techniques are available to correct eye defects. **Laser surgery** is used to change the shape of the cornea. The eye normally recovers in a few days.

With serious refractive error or age-related problems, where the problem is with the lens itself, an artificial lens is inserted. Some modern lenses are multifocal, so the patient will regain their ability to see objects at different distances.

5 Describe the difference between a hard and soft contact lens.

6 Explain the principle of laser surgery.

7 It is possible to get replacement lens surgery. Suggest a potential advantage and disadvantage of having this surgery to correct an eye defect such as myopia or hyperopia.

ADVICE

You need to be able to interpret ray diagrams demonstrating how spectacle lenses can correct myopia and hyperopia.

Controlling body temperature

Learning objectives:

- understand the mechanisms by which body temperature is controlled when too hot or cold

- explain how body temperature can be controlled in a specific context.

KEY WORDS

cooling
evaporation
negative feedback
thermoregulatory centre
vasoconstriction
vasodilation

As Huan cycles, the increased respiration in his cells releases heat. His body temperature rises.

The body temperature rises

Huan is competing in a cycle race. He is producing more heat as a result of his increased respiration.

His increased body temperature is monitored and controlled by the **thermoregulatory centre** in his brain. This detects a change in body temperature in two ways:

- The centre has receptors that are sensitive to a change in the temperature of the blood circulating through it.
- The skin has temperature receptors that send nervous impulses to the thermoregulatory centre.

Huan's body temperature is too high. Blood vessels in his skin become wider, or *dilate* (**vasodilation**). His sweating also increases. These two activities mean that he loses more heat through his skin.

1. Where is body temperature monitored and controlled?

2. Give two actions that the body takes when body temperature increases.

Figure 5.24 A thermal image of Huan cycling. The image shows the heat from his muscles – and his bicycle tyres. False colours are used to represent the different temperatures.

DID YOU KNOW?

The cooling effect of sweating makes use of the properties of water. It has a very large latent heat of vaporisation.
Even when we are unaware of sweating, we lose 0.6 kg of moisture from the skin per day.

The body temperature decreases

After the race, Huan's body temperature drops. He is given a thermal blanket.

A fall in body temperature is again monitored and controlled by the thermoregulatory centre. This time:

- Blood vessels in the skin become narrower, or *constrict* (**vasoconstriction**).
- Sweating is reduced or stopped.
- Skeletal muscles contract and the body shivers.

3. What is vasoconstriction?

4. Give one other mechanism that the body uses to control its temperature.

ADVICE

You need to be able to explain how mechanisms that are used to control body temperature operate in different contexts.

Mechanisms for body temperature regulation

Figure 5.25 shows how human body temperature is restored to normal, if it becomes changed.

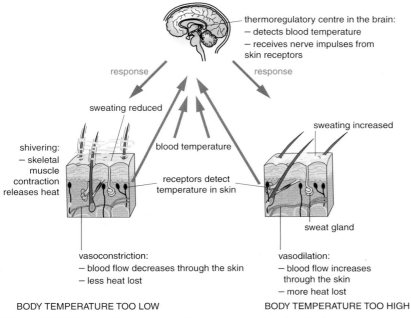

Figure 5.25 Maintaining a constant body temperature is called thermoregulation

Vasodilation, vasoconstriction and sweating are all processes that change the rate of energy transfer from the skin to the environment.

When blood flows through our skin, heat is lost through its surface. When we are warm or cold the arterioles and venules lower in the skin dilate or constrict, allowing more or less blood to flow near the surface in the capillaries.

Sweating works by **cooling** through **evaporation**. As water molecules in sweat evaporate, they take heat from the skin away.

Another response to getting cold is shivering. Shivering is repeated involuntary contraction of our muscles. Respiration is not 100% efficient, so the heat released as muscles contract helps to warm our bodies.

An important part of homeostasis is being able to reverse changes to the body when they have happened – this is known as **negative feedback**.

⑤ **Describe how heat transfer to the environment is increased when we get hot.**

HIGHER TIER ONLY

⑥ **Jenny is out with friends on a cold winter's evening. She is not wearing a coat. Explain fully the mechanisms in her body to keep warm.**

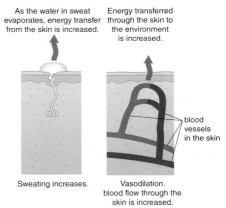

As the water in sweat evaporates, energy transfer from the skin is increased.

Energy transferred through the skin to the environment is increased.

Sweating increases.

Vasodilation. blood flow through the skin is increased.

blood vessels in the skin

THE BODY IS TOO HOT

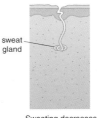

As the water in sweat evaporates, energy transfer from the skin is decreased.

Energy transferred through the skin to the environment is decreased.

sweat gland

Sweating decreases or stops.

Vasoconstriction: blood flow through the skin is decreased.

THE BODY IS TOO COLD

Figure 5.26 Skin blood flow during overheating and overcooling

The endocrine system

Learning objectives:

- recall that the endocrine system is made up of glands that secrete hormones into the blood
- know the location of the major endocrine glands
- understand why the pituitary gland is the 'master gland'.

KEY WORDS

endocrine gland
endocrine system
hormone

The world's tallest man was Robert Pershing Wadlow (1918–1940). He was 2.72 m when he died at the age of 22.

His height was caused by an excess of growth hormone, produced by the pituitary gland.

The endocrine system

Hormones are produced by glands of the **endocrine system**. **Endocrine glands** secrete hormones directly into the blood.

Hormones are often described as *chemical messengers*. They circulate in the blood and produce an effect on target organs. Many hormones are large molecules.

Like the nervous system, hormones work on effectors. But, unlike the nervous system, the effects they produce don't take milliseconds. With the exception of a hormone called adrenaline, effects of hormones take minutes or hours to occur, and in the case of hormones involved in our development, they can act for years.

Figure 5.27 We all produce growth hormone. Maximum production is during our youth

1 **What is the function of the endocrine system?**

2 **On what does the endocrine system act?**

Location of the endocrine glands

There are a number of endocrine glands situated in different parts of the body. Some of these are gender specific and some produce enzymes as well as hormones.

3 **Which organ secretes hormones and also digestive enzymes?**

4 **Which endocrine glands are found in the female only?**

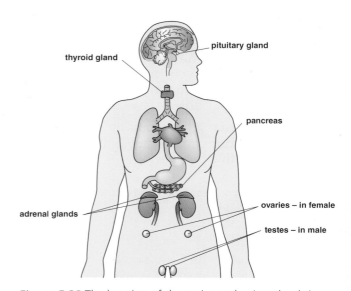

Figure 5.28 The location of the major endocrine glands in the body

The master gland

The pituitary gland is an outgrowth from the base of the brain.

Some of the hormones that it secretes, such as growth hormone, have a direct effect on their target organs. Other hormones have an indirect effect; they cause other glands to secrete hormones. It is, therefore, called the master gland as it regulates the secretion of other endocrine glands.

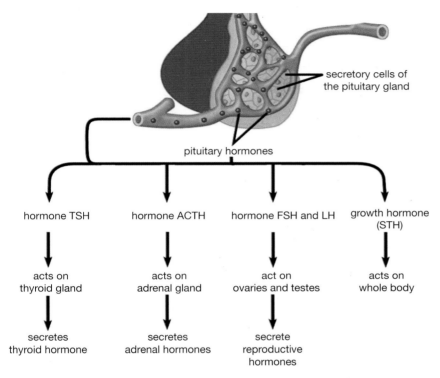

Figure 5.29 Some of the hormones secreted by the pituitary gland

5 **Explain why the pituitary gland is called the master gland.**

6 **Name one hormone that exerts its effects over the whole body.**

ADVICE

You need to be able to label the major endocrine glands on a diagram of the human body.

Controlling blood glucose

KEY WORDS

glucagon
insulin

Learning objectives:

- recall that blood glucose is monitored and controlled by the pancreas
- understand how insulin controls the blood glucose level
- understand how insulin works with another hormone – glucagon – to control blood sugar level.

Pao cannot produce enough of the hormone insulin to control her blood sugar. She has diabetes. She needs injections of insulin to stay alive.

Controlling blood sugar

The pancreas secretes enzymes that digest carbohydrates, proteins and lipids.

The pancreas also has another function. It produces hormones that control the concentration of our blood glucose. Glucose is needed by all cells for respiration to release energy. It is carried to cells in our blood. But its concentration must be strictly controlled within certain limits.

After a meal, the concentration of glucose in our blood increases. The hormone **insulin** causes glucose in the blood to move into our body's cells. Here, this glucose can be used for respiration. In cells of the liver and muscle, the glucose is also converted into glycogen so that it can be stored.

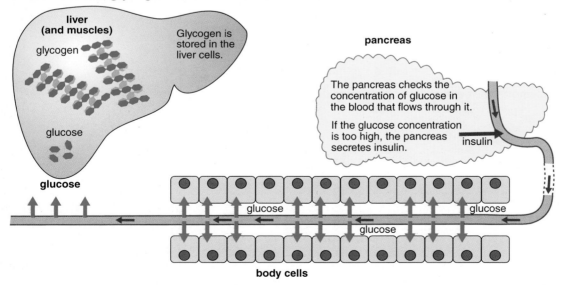

Figure 5.30 Insulin secreted into the bloodstream causes our body's cells to take up glucose

1 **Name a hormone that controls blood glucose.**

2 **What effect does this hormone have on our body's cells?**

Blood glucose concentration

Insulin restores the blood glucose concentration to its normal level. A little insulin is produced as you first smell or chew food. As you eat food and it is digested, the blood glucose level rises, which causes a surge in insulin. The insulin level reaches a peak, then gradually falls.

Figure 5.31 shows what happens to blood glucose on eating a meal and then with the effect of insulin.

3 **Describe how a person's blood glucose concentration changes after a meal.**

4 **How long after having a meal did this person's blood glucose level start to fall?**

HIGHER TIER ONLY

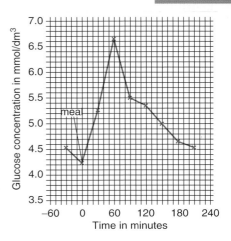

Figure 5.31 Blood glucose concentrations can rise to up to 8 mmol/dm³ after a meal or snack, before falling as insulin is secreted.

Glucagon

Blood glucose is normally regulated at between 4.5–7.5 mmol per dm³ of blood. To obtain this fine control, another hormone that is produced by the pancreas is also involved – **glucagon**.

Insulin and glucagon achieve this fine control by balancing glucose with carbohydrate stored as glycogen. This balance is maintained via a negative feedback cycle.

Insulin promotes the uptake of glucose by cells, and its conversion into glycogen in the liver and muscles.

Glucagon, which is secreted in response to low blood glucose concentration, promotes the conversion of stored glycogen into glucose, which is released into the bloodstream.

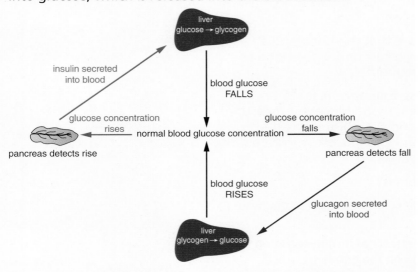

Figure 5.32 The control of blood glucose concentration by insulin and glucagon

5 **What is normal concentration of blood glucose? Why is it important to maintain this level of blood glucose?**

6 **Explain how a constant level of blood glucose concentration is maintained.**

DID YOU KNOW?

Until the discovery of insulin in 1921 by Frederick Banting and Charles Best, diabetes wasn't treatable. Banting and Best carried out experiments on dogs and, later, themselves, before testing purified insulin on a teenager with diabetes in 1922.

ADVICE

Remember: don't confuse the storage compound glycogen with the hormone glucagon.

Diabetes

Learning objectives:

- understand the causes of Type 1 and Type 2 diabetes
- compare Type 1 and Type 2 diabetes
- evaluate information on the relationship between obesity and diabetes, and make appropriate recommendations.

KEY WORDS

body mass index
glucose tolerance
 test
Type 1 diabetes
Type 2 diabetes

Diabetes, if poorly controlled, can damage circulation, nerves, and muscles in the feet and legs, which may then need amputating.

Type 1 diabetes

In **Type 1 diabetes**, the pancreas is unable to produce enough, or any, insulin.

Without insulin, the body's cells are unable to take up glucose. The blood glucose level becomes uncontrollably high, and glucose is excreted in the urine.

Without glucose, cells must use alternative energy sources. Fat and protein are used. The person will lose weight. If the condition is not controlled, kidney failure and death will result.

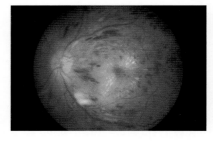

Figure 5.33 Diabetes causes damage to the retina of the eye. If left untreated, it may lead to blindness

1 **What is the cause of Type 1 diabetes?**

2 **Describe the sequence of events that will occur if a person has diabetes that isn't controlled.**

Type 2 diabetes

The main cause of **Type 2 diabetes** is that the body's cells lose their sensitivity – they no longer respond, or respond as effectively – to the insulin being produced.

One test for diabetes is the **glucose tolerance test**. After 8–12 hours of no eating or drinking, blood glucose is measured. The person is then given glucose and their blood retested 2 hours later. If the person's tolerance to glucose is lowered, the glucose will be above a certain level when retested.

3 **What is the cause of Type 2 diabetes?**

4 **Describe how the glucose tolerance test can help to diagnose Type 2 diabetes.**

5 **Use Figure 5.34 to describe the effects of insulin secretion in the three examples of people shown on the graph.**

What causes diabetes?

Only 10% of people with diabetes have Type 1. The cells that produce insulin have been destroyed. This can be an autoimmune condition – in which the immune system attacks the person's own body.

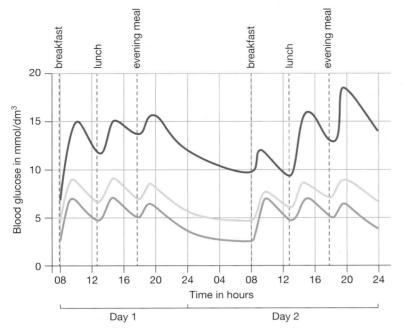

Figure 5.34 Blood glucose increases after a meal in a person with Type 2 diabetes and is not brought back to normal

Type 2 diabetes tends to cluster in families. But it's also associated with a Western lifestyle – one that includes high energy 'fast' food and an inactive life. There's a clear link with obesity.

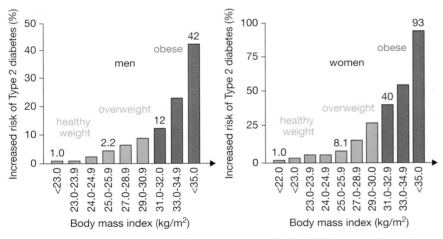

Figure 5.35 There is a correlation between increased risk of diabetes and obesity, as indicated by **body mass index** (BMI)

Type 2 diabetes is now emerging in young people. The first cases were in 2000, in children aged 9–16. Around three in ten children are now obese.

6 Discuss links between genetics and the environment with Type 2 diabetes.

7 Compare the correlation between obesity and Type 2 diabetes in men and women.

Diabetes recommendations

Learning objectives:

- understand the causes of Type 1 and Type 2 diabetes
- compare Type 1 and Type 2 diabetes
- evaluate information on the relationship between obesity and diabetes, and make appropriate recommendations.

KEY WORDS

Type 1 diabetes
Type 2 diabetes

In 2014, 3.3 million people in the UK had been diagnosed with diabetes. Diabetic patients are currently costing the NHS about £10 billion a year. *That's £1 million per hour.*

Treatment

There's currently no cure for diabetes. **Type 1 diabetes** cannot be prevented. Patients with Type 1 diabetes *control* it with insulin injections.

For **Type 2 diabetes**, obesity accounts for 80–85% of the risk of developing it; so, it is preventable. The most important way of managing the condition is by modification of lifestyle. This involves exercise and diet. Carbohydrate-controlled diets are important for all diabetics. Foods that rapidly affect blood sugar level should be avoided.

1 **How can Type 1 diabetics control the condition?**

2 **What recommendations should be given to Type 2 diabetics?**

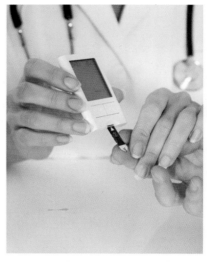

Figure 5.36 It's within our hands to control one type of diabetes – Type 2

Ethical considerations

Public guidelines and national programmes of weight management have been designed to help to prevent Type 2 diabetes.

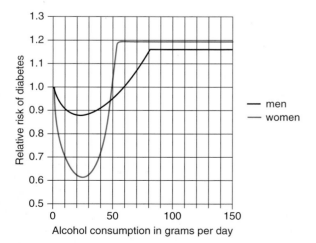

Figure 5.37 The intake of alcohol is one factor that affects the risk of developing diabetes. The risk is relative to that for people who have never drunk alcohol.

DID YOU KNOW?

There are '4Ts' of Type 1 diabetes symptoms – Toilet, Thirsty, Tired and Thinner.
Only 9% of parents are aware of this – a quarter of the 2000 children who develop diabetes every year are not diagnosed until they are seriously ill.

It is clear that a change in the lifestyle of many people would reduce the incidence of Type 2 diabetes. However, we cannot *insist* that people do this. In any case, do we have a right to? Factors, such as a person's ethnicity also affect the prevalence of the condition. Is it ethical for the authorities to make this type of distinction when giving guidance or making policy?

The sugary drinks and snacks produced by food manufacturers contribute to obesity. Do they have an ethical responsibility to produce healthier foods? Must restaurants and school canteens be responsible for providing healthy foods?

(3) **What measures could be taken to reduce Type 2 diabetes?**

(4) **What advice should be given to anyone who drinks alcohol about the risk of developing diabetes?**

Social considerations

A professor of food policy considers that some snacks, that are just sugar and flavouring, should not be termed 'foods' at all.

One possible solution is a 'sugar tax' on the cost of sugary foods, or on food companies that continue to produce foods that contribute to poor health. However, a tax might affect most the people who could least afford it. A Health Survey for England report found that the risk of diabetes was highest among the most deprived of the population.

Typical values	100g contains	45g serving contains
Energy	1570kJ	
	375kcal	710kJ
Protein	10.3g	170kcal
Carbohydrate	73.8g	4.6g
of which sugars	15.0g	33.2g
Fat	2.0g	6.8g
of which saturates	0.3g	0.9g
Fibre‡‡	8.2g	0.1g
Sodium	0.2g	3.7g
Salt equivalent	0.6g	0.1g

Figure 5.38 A 'traffic light system' on food labels informs consumers of its nutritional content

ADVICE

Be prepared to make recommendations for control of diabetes, considering social and ethical issues and consequences.

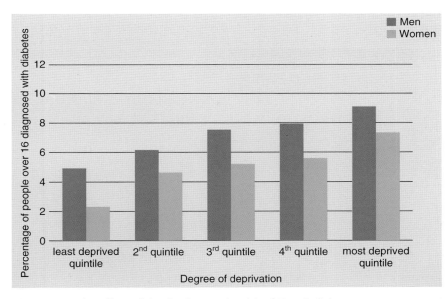

Figure 5.39 The effect of deprivation on the risk of Type 2 diabetes

(5) **Discuss what food manufacturers and caterers could do to reduce diabetes.**

(6) **Suggest explanations for the trend in the graph in Figure 5.39.**

KEY INFORMATION

Quintiles refer to any of five equal groups into which a population can be divided, according to the distribution of values; in this case, according to how deprived they were.

Water balance

Learning objectives:

- recall the ways in which the body loses water
- explain why cells do not function efficiently if they lose or gain too much water
- explain how excess protein is converted to urea for excretion.

It's crucial that patients in hospital – particularly elderly ones – don't become dehydrated.

Water balance

Our daily water intake can vary enormously. The amount in our body must stay constant. Maintaining the correct water balance is essential for good health.

We lose most water via our urine. We also lose ions that are not required by the body, and excretory products such as **urea**, in this way. Most of the other water we lose is from:

- the warm, moist surfaces inside our lungs as we breathe out
- sweating; sweat also contains ions and urea.

We have no control over the water lost from the lungs, and we lose water continually through our skin.

Water gain		Water loss	
Source	**Volume/dm³**	**Route**	**Volume/dm³**
food: present in food	1.0	urine	1.5
food: from respiration	0.4	faeces	0.1
drink	1.5	lungs	0.4
		skin	0.9
Total	2.9	**Total**	2.9

Typical values of a person's 'water balance sheet'

1 What is the typical volume of water lost through a person's lungs?

2 From which parts of our body do we have no, or limited, control over water loss?

DID YOU KNOW?

To function properly, the body's water must be kept to within around 1% of its normal level.

Why is water balance important?

Body cells lose and gain water by osmosis. But the amount in our bodies must be kept reasonably constant.

The cell's water content is important for the chemical reactions that go on there. Our cells will not function efficiently with too little water, and too much would cause them to burst.

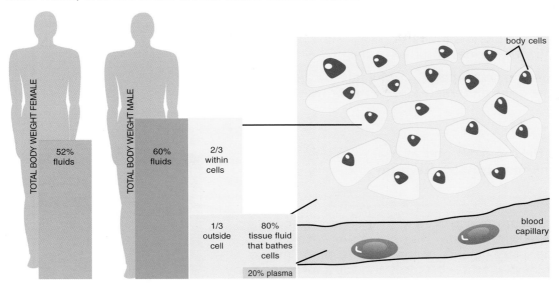

Figure 5.40 The location of the fluid in the human body

Our cells are affected by improper water balance, but our body shows symptoms of inadequate amounts of water, or **dehydration**. As little as 2% water loss by body weight can make us feel very thirsty and uncomfortable; at 5% we lose our ability to work; 10–20% is life threatening

3 **What effect does too little water have on the body's cells?**

4 **Make and justify a recommendation to a company about the level of hydration of their workforce.**

HIGHER TIER ONLY

Removing excess amino acids

Even when we are dehydrated, we will continue to urinate. This is because we must remove excretory products (and some excess ions) from our bodies.

Unlike carbohydrate and fat, we do not store excess amino acids in our bodies. Excess protein or amino acids in our diet must be excreted.

The excess amino acids are taken to the liver. Their amino groups are removed as ammonia. This process is called **deamination**.

Ammonia is toxic, so it is quickly converted to urea, a less toxic compound. Urea is excreted from our bodies by our kidneys.

5 **Explain why amino acids are transported to the liver for processing.**

6 **Describe how amino acids are converted into urea.**

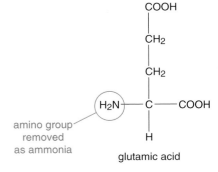

Figure 5.41 Deamination

> **REMEMBER!**
>
> The data shown here are to help you understand the principles.

The kidneys

Learning objectives:

- recall that excess water, ions and urea are removed from the body by the kidneys in urine
- describe how the kidneys produce urine
- explain how the hormone ADH regulates the amount of water in the urine and, therefore, in the body.

KEY WORDS

ADH
aorta
filtrate

negative
feedback
selective
reabsorption

Esteban is suffering from abdominal pain. His doctor arranges a hospital appointment for a kidney scan.

The kidneys

Our kidneys are bean-shaped organs, located in the abdomen, one on either side of the **aorta**. They are important in maintaining the water balance of the body. As blood moves through the kidneys, waste substances and substances the body doesn't require are removed.

As blood enters the kidneys, small molecules and ions are filtered out from the blood, into the small tubes, or tubules, running through the kidneys. These molecules include water, glucose, urea and ions.

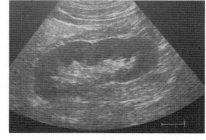

Figure 5.42 An ultrasound scan – using high-frequency sound waves – reveals that both kidneys are normal

1. What is the role of the kidneys?

2. Which substances pass from the blood into the kidney tubules?

Selective reabsorption

All types of *small* dissolved molecules and ions pass into the kidney tubules. Large molecules in the blood, such as proteins, do not.

Some of the small molecules in the tubules, such as glucose and amino acids, are useful to the body. The body can't allow them to be lost. These are reabsorbed back into the bloodstream. This is **selective reabsorption**.

Depending on the body's levels and requirements, molecules may be excreted *or* reabsorbed. You'll reabsorb more sodium and chloride ions, for instance, if you've been sweating all day, compared with after a salty meal.

There may also be some reabsorption of water, again depending on the level in the body, and the concentration of urine will reflect this.

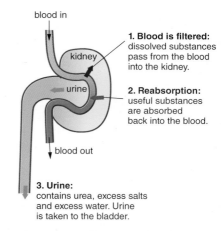

Figure 5.43 Filtration and reabsorption of substances by the kidney

Substance	Concentration in blood/ g per 100 cm³	Concentration in kidney tubule filtrate/g per 100 cm³	Concentration in urine/ g per 100 cm³
water	91.00	99.00	96.00
proteins	8.00	0.00	0.00
glucose	0.10	0.10	0.00
ions	0.75	0.75	1.50
urea	0.03	0.03	2.00

An example of concentrations of different substances in the blood, the **filtrate** in the kidney tubules, and in urine

3 Which substances are not filtered into the kidney tubules?

4 In this example, which substance, or substances, are selectively reabsorbed into the blood?

HIGHER TIER ONLY

ADH

The water level in the body is controlled by the hormone, **ADH**, short for anti-diuretic hormone. ADH is released by the pituitary gland in response to changes in the concentration of blood plasma.

ADH acts on the kidney tubules, increasing the reabsorption of water by increasing their permeability. It decreases the volume of water in the urine.

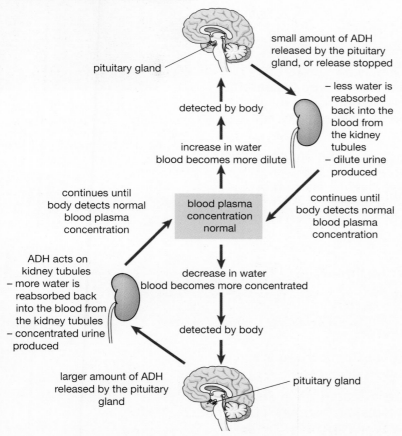

Figure 5.44 ADH controls the body's water balance

This is an example of **negative feedback**. As the water in the blood *falls*, control by negative feedback means that the amount of ADH *rises*. And vice versa.

5 Explain what happens when the concentration of water in the blood falls.

6 Alcohol decreases the release of ADH. Explain how this results in more dilute urine.

Negative feedback

Learning objectives:

- explain the role of thyroxine in the body
- understand the principles of negative feedback, as applied to thyroxine.

KEY WORDS

basal metabolic rate
negative feedback
pituitary gland
thyroxine

Priya is tired and sluggish and is putting on weight. Her heart rate feels slow. She also feels the cold. Priya could have a problem with her thyroid gland.

HIGHER TIER ONLY

The thyroid gland

The thyroid gland affects our activity by producing the hormone **thyroxine**. Thyroxine stimulates the body's **basal metabolic rate** – it increases the metabolism of all the body's cells. Priya has an underactive thyroid. That explains her symptoms.

Under- or over-active thyroid glands are common. As adults, these conditions are fairly easily treated. But thyroxine also controls our growth and development – starting in the uterus. In the embryo, infant or child, mental and physical development is severely retarded by insufficient levels of thyroxine.

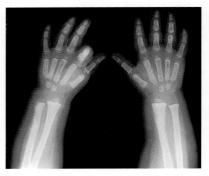

Figure 5.45 This 6-year-old's finger shows stunted development owing to lack of thyroxine. Children in the UK are tested at birth using a drop of blood from their heel.

1. Suggest a symptom of an *over*active thyroid.

2. What is the function of thyroxine?

Control of thyroxine secretion

The **pituitary gland** controls secretions from other glands. One such gland is the thyroid gland.

Levels of thyroxine in the blood are increased when the pituitary gland secretes thyroid-stimulating hormone (TSH). The thyroid gland responds by secreting thyroxine. If levels of thyroxine become too high, TSH secretion is blocked.

Thyroxine increases the respiration rate of cells. In the process, thermal energy is released. The body will secrete thyroxine when its temperature falls. The same is true of adrenaline, the hormone that is released when we are frightened or excited. Both hormones raise our body temperature, particularly with thermal energy released by the liver. If our body temperature becomes too high, the thermoregulatory centre in the brain detects the temperature rise, and the hormones' secretion is blocked.

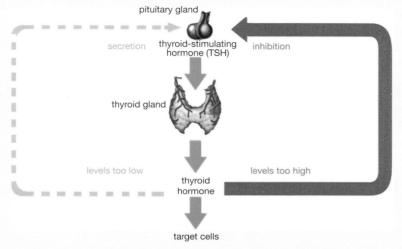

Figure 5.46 A simplified version of the negative-feedback system that controls thyroxine secretion

3 Which hormone regulates thyroxine production?

4 Which hormones help to control our body temperature?

The principles of negative feedback

The endocrine system keeps the conditions in the body constant using feedback systems.

A simple *negative*-feedback system is the central-heating system in your home. If the temperature falls in your living room, the thermostat detects this and switches on the heating. The room warms up.

When the thermostat temperature is reached, it detects this, and turns off the heating. This is **negative feedback** because when the desired effect is reached, the system is switched off.

Negative feedback within the endocrine system prevents a system from becoming overactive. The system is inhibited by its own products.

5 Give a definition of a negative feedback system.

6 Figure 5.47 illustrates a *simple* negative feedback system, but it's not a perfect comparison with thyroid secretion. Explain why.

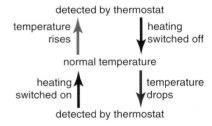

Figure 5.47 Negative feedback in a central-heating system

Kidney failure

Learning objectives:

- recall that people who suffer from kidney failure can be treated by dialysis or kidney transplant
- understand the principles of dialysis
- evaluate the advantages and disadvantages of treating organ failure using a mechanical device or transplant.

Anthony has lost most of his kidney function. He needs a kidney transplant. Anthony receives dialysis until a donated kidney becomes available.

Dialysis

The principle of **dialysis** is simple. Blood is removed from the patient's arm and circulated through the kidney machine (or dialysis machine). Wastes filter out from the patient's blood through a partially permeable membrane. The blood is then returned to their arm.

The concentration of glucose and ions in the dialysis fluid is carefully controlled to maintain the appropriate concentration in the blood.

A patient needs dialysis on 3 days a week, for around 4 hours each time.

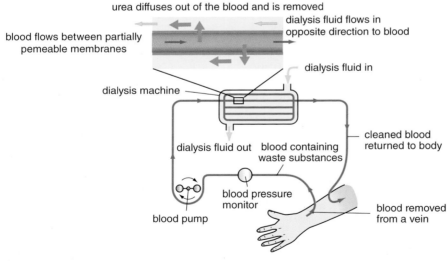

Figure 5.48 Dialysis

1 What is dialysis?

2 Why would a person receive dialysis?

A kidney transplant

A single kidney can do the work previously done by the person's two kidneys. The kidney will be found from a **donor**.

It's best if the donor is a close relative. That way, there's likely to be a better match between the patient's and the donor's tissue types and blood groups. Alternatively, a donor might be very recently deceased. But transplants from living donors have a greater chance of success.

During surgery, the recipient's abdomen is opened below the navel. The new kidney is connected to:

- an artery and vein that enable it to function
- the tube that leads to the organ where urine is collected – the bladder.

Figure 5.49 On the Organ Donor Register, your tissue type and blood group are kept on record

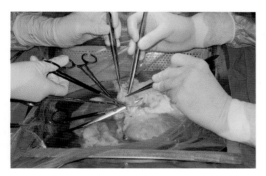

Figure 5.50 The transplanted kidney. Provided the patient's own kidneys aren't causing a problem, they can be left in place

The new kidney usually starts to work straightaway, but can take up to several weeks to work properly.

3 Which kidney donors are most appropriate?

4 Suggest how being on the Organ Donor Register speeds up the process of finding a donor.

After the operation

The operation is complex and takes 2–3 hours. A major risk is of rejection of the kidney by the patient's body. The patient will need immunosuppressant drugs for the rest of their life. These tone down the patient's immune system and reduce the chances of rejection. However, with these drugs, there's an increased risk of infection and cancer. A balance must be found between these risks and preventing rejection.

If a transplant fails, or a transplanted kidney is at the end of its lifetime, a new kidney must be found.

5 What are the risks of the operation?

6 How is the risk of rejection reduced?

> **DID YOU KNOW?**
>
> If the donor is deceased, the kidney must be used as quickly as possible. It can be kept alive in cold saline – a dilute solution of salt – for up to 48 hours.

> **REMEMBER!**
>
> Use this and the next spread to discuss the advantages and disadvantages of kidney transplants, compared with alternative procedures.

Dialysis or transplant?

Learning objectives:

- recall that people who suffer from kidney failure can be treated by dialysis or kidney transplant
- evaluate the advantages and disadvantages of treating organ failure using a mechanical device or transplant.

KEY WORDS

haemodialysis
peritoneal
  dialysis

Niccolò uses a different type of dialysis. He pumps fluid into his abdomen. Wastes diffuse into it across a body membrane called the peritoneum.

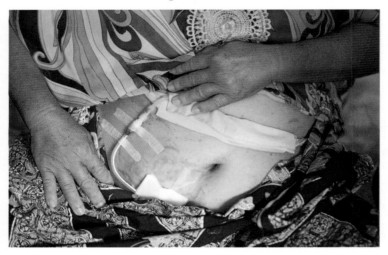

Figure 5.51 **Peritoneal dialysis** is carried out daily, in several sessions or overnight

Dialysis machine or transplant?

The most suitable type of dialysis treatment depends on the individual. For the elderly, dialysis may *reduce* life expectancy because of complications.

Both types of dialysis *can* be done at home. For **haemodialysis** (covered in the previous spread), most patients opt for a hospital or other medical centre.

Dialysis has improved over the years, but there can be complications. Patients can experience a sudden fall in blood pressure. And they must watch out for signs of infection.

Dialysis patients usually need treatment for the rest of their lives, and many progress to a transplant. Transplants provide greater independence and a better quality of life. Some people are too weak for a transplant. Other conditions such as heart disease may prevent them from having the operation.

1. Give one complication arising from dialysis.
2. Why aren't transplants suitable for everyone?

Long-term consequences of transplants

Immunosuppressant drugs increase the risk of infection, cancer and diabetes.

Most people will be able to have a much more varied diet after the transplant, but must make lifestyle changes to minimise risk. Patients should stop smoking, limit alcohol and restrict their weight.

A number of factors affects how long the transplanted kidney lasts.

Average kidney survival time in years	Percentage of patients
1	85–95
5	70–80
15	50–60

The probability of survival of a transplanted kidney. The range reflects that survival time is shorter if the donor was deceased

3 **Give two long-term effects of transplants.**

4 **What is the probability of survival of a kidney for 15 years?**

Availability of organs

More organ donors are needed. In the UK, over 7500 people are waiting for a transplant. For a kidney, the waiting time is around 3 years.

Organ donation raises ethical issues. Only around a third of people are actually on the Organ Donation Register. Relatives of 10% of people carrying donor cards have overruled their wishes after the person has died.

Some people who are short of money have advertised a kidney for sale. It's illegal, but it's suggested by some people that organ sales could save lives.

Biotechnology could have some answers. Transplanting organs from other animals is a possibility. But there are concerns about cross-species infection. Breeding of animals *specifically* for organs raises serious animal rights issues.

In 2014, scientists were able to grow a thymus gland from reprogrammed body cells. In the not too distant future, scientists may be able to grow *complex* organs.

5 **Discuss one ethical issue of organ transplants.**

6 **Suggest why some scientists have suggested that *genetically modified* pigs could be suitable organ donors.**

7 **How could a patient change their lifestyle to improve their kidney survival time after receiving a transplant?**

DID YOU KNOW?

Transplanting organs from other animals was attempted in the early 1900s. At that time, no-one imagined how *human* kidneys could be obtained ethically.

REMEMBER!

Many of the advantages and disadvantages of kidney transplants apply to other types of transplant. Be prepared to discuss these.

Human reproduction

Learning objectives:

- describe the roles of hormones in sexual reproduction
- explain how hormones interact in the menstrual cycle.

KEY WORDS

FSH
LH
oestrogen
progesterone

secondary
sex
characteristics
testosterone
follicle

Our first exposure to sex hormones is not at puberty, but as a foetus. In the presence of male sex hormones (which needs a Y chromosome), male characteristics develop; in their absence, the foetus becomes female.

Reproductive hormones

Secondary sex characteristics develop as our bodies produce reproductive hormones at puberty.

Oestrogen is the main female reproductive hormone that is produced at this time. It is produced by the ovaries. Eggs start to mature in the ovaries and are released, at approximately one every 28 days. This process is called ovulation.

In the male, **testosterone** is the main reproductive hormone. It is produced by the testes. It stimulates sperm production.

DID YOU KNOW?

Oestrogen, progesterone and testosterone are steroid hormones. Steroid hormones also help to control glucose and protein metabolism, and water balance.

1 **Give the main reproductive hormones in males and females.**

2 **What happens in males and females under the influence of these hormones?**

The menstrual cycle

The menstrual cycle is the reproductive cycle in women, which – by convention – starts with a period (menstruation), if the woman is not pregnant.

Four hormones control the menstrual cycle:

- follicle stimulating hormone (**FSH**) causes eggs to mature in the ovaries
- luteinising hormone (**LH**) stimulates the release of an egg from an ovary
- oestrogen and **progesterone** maintain the lining of the uterus.

3 **Which four hormones control the menstrual cycle?**

4 **Which hormones maintain the lining of the uterus?**

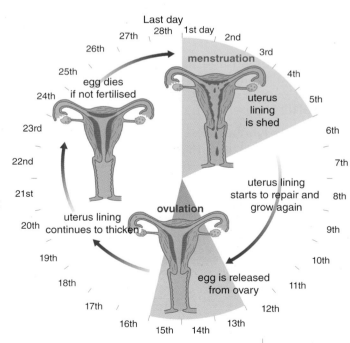

Figure 5.52 The menstrual cycle lasts approximately 28 days, but this is highly variable

Menstrual cycle hormones

The menstrual cycle is concerned with the maturation of an egg every month and preparing the uterus to receive that egg if it is fertilised. This could result in pregnancy.

These hormones interact with each other during the menstrual cycle.

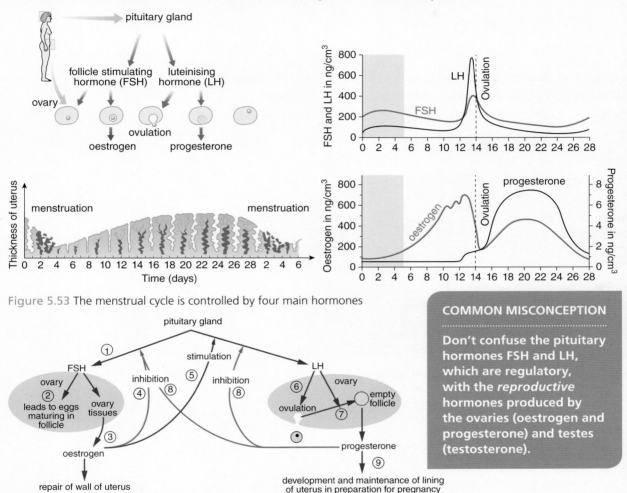

Figure 5.53 The menstrual cycle is controlled by four main hormones

Figure 5.54 The roles of the hormones as the cycle progresses are:

1 FSH is secreted by the pituitary gland.

2 FSH causes the eggs to mature in the ovaries.

3 FSH stimulates the ovaries to produce oestrogen.

4 & 5 Oestrogen inhibits further release of FSH and stimulates release of LH.

6 LH triggers ovulation – the release of the mature egg from the ovary – and …

7 … leads to the secretion of progesterone by the empty **follicle** that contained the egg.

8 Progesterone inhibits the release of LH and FSH.

9 Progesterone maintains the lining of the uterus during the second half of the menstrual cycle, in readiness for receiving a fertilised egg.

COMMON MISCONCEPTION

Don't confuse the pituitary hormones FSH and LH, which are regulatory, with the *reproductive* hormones produced by the ovaries (oestrogen and progesterone) and testes (testosterone).

5 Which two hormones repair the uterus after menstruation and encourage its growth?

6 Suggest which hormones are involved in negative feedback.

IVF

Learning objectives:

- explain the use of hormones in technologies to treat infertility
- describe the technique of *in-vitro* fertilisation
- evaluate the scientific, emotional, social and ethical issues of *in-vitro* fertilisation.

KEY WORDS

fertility drug
in-vitro
 fertilisation
IVF cycle

Louise Joy Brown was born on 25 July 1978 at Oldham General Hospital. She was the first 'test tube baby'.

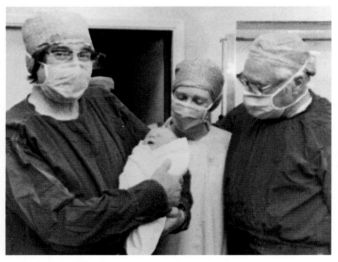

Figure 5.55 Louise was born following the pioneering technique of physiologist Dr Robert Edwards and gynaecologist Dr Patrick Steptoe

HIGHER TIER ONLY

Treating infertility

The NHS recommends that if a couple has been trying to conceive for a year – 6 months if the woman is over 35 – but with no success, it may be time to investigate.

For women whose levels of FSH are too low to conceive, combined hormones can be given as a **'fertility drug'**.

After treatment with a fertility drug, many women ovulate and become pregnant. They are warned of the possibility of multiple births.

If treatment is successful, it's usually within the first 3 months.

Figure 5.56 The women's ovaries are monitored with ultrasound to look at the number and size of developing follicles

① **Explain why FSH and LH are given as fertility treatments.**

② **Why are the ovaries monitored after treatment?**

In-vitro fertilisation

In-vitro **fertilisation** (IVF) may be an option to treat infertility. Here, eggs are fertilised outside the body. '*In vitro*' means, literally, 'in glass'. Eggs are removed from the mother and are fertilised, in the laboratory, with sperm collected from the father.

The technique is more successful if the woman:

- is younger
- has previously been pregnant
- has a BMI within the range of 19–30
- has low alcohol and caffeine intake, and does not smoke.

Counselling is important at this stage. The couple must be optimistic, yet prepared for failure.

3 What is *in-vitro* fertilisation?

4 Give two criteria that increase the possibility of successful IVF.

Stages of the process

The stages of IVF:

- The woman is given FSH and LH to stimulate the production of more eggs than normal in her ovaries.
- Eggs are then collected. The woman is sedated but conscious.
- Eggs are mixed with the father's sperm in the lab for 16–20 hours. They are monitored microscopically for fertilisation.
- Any embryos are allowed to develop for 5 days. They will contain around 100 cells.
- One or two embryos are selected and placed in the mother's uterus.

A number of **IVF cycles** – from stimulation of the ovaries to implantation – can be attempted. If a cycle is unsuccessful, a gap of 2 months is usually left as the treatment is emotionally and physically stressful.

5 If the father's sperm count is low, explain how the procedure is sometimes modified.

6 What is a cycle of IVF treatment?

7 Compare and contrast fertility treatment with IVF.

DID YOU KNOW?

Cryopreservation – or freezing – of eggs can be an option for someone who is undergoing a harmful cancer therapy and yet wants to be able to conceive later on.

Figure 5.57 If the man's sperm count is low, a *single* sperm is sometimes selected and injected into the egg. The procedure is carried out microscopically

COMMON MISCONCEPTION

Don't confuse *fertility treatment* and *IVF*. IVF is just one type of fertility treatment.

IVF evaluation

Learning objectives:

- describe the technique of *in-vitro* fertilisation
- evaluate the scientific, emotional, social and ethical issues of *in-vitro* fertilisation.

Scientists in Oxford have devised a test that could improve the success rate of IVF. A clinical trial resulted in a pregnancy rate of 80%.

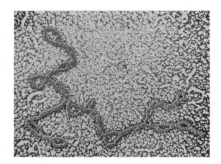

Figure 5.58 The test is based on mitochondrial DNA. Many IVF embryos had unusually high levels that prevented their implantation

HIGHER TIER ONLY

The couple's perspective

IVF has been available on the NHS – provided certain criteria are fulfilled. The National Institute for Health and Care Excellence (NICE) produces guidelines, but the final decision is made by the woman's local NHS.

The process begins with counselling. This prepares the couple for the success, or the chance of failure. Emotional support is continuous through the process.

1. **Who determines the selection procedure for IVF on the NHS?**

2. **Why is counselling important for potential IVF couples?**

Medical and scientific evaluation

Around half the embryos produced by **IVF** have an incorrect number of chromosomes. One-third of the normal embryos actually selected will not implant in the uterus.

Age of woman in years	Proportion of live births in %
Under 35	32.2
35–37	27.7
38–39	20.8
40–42	13.6
43–44	5.0
Over 44	1.9

The woman's age is one factor affecting success

Figure 5.59 Eggs or embryos are stored cryogenically. That way, they can be used later on. But does freezing and then thawing affect them?

The success rate is not high, but it *is* rising – by 1% per year. There are reports of higher incidences, among IVF babies, of premature births, stillbirths, low birth-weights and infant deaths. But rates are higher still in babies born to couples with infertility problems who eventually manage to conceive naturally.

One possible drawback of successful IVF is an increased possibility of multiple births, as more than one embryo is implanted. This increases the risk to the mother and babies.

3 The data in the table are based on NHS 2010 statistics. Estimate the *current* success rate for a woman under 35.

4 Describe three medical issues of IVF.

Ethical issues

Many people object to the technique and the treatment of embryos:

- One argument says that fertility treatments are just removing natural obstacles to fertility, but IVF is not natural. It's replacing the physical and emotional relationship of conceiving with a laboratory technique.
- Embryos that are not transplanted are eventually destroyed. Is a human embryo a mass of cells to be used, selected and discarded? Should we accept these losses as the price for success? Or does an embryo demand the *unconditional* moral respect given to any human being?

Modern microscopical and genetic techniques have enabled embryos to be screened for abnormalities. One serious concern is that couples with no fertility problems may use IVF as a technique to select a child, possibly using dubious selection criteria – a form of **eugenics**.

5 Discuss one ethical issue of IVF.

6 How could the screening of embryos be misused?

DID YOU KNOW?

Nobel Prize winner Professor, later Sir, Robert Edwards was aware that his work was controversial. After publication of a paper on producing embryos, he stopped research for 2 years while considering whether it was right to continue.

REMEMBER!

Use this topic to help you to evaluate the range of issues associated with IVF.

KEY CONCEPT

Systems working together

Learning objectives:

- describe the effects of adrenaline
- understand that automatic control systems may involve nervous responses and chemical responses
- understand that combinations of hormones work to produce a response.

Harry is frightened of having of having a flu vaccination. He screams. His heart begins to race. His skin goes pale.

HIGHER TIER ONLY

Figure 5.60 These effects are the result of the release of a hormone called adrenaline

Adrenaline

Adrenaline prepares us – very quickly when we get a fright – for emergency action. It has a number of effects on the body.

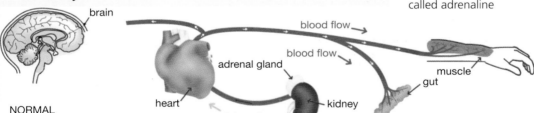

NORMAL

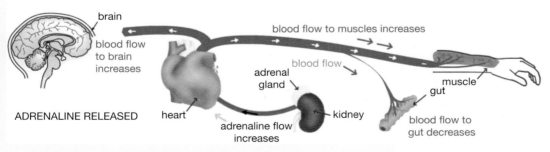

ADRENALINE RELEASED

Figure 5.61 As the heart pumps more rapidly, more blood, with glucose and oxygen, is delivered to our muscles and brain

Our muscles are being prepared for us to run, or maybe to fight. Adrenaline is called the 'flight-or-fight' hormone.

1. **Which hormone is released when we become frightened?**

2. **There is only a set volume of blood in the body. Explain how more blood can be pumped to the brain and muscles.**

HIGHER TIER ONLY

The nervous and endocrine systems interact

The release of adrenaline is a good example of how the nervous and endocrine systems interact.

Connections to and in the brain and to the adrenal glands are *nervous*. The part of the adrenal gland called the **adrenal medulla** responds to nervous stimulation by releasing the hormone adrenaline.

The adrenaline that is released – as you have seen previously – acts on various parts of the body, including the liver cells. In the liver, it promotes the breakdown of glycogen into glucose, and its release into the bloodstream.

Most of us don't use the flight-or-fight response on a regular basis. But it's there if we need it.

3 **Explain how the nervous and endocrine systems work together in times of stress.**

4 **Which other hormone stimulates the breakdown of glycogen in the liver?**

Hormones working together

You will now have realised that *several* hormones are involved in glucose metabolism – insulin and glucagon (topic 5.13), thyroxine (topic 5.16), and now adrenaline (and also cortisol).

Likewise, several hormones, in addition to ADH (topic 5.15), act together to control water balance.

And a number of hormones work together to control our development and reproduction.

These hormones and many others work together in our bodies to ensure that our bodies *do* work as a coordinated whole.

5 **Describe the action of three hormones on glucose metabolism.**

6 **Describe how two hormones affect our development.**

7 **Use your knowledge of the endocrine system to create a concept map which links key words covered in the topic and explains how the words are linked.**

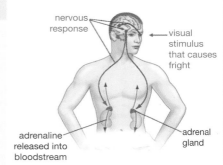

nervous response

visual stimulus that causes fright

adrenaline released into bloodstream

adrenal gland

Figure 5.62 The nervous and endocrine systems produce our reaction to a fright

DID YOU KNOW?

The brain's response to fear also leads to the release of a *hormone* by the pituitary gland. This hormone acts on a different part of the adrenal gland. A hormone called cortisol is released. Its release isn't as quick as adrenaline – it sustains our response to possible danger.

ADVICE

This topic provides an illustration of how the body systems work together to produce a coordinated response.

Contraception

Learning objectives:

- understand that fertility can be controlled by different hormonal and non-hormonal methods of contraception
- evaluate the different methods of contraception.

KEY WORDS

cervix
combined
  contraceptive pill
progesterone
progestogen-
  only pill

If the population continues to increase at the current rate, human life on Earth will not be sustainable. Efforts to bring down the birth rate need to continue.

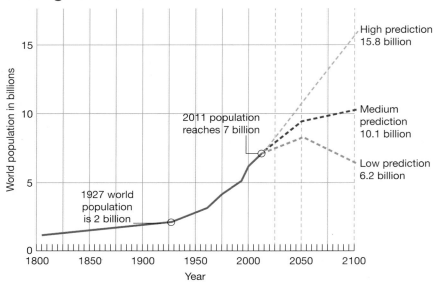

Figure 5.63 One set of predictions for world population growth

Contraceptive methods

Most contraceptive methods fall into one of two types:

- methods that use hormones
- non-hormonal methods, or barrier methods.

Some people choose *natural* planning methods. A woman's time of ovulation is linked with:

- her menstrual cycle, occurring at around 14 days
- a *slight* increase in body temperature
- thinning of mucus secreted from the **cervix**.

By estimating when ovulation occurs, it's possible to avoid having sexual intercourse when an egg might be in the oviduct. But eggs and sperm can live for several days, and women's cycles can be irregular.

Some people choose to have surgery. In the woman, the oviducts are cut, sealed or blocked by an operation. In the man, the sperm ducts are cut, sealed or tied. Surgical methods are designed to be permanent.

DID YOU KNOW?

Oral contraceptives were first approved for use in Britain in 1961. The hormone content was equivalent to seven of today's pills.

KEY SKILL

You may be provided with new information or data on contraceptives that you have to analyse and interpret.

1. How can contraceptives be divided into broad categories?
2. List three indicators of ovulation.

Barrier methods

Barrier methods prevent sperm from reaching an egg.

Condom (male)	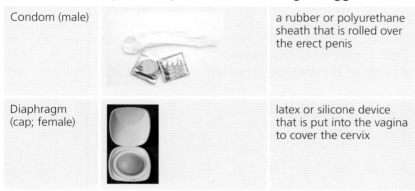	a rubber or polyurethane sheath that is rolled over the erect penis
Diaphragm (cap; female)		latex or silicone device that is put into the vagina to cover the cervix

Two commonly used barrier methods

A spermicidal cream is toxic to sperm. It can help with the effectiveness of other contraceptives, such as diaphragms. It should not be used on its own, or with condoms.

3. Give two examples of barrier methods of contraception.
4. When should a spermicidal cream be used?

Hormonal methods

Hormonal methods of contraception use reproductive hormones to prevent pregnancy. The **combined contraceptive pill** contains a synthetic oestrogen and **progesterone**.

For women for whom the combined pill isn't suitable, for instance, if they're older or have high blood pressure, there's also a **progestogen-only pill** (POP, mini pill). This contains a synthetic version of progesterone.

These contraceptives inhibit the release of the pituitary hormones that control egg maturation and release. They also thicken cervical mucus which helps to prevent sperm reaching an egg. The combined pill is taken for 21 days, allowing periods to occur. The POP is taken every day.

Some other types of contraceptive can also include hormones. Intrauterine devices (IUDs) are placed in the uterus. They prevent a fertilised egg from implanting in the uterus.

Plastic IUDs also release progestogen. Copper versions have copper wound around plastic. Copper is toxic to sperm.

5. How do oral contraceptives work?
6. What is an IUD?
7. **IUDs could be considered unethical because they do not prevent fertilisation of an egg. What do you think? Give reasons to justify your answer.**

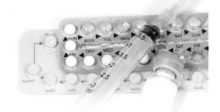

Injection: an injection of a progestogen is given.

Implant: a small flexible rod is implanted under skin of the upper arm. Progestogen is released slowly.

Patch: A sticky patch is put on the skin. It releases oestrogen and progestogen slowly.

Figure 5.64 Hormonal contraceptives can also be used as injections, implants or patches

Figure 5.65 An IUD.

Which contraceptive?

KEY WORDS

condom
IUD

Learning objectives:

- understand that fertility can be controlled by different hormonal and non-hormonal methods of contraception
- evaluate the different methods of contraception.

A microchip implant has been developed that can store and release regular, controlled doses of a contraceptive for up to 16 years.

The chip can be controlled wirelessly. It could be turned off at any time to start a family.

Reliability

People who are considering using contraceptives should look at their success rate.

Successful surgery is 100% effective but irreversible. Natural family planning methods *can* be effective if used correctly, with guidance and teaching. Computerised devices that monitor hormone changes improve reliability of natural methods.

1 **How can the reliability of natural family planning methods be improved?**

2 **Which method(s) is, or are, the most effective contraceptive?**

Advantages and disadvantages of different methods

With surgery, couples no longer need to think about contraception at all. But the decision should be carefully considered.

Hormonal contraceptives are convenient, although POPs do need to be taken at the same time every day to be effective. Implants last for 3 years, injections 12–13 weeks and patches 1 week. With injections, fertility may not return to normal straightaway.

With the combined pill, implant and patch, periods may be more regular, lighter and less painful. Oral contraceptives also reduce the risk of certain cancers, though those containing oestrogen slightly increase the risk of blood clots, and breast and cervical cancer.

Type of contraception	Percentage of pregnancies prevented
condoms	98
diaphragm	92–96
implants	99
IUD	>99
oral contraceptives	>99

The effectiveness of contraceptives when used *perfectly*

Method	Advantages	Disadvantages
condoms	• widely available • can protect against transmitted infections, e.g. HIV	• may slip off • must withdraw after ejaculation and not spill semen
diaphragm	• put in just before sex • no health risks	• needs to be left in for several hours after sex • some people are sensitive to spermicide
IUD	• works immediately • can stay in place for 10 years (copper); 3–5 years (hormonal)	• insertion may be uncomfortable periods may be longer or more painful

Advantages and disadvantages of other methods

3 **Suggest why oral contraceptives are affected by vomiting or diarrhoea, but injections, implants and patches are not.**

4 **Give one advantage and one disadvantage of using an IUD.**

Wider issues

People must decide whether to use contraception or not. This could be influenced by religious factors.

If people do use contraception, they must choose a method.

Health factors are important in family planning, but for many people the decision to use contraception is *economic*, or can be a *lifestyle* choice:

- Can a couple afford to start a family?
- Is contraception likely to be needed every day?
- How soon would the woman like to become pregnant *after* using contraception?

A wider issue associated with contraception is one of world population and sustainability, although there are ethical issues with its use in population control. It can be seen as interfering with human rights and reproductive freedom.

5 **Name the most widely used contraceptive method in the UK.**

6 **Discuss non-scientific questions around a need for contraception.**

7 **Some governments have considered compulsory sterilisation to control population growth. Evaluate the advantages and disadvantages of sterilisation.**

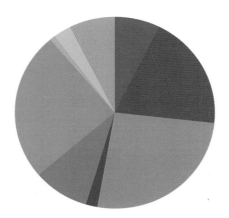

- ▅ sterilisation: female
- ▅ sterilisation: male
- ▅ contraceptive pill
- ▅ other methods, including emergency pill
- ▅ intrauterine devices (IUDs)
- ▅ male condom
- ▅ vaginal barrier, including spermicides
- ▅ hormonal injections and implants
- ▅ traditional methods, e.g. abstaining and withdrawal

Figure 5.66 Contraceptive use in the UK, 2008–9

DID YOU KNOW?

Fifteen types of contraceptive are used in the UK. Two are designed *specifically* for men; 13 for women.

KEY SKILL

Be prepared to evaluate data on the use of different types of contraception.

Auxins

Learning objectives:

- recall that plants produce hormones to coordinate and control growth, and responses to light and gravity
- describe how unequal distributions of auxins cause unequal growth rates in plant shoots and roots.

Angus has bought a house plant. He places on the windowsill. After a couple of weeks, the plant's stems have bent and grown towards the light.

Tropisms

Like animals, plants detect and respond to stimuli. Plants produce hormones to coordinate their responses.

Plants will respond to light and gravity. The response of a plant by growing towards or away from a stimulus is called a **tropism**.

The response to light is called **phototropism**. The shoots of most plants respond to light by growing *towards* it. They are, therefore, said to be *positively* phototropic.

1. What is a plant growth movement called?
2. How do plant shoots respond to light?

Gravitropism

The growth response of a plant to gravity is called **gravitropism** (also called geotropism).

- Roots grow downwards, so are *positively* gravitropic.
- Shoots grow upwards: they are *negatively* gravitropic.

If a plant is placed on its side, then its roots soon start to change direction and grow downwards. The shoots will grow upwards.

3. How do plant shoots respond to gravity?
4. What happens when a seed is planted upside down?

Figure 5.67 The maize seed in the middle was planted the right way up. The seed on the left was planted upside down; the one on the right, on its side. The roots still grew downwards, and the shoots upwards

Auxins

Tropisms are caused by unequal distributions of plant hormones. The type of hormone involved is called an **auxin**.

Auxins are produced in the *tips* of plant shoots and roots. They diffuse away from the tip. Lower down the stem, auxins promote cell division and cause the cells to elongate.

In phototropism, the distribution of auxins is affected by light. Auxins synthesised in the tip of a shoot move *away* from the light.

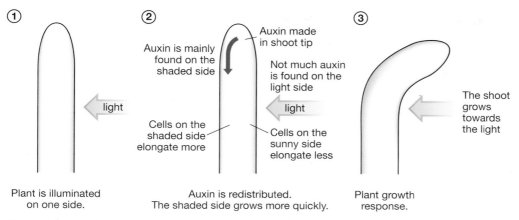

Figure 5.68 In a shoot tip, auxins lead to more rapid growth on the shaded side. The plant grows *towards* the light

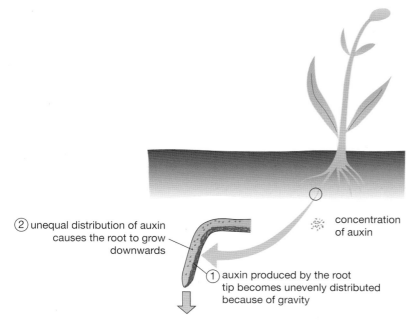

Figure 5.69 In roots, unequal distribution of auxins cause roots to grow downwards.

DID YOU KNOW?

Charles Darwin carried out some of the earliest work on tropisms. By removing the tip of oat shoots, or covering them up, he showed that it was shoot tips that detected the light.

REMEMBER!

Be able to give a precise definition of a tropism. It's a plant growth movement in response to a directional stimulus. Remember that the mechanism is by the redistribution of auxins.

5 How do auxins cause plants to bend towards the light?

6 Explain the negatively gravitropic response by plant shoots. How does this response help the plant survive?

Applications of auxins

Learning objectives:

- explain how auxins coordinate and control responses to light and gravity
- explain that auxins act on 'stem cells' in plants called meristems
- describe some applications of auxins.

Farmers are discovering that weeds are becoming resistant to the herbicide, glyphosate.

In response, biotechnology companies are developing crop varieties that are resistant to a synthetic auxin called 2,4-D.

Meristems

Like animals, plants have undifferentiated cells that can divide to produce new cells. In animals, these cells are called **stem cells**. In plants, these cells are found in regions called **meristems**.

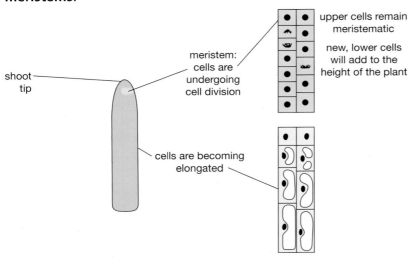

upper cells remain meristematic

new, lower cells will add to the height of the plant

meristem: cells are undergoing cell division

shoot tip

cells are becoming elongated

Figure 5.70 One meristem in plants is found towards the tip of the shoot; one towards the tip of the root

When cells from the meristems of plants are removed, scientists are able to grow them in **tissue culture**. Tissue culture is used to produce clones of plants quickly. It means that plants that are useful to us, and are genetically identical, can be grown economically.

1 Where are stem cells found in plants?

2 How is the tissue culture of plant cells used?

DID YOU KNOW?

In the Vietnam War, the U.S. Air Force sprayed the jungle with auxins. The leaves fell off trees and the enemy could be seen. The mixture of auxins, 2,4-D and 2,4,5-T, was called Agent Orange because of the orange stripe on the containers.
Agent Orange caused birth defects and cancers in the Vietnamese people because it was contaminated with chemicals called dioxins.

Promoting cell division

When plants are grown in tissue culture, auxins – usually in combination with other plant hormones – are added to the culture medium. They promote growth through the division of cells in meristems and enlargement of the daughter cells. The culture medium also contains nutrients, along with agar to support the growth of the plantlets.

We can also make use of the action of auxins when growing **plant cuttings** (see Chapter 7). Hormone rooting powders contain another type of synthetic auxin, called NAA. In the amounts used, it promotes the growth of new roots from the cutting.

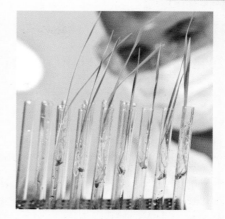

Figure 5.71 These cereal plants are being grown in tissue culture

3 What are the contents of a tissue culture medium?

4 Explain why plant hormones are included in tissue culture media.

Weedkillers

In higher concentrations, auxins disrupt cell metabolism and kill plants. Synthetic auxins are therefore used as weedkillers or herbicides.

One type of weedkiller is called a **selective weedkiller**. Plants with broad leaves are more sensitive to the auxins in weedkillers than plants with narrow-leaved plants, such as grass and cereal crops. So, farmers could kill weeds in a field of wheat or barley without harming the crop. Gardeners can also use them to kill weeds such as dandelions in their lawns.

5 Why can auxins be used as weedkiller?

6 What is the advantage of using auxins as a weedkiller?

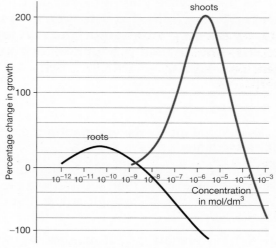

Figure 5.72 When certain concentrations are reached, natural auxins inhibit the growth of plants

REMEMBER!

Know the uses of auxins in agriculture and horticulture.

REQUIRED PRACTICAL

The effect of light and gravity on the growth of newly germinated seedlings

KEY WORD

phototropism

Learning objectives:

- describe how an experiment is planned for a specific purpose
- make and record observations and translate data from one form to another
- interpret observations and other data, identifying patterns and trends, make inferences and draw conclusions.

These pages are designed ❗ to help you think about aspects of the investigation rather than to guide you through it step by step.

NASA is studying plant growth on the International Space Station (ISS). Their response to radiation and gravity is being monitored. The research is essential for sustainable food production for future long-duration space missions.

Responding to gravity

A class of students was investigating the effect of gravity on maize seedlings.

Some maize seeds were planted and allowed to germinate. Each of the seeds was then secured in a separate Petri dish. Each Petri dish was placed upright so that the emerging root was horizontal.

The dishes were left for 200 minutes. At certain intervals, a photograph was taken of the roots of the seedlings.

Figure 5.73a One of the seeds at the beginning of the experiment

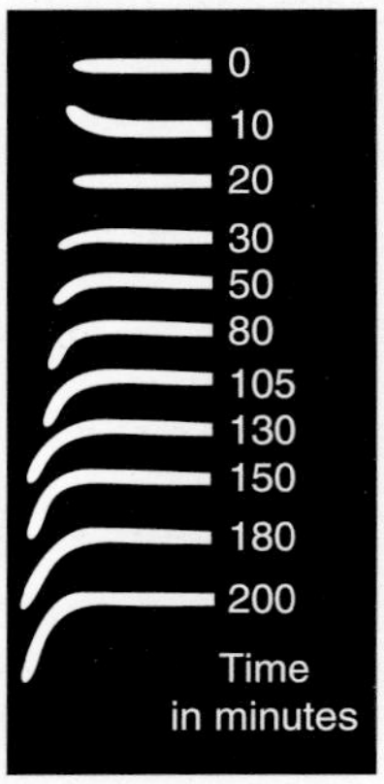

1. Make a list of the variables that needed to be controlled in this experiment.

2. One of the results might be anomalous. Suggest which one and state why.

3. Name the process by which plants respond to gravity.

4. When did this plant's response to gravity occur?

Figure 5.73b The results of the experiment

Phototropism

Some newly germinated, upright seedlings in the school lab were placed on a bench close to a window. They began to grow towards the light.

5. Draw an accurate, scale diagram of the seedling in Figure 5.74.

6. Two students decide to test whether light, that leads to the response, is detected in the tip of the shoot.

They each use 20 oat seedlings, that haven't yet developed leaves. George suggests cutting off the tips of his plants, but Thomas says it's better to place a foil cap over each tip.

Discuss which method is better.

Investigating phototropism

A small seedling was placed in a black film canister lined with floral foam. A hole had been pierced in the side of the canister to allow light to pass through. The hole was covered with a blue filter.

The seedling had been attached carefully to the floral foam. It was kept moist with water dripped onto it.

The plant was illuminated.

Every 30 minutes the canister was opened and the angle of the stem measured using a protractor.

Figure 5.74 The response of a plant to one-directional light. Actual size.

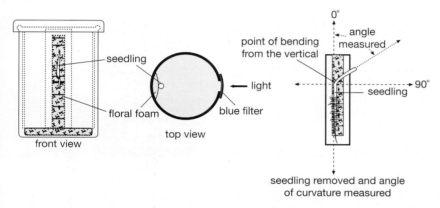

Figure 5.75 The technique used to investigate phototropism.

Time in minutes	0	30	60	90	120	150	180	210	240
Mean angle of curvature in degrees	5	7	10	13	20	40	53	61	70

The students' results of their phototropism experiment.

7 **Draw a graph of the students' results.**

8 **Describe the trend in the data.**

9 **Explain fully the results of the investigation.**

10 **Results of experiments on the ISS suggest that plant responses to light and gravity are interconnected.**

Devise an experiment to test the effect of blue and red light on gravitropism. You are provided with sheets of coloured filters – red, green and blue.

DID YOU KNOW?

Experiments on the ISS have been invaluable in studying phototropism in an environment without a significant gravitational force.

HINTS & TIPS

A skill you need to develop is to be able to make drawings of biological specimens.

Other plant hormones

Learning objectives:

- recall that gibberellins are important in seed germination, and ethene in cell division and ripening of fruit
- explain the application of the plant hormones gibberellins and ethene.

In brewing, the germination of barley, or malting process, is necessary to produce the sugars and amino acids that are required by yeast. If water containing a gibberellin is sprayed on the barley grains, they germinate quickly.

HIGHER TIER ONLY

Gibberellins

Gibberellins are produced naturally by barley and other seeds. These hormones initiate seed germination. But spraying seeds with a very low concentration of gibberellins speeds up the process.

Seeds such as barley germinate straight away, but gibberellins also promotes germination in seeds that have become **dormant**.

Gibberellin spraying is also used to promote:

- flowering, for example, in some apple varieties in a poor year
- fruit growth, such as to increase the size of some cherry varieties, which are sprayed 4–6 weeks before harvest.

Concentration of gibberellin sprayed / ppm	Average mass of grape / g
0	1.59
5	1.91
20	2.71
50	3.15

The effects of gibberellins sprayed on seedless grapes

1 Give three uses of gibberellins.

2 Use data in the table to describe the effect of gibberellins on seedless grapes.

Ethene

Have you ever noticed how, if you leave a bunch of bananas in a plastic bag, they will ripen?

This is because the bananas release **ethene**, a **hydrocarbon** gas. Many plants increase their ethene production and rate of respiration as they begin to ripen.

Ethene regulates many aspects of plant development, including cell division.

DID YOU KNOW?

Mendel's dwarf pea variety is thought to have been the result of a mutation, making it deficient in gibberellins.

KEY INFORMATION

Know the applications of gibberellins and ethene.

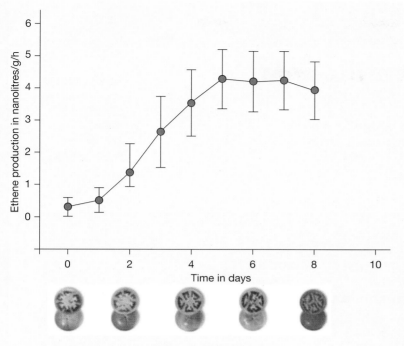

Figure 5.76 Ethene production by ripening tomatoes

Bananas are picked when green.

❸ **Explain why a tomato, when placed in the same bag as a ripe banana, will ripen quickly.**

❹ **Using Figure 5.76, describe the production of ethene by a ripening tomato.**

Regulating fruit ripening

As a gas, ethene is difficult to use on fruit crops. But compounds that release ethene, including a product called Ethrel, *do* ripen fruit in the field.

Some fruit will *only* ripen on the tree. But many will ripen *after* they're harvested. These include apples, bananas and tomatoes. Such fruit can be harvested and transported unripe. The fruit are then ripened in special rooms where ethene concentration, temperature and humidity are carefully controlled. The fruit reach the supermarket at a perfect degree of ripeness.

Sometimes it's desirable to *reduce* ethene synthesis, by chilling or spraying with a chemical inhibitor, when fruit are being transported or stored.

❺ **Why are bananas ripened in the UK, and not in their country of origin?**

❻ **Ethene is flammable. Carbon dioxide produced by ripening fruit is toxic. Suggest how hazards and risks are managed in ripening rooms.**

❼ **Suggest why fruit ripen more quickly when placed next to bananas.**

Bananas are ripened in the UK in ripening rooms using ethene.

Figure 5.77 By ripening them in the UK, bananas are not overripe when they reach the supermarket

Figure 5.78 Some supermarkets use a sheet in their packaging that removes ethene, preventing over-ripening of the fruit

MATHS SKILLS

The spread of scientific data

Learning objectives:

- be able to calculate means and ranges of data
- understand how to estimate uncertainty from a set of measurements.

KEY WORDS

estimate
mean
range
repeatability
uncertainty

When results are collected, how they are spread out is important. It helps us to make judgements about the quality of the data we have collected. This is important when attempting to identify trends in data.

The spread of data

A person's blood glucose level was measured. The measurement was repeated three times on the same blood sample. The following values were obtained:

6.2 mmol/dm³   6.1 mmol/dm³   6.0 mmol/dm³

If you carry out *any* experiment and then do it again, you often get a slightly different result. This may not be because you've used the equipment wrongly. In any measurement there are always random errors that cause measurements to be spread around the true value. This is why it's best to repeat measurements and find the mean.

A **mean** reduces the effect of random errors and gives you the best **estimate** of the true value.

The mean value of a set of measurements is the sum of the values divided by the number of values:

$$\text{mean} = \frac{6.2 + 6.1 + 6.0}{3} = 6.1 \text{ mmol glucose/dm}^3$$

The **range** is a measure of spread. It is calculated as the difference between the largest and smallest values

6.0 – 6.2 mmol/dm³, or 0.2 mmol/dm³.

The spread of data on graphs

Data that are consistent are said to be repeatable; the narrower the range of a set of data, the higher the degree of **repeatability**. Vertical range bars are sometimes added to points on a graph to show the spread of the measured data about the mean.

1 Another set of readings, using a blood sample from a different person, were:

9.6 mmol/dm³
9.5 mmol/dm³
9.8 mmol/dm³

What is the mean of these values?

2 What is the range of this set of values?

Estimating uncertainty in data

The best estimate of the true value of a quantity is the mean of repeated measurements. When calculating a mean, include all the values for data you have collected, unless you have any anomalous results (measurements that do not fit into the pattern of the other results). You could check if a result was an anomaly by repeating it. The more repeated readings you take, the better estimate you'll get of the true value. Three to five repeats are often suggested.

The table shows the data collected on the effect of the hormone thyroxine on heart muscle tissue:

Time in minutes	Oxygen uptake, in cm³ oxygen/g of heart muscle tissue					
	Experiment 1	Experiment 2	Experiment 3	Experiment 4	Experiment 5	Mean
10	3.3	3.9	3.6	3.8	3.9	3.7
20	7.8	8.3	8.0	7.8	8.1	8.0
30	11.7	11.2	11.5	11.6	11.5	11.5
40	14.6	14.8	15.1	14.9	14.6	14.8
50	17.7	17.2	17.5	17.6	17.5	17.5
60	26.7	26.0	26.5	27.4	26.9	

The effect of thyroxine on heart muscle was measured by the oxygen uptake by the cell.

For the data collected after ten minutes:

The mean is 3.7 cm³ oxygen/g of heart muscle.

The range *about* the *mean* gives an estimate of the level of **uncertainty** in the data collected

For this set of data, the upper limit is 3.9 cm³/g; the lower limit 3.3 cm³/g.

The range is 0.6 cm³/g. So, according to our data, the true value could be up to 0.3 cm³/g above the mean, or 0.3 cm³/g below the mean.

Uncertainty is therefore calculated by:

$$\text{uncertainty} = \frac{\text{upper limit of range} - \text{lower limit of range}}{2}$$

So, as we have seen above, uncertainty $= \frac{3.9 - 3.3}{2} = \frac{0.6}{2} = 0.3$

Uncertainty is written next to the mean. In this instance, it is:

3.7 cm³/g ± 0.3 cm³/g

Note that the units are written *both* after the mean and value of uncertainty. The value of uncertainty has the same units and number of decimal places as the measurements.

3 Calculate the mean for the set of data collected at 60 minutes.

4 Calculate the uncertainty in these measurements.

Check your progress

You should be able to:

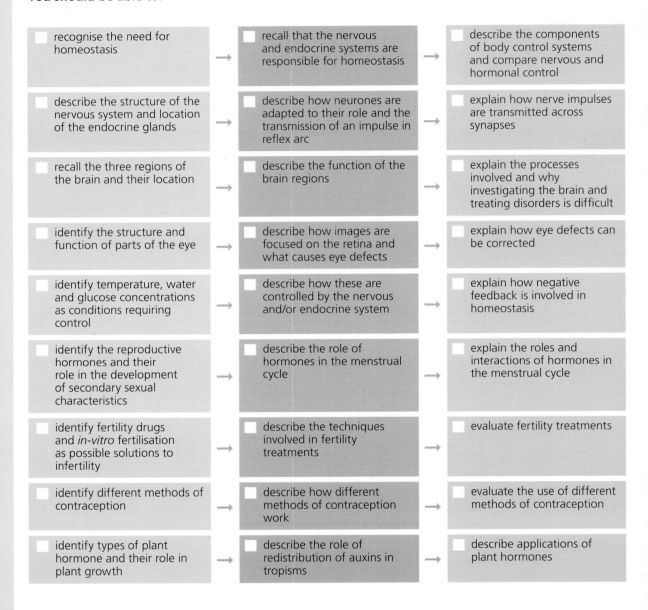

recognise the need for homeostasis → recall that the nervous and endocrine systems are responsible for homeostasis → describe the components of body control systems and compare nervous and hormonal control

describe the structure of the nervous system and location of the endocrine glands → describe how neurones are adapted to their role and the transmission of an impulse in reflex arc → explain how nerve impulses are transmitted across synapses

recall the three regions of the brain and their location → describe the function of the brain regions → explain the processes involved and why investigating the brain and treating disorders is difficult

identify the structure and function of parts of the eye → describe how images are focused on the retina and what causes eye defects → explain how eye defects can be corrected

identify temperature, water and glucose concentrations as conditions requiring control → describe how these are controlled by the nervous and/or endocrine system → explain how negative feedback is involved in homeostasis

identify the reproductive hormones and their role in the development of secondary sexual characteristics → describe the role of hormones in the menstrual cycle → explain the roles and interactions of hormones in the menstrual cycle

identify fertility drugs and *in-vitro* fertilisation as possible solutions to infertility → describe the techniques involved in fertility treatments → evaluate fertility treatments

identify different methods of contraception → describe how different methods of contraception work → evaluate the use of different methods of contraception

identify types of plant hormone and their role in plant growth → describe the role of redistribution of auxins in tropisms → describe applications of plant hormones

Worked example

1 **Describe how the kidneys produce urine.**

The kidneys produce urine by filtering waste substances from the blood. If there is more water than we need in the body, the urine will be dilute. If there is not enough, it will be concentrated.

This is a part answer to the question. The filtration and reabsorption processes should be described in more detail.

During filtration of the blood, small molecules, such as glucose and urea, and ions, pass into the kidney tubule. Useful substances, such as glucose, are reab-sorbed back into the blood, along with water and some ions. As the student answer says, the composition of urine as it leaves the kidney will depend on amount of water and ions that the body needs to reabsorb.

As a general point, while a lead-in to the answer is important, the student has simply repeated part of the question in the answer, which reduces space for writing the answer.

2 **Explain how the body uses the hormone ADH to regulate its water level.**

A hormone called ADH controls the amount of water in the body. If the body is short of water, the pituitary gland produces more ADH. It acts on the kidney to reduce water lost.

This answer could be im-proved. The student could have described the condition that brings about ADH release as being the blood becoming *more concentrated*, rather than the body being 'short of water'. The student could have been more specific in terms of ADH promoting reabsorption of water by the kidney *tubules*, and the reverse of this, when water levels in the body become too high. The student has omitted a key term – negative feedback.

A scientific point is that ADH is only released from the pituitary gland. It is *produced* elsewhere.

End of chapter questions

Getting started

1 The following diagram shows the location of the major endocrine glands.

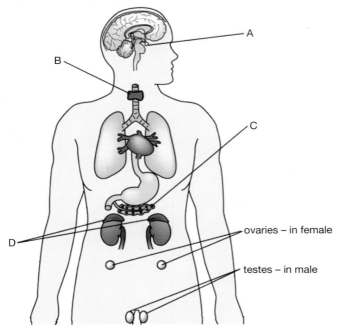

    a Which letter labels the thyroid gland?

1 Mark

    b Which letter represents the gland where, if it produces insufficient
       hormone, diabetes results?

1 Mark

2 What is the ruler-drop test used to measure?

1 Mark

3 Describe the recommendations a doctor would make to a patient diagnosed
with Type 2 diabetes.

2 Marks

4 How should the person being tested hold the ruler at the beginning
of the test?

1 Mark

5 The table below shows some results from the ruler-drop test:

Test number	Distance ruler dropped in cm
1	8
2	4
3	3
4	6
5	5
6	7
7	4
Mean	

    a Calculate the mean distance dropped by the ruler.

1 Mark

b What is the median result? 1 Mark

> i 6
> ii 5
> iii 4
> iv 8

6 **Look at the photograph of the plant. Explain why it is growing in this way.** 2 Marks

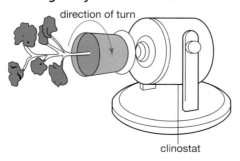

Going further

7 **Compare the nervous and endocrine systems according to their:**

- **type of message**

- **speed of response.** 2 Marks

8 **Which hormone leads to oestrogen production by the ovaries?** 1 Mark

9 **Short-sightedness is a condition when people struggle to see objects at a distance.**

Give a cause of short-sightedness. 1 Mark

10 **Scientists investigating the effect of gravity on a plant placed it on an instrument called a clinostat.**

A clinostat rotates, so the plant is exposed to gravity equally from all directions. The effects of gravity are neutralised.

direction of turn

clinostat

Describe and explain what would happen to the plant after several days on the clinostat. 3 Marks

11 Scientists have researched the relationship between the number of sugar-sweetened soft drinks consumed by a group of women and the relative risk of diabetes.

Their results are shown here.

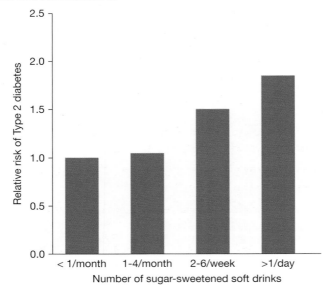

Describe the pattern shown by the graph and explain what this suggests about the cause of Type 2 diabetes.

`2 Marks`

12 A mother places her finger on her baby daughter's palm. The baby grasps it. It is a reflex action.

Describe the process by which this reflex action occurs. You can use a diagram to help with your description.

`6 Marks`

More challenging

13 Explain how hormonal methods of contraception work.

`2 Marks`

14 Describe two applications of auxins.

`2 Marks`

Most demanding

15 Explain how adrenaline and thyroxine are involved in a negative-feedback system to control body temperature.

`2 Marks`

16 Myopia affects approximately 34% of the UK population. In a school of 1456 students how many would you expect to be short sighted?

`1 Mark`

17 Baljit measured the length of four individual roots from plants treated with rooting powder.

Her results are shown in the table.

Percentage rooting powder (%)	Root length / cm				
	1	2	3	4	Mean
2	0.9		1.0	1.1	1.1
5	2.3	2.5	1.3	3.1	2.3

Calculate the missing value.

`1 Mark`

18 Endometriosis is a condition where the type of cells that normally line the uterus are found 'trapped' in the pelvic area and lower tummy.

During the menstrual cycle the cells that have moved can react in the normal way to the hormones controlling menstruation.

This can cause a number of problems including abdominal pain and painful periods.

Using the information from the graph below, as well as your own knowledge, explain what happens to the cells that have moved and how doctors could treat the condition using sex hormones.

`6 Marks`

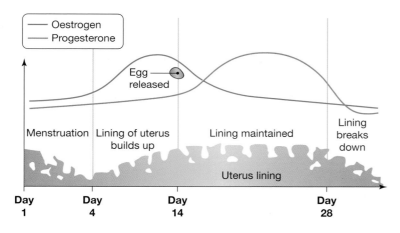

`Total: 40 Marks`

GENETICS

OUR CELLS CONTAIN GENETIC INFORMATION.

- Genetic information is contained within genes on our chromosomes.
- Our chromosomes and genes are found in the nucleus of our cells.
- Our knowledge of the structure of DNA comes from the work of Crick, Watson, Franklin and Wilkins.
- A change to our DNA, called a mutation, can lead to cancer.

IN MULTICELLULAR ORGANISMS, CELLS HAVE TO DIVIDE.

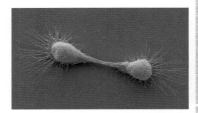

- Cells divide as we're growing, and to replace cells that are injured, worn out or have died.
- This type of cell division is called mitosis.
- When a cell divides by mitosis, two daughter cells are produced, each with an identical number of chromosomes and identical DNA.

OUR CHARACTERISTICS ARE INHERITED AND PASSED ON.

- Passing on genetic information is called heredity.
- Sexual reproduction in humans leads to similarities and variation between individuals.

IN THIS CHAPTER YOU WILL FIND OUT ABOUT:

OUR UNDERSTANDING OF DNA AND THE WAY GENES WORK.

- DNA is a polymer made up of units called nucleotides.
- A gene is a short section of DNA that codes for the production of a particular protein.
- The code for the synthesis of each protein is carried by a sequence of chemicals called bases.
- Proteins are synthesised from the DNA template.
- The unique structure of each protein enables it to do a particular job.
- Mutations occur when errors are made copying DNA.
- The Human Genome Project (HGP) is increasing our understanding of DNA which will lead, in the future, to personalised medicine.
- Because of the HGP, we now know that not all genes that are present control protein synthesis.

PRODUCTION OF SEX CELLS FOR REPRODUCTION.

- In asexual reproduction, only one parent is involved. *No* sex cells are produced and cells divide by mitosis.
- During sexual reproduction, a cell divides by meiosis to produce four gametes, each with half the number of chromosomes.
- Meiosis ensures that we keep our chromosome number constant – 46, or 23 pairs – in each generation.
- Meiosis also produces gametes that are genetically unique, leading to variation between individuals.
- Our sex is determined by the 23rd pair of chromosomes.
- The way chromosomes are inherited means that the number of boys who are born is roughly same as the number of girls.

CHARACTERISTICS ARE INHERITED FROM ONE GENERATION TO THE NEXT.

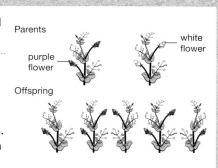

Parents

purple flower

white flower

Offspring

- The science of genetics was established by the work of Austrian monk, Gregor Mendel.
- Genetics allows us to understand the inheritance of certain characteristics in humans and in many other organisms.
- We can use genetic terms and predict the outcome of crosses.
- Genetics enables us to track the inheritance of certain human diseases, although the inheritance of most is more complex.

DNA and genes

Learning objectives:

- describe the structure of DNA
- describe a gene as a small section of DNA that codes for a protein.

KEY WORDS

gene
genome
chromosome

DNA, deoxyribonucleic acid, the molecule that carries our inheritance, was first isolated by the Swiss scientist, Friedrich Miescher, from white blood cells in 1869. He called it nuclein, as it was from the cells' nuclei.

Figure 6.1 You may have extracted DNA in the lab from fruit or vegetables, such as kiwi, strawberries or peas

The genome of an organism

The **genome** of an organism is the entire genetic material of that organism.

The genetic information is carried by a chemical called deoxyribonucleic acid (DNA). DNA is a polymer of nucleotides made up of two strands that form a double helix. The DNA is found within structures called **chromosomes**.

A **gene** is a short section of DNA that contains the instructions for one characteristic of an organism. Genes control all our inherited characteristics. These characteristics range from things like our blood group, hair colour and eye colour, down to the minutest detail of our cells.

Genes work by providing the code for the production of a particular protein.

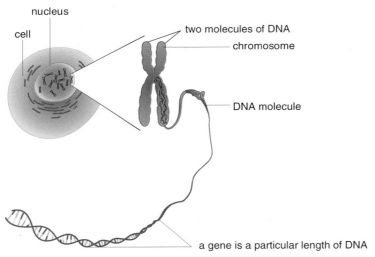

Figure 6.2 Each half of a double chromosome is a molecule of DNA

1. What is the chemical called that makes up our chromosomes?

2. What is a gene?

Genes

Our genes are distributed across our chromosomes.

The chromosomes in a chromosome pair have the same types of genes, in the same order, along their length.

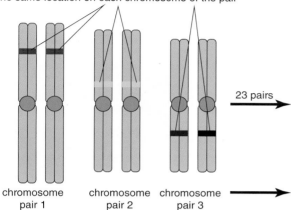

the genes controlling a certain characteristic are at the same location on each chromosome of the pair

23 pairs

chromosome pair 1

chromosome pair 2

chromosome pair 3

Figure 6.3 The chromosomes of an organism contain our genes. One gene pair per chromosome is shown here. In reality, there are thousands on each chromosome

Every person has two copies of each gene – one inherited from each parent.

3 **Where are genes located on a chromosome pair?**

4 **How many copies of a gene does an organism have?**

Figure 6.4 The female jumper ant has just one pair of chromosomes; the male just one chromosome. One species of fern has over a thousand

DNA, genes and chromosomes

Human cells have 46 chromosomes. This number varies from one organism to another. It's the chromosomes contained within a cell and their genetic make-up that determine the type of organism.

The DNA has proteins associated with it that help it to condense into chromosomes.

Certain other parts of the cell also contain DNA – for example, mitochondria.

The plasmids of bacteria are made from DNA. They're not part of the chromosome, so do not contribute to the running of the cell, but they have genes that help the survival of the organism, for instance, in antibiotic resistance.

5 **Name two structures of human cells that contain DNA.**

6 **Give an advantage to a bacterium of having one or more plasmids.**

7 **Why do plasmids not contribute to the species of the bacterium?**

8 **What is the difference between a gene, DNA and a chromosome?**

COMMON MISCONCEPTIONS

Don't think that the amount of DNA or number of genes in a cell is linked with complexity of an organism.

DID YOU KNOW?

The marbled lungfish has more than 40 times the amount of DNA per cell as humans. And the number of protein-coding genes of a chicken is similar to that of a human.

The human genome

Learning objectives:

- describe a gene as a small section of DNA that codes for a protein
- explain the importance of understanding the human genome.

KEY WORDS

gene expression
gene therapy
genome
genomics
genome editing

The Human Genome Project (HGP) was a study to map all the genetic information on the chromosomes of a human being. The work began in 1990 and the research was published in 2003.

Work is continuing. Now a new science – *genomics* – has developed with the aim of increasing our understanding of human DNA.

Increasing our understanding of genes

The HGP has helped scientists to understand more about human diseases, such as diabetes, cancer and heart disease. So far, around 4000 genes have been shown to be involved in human diseases.

The HGP has also helped scientists to understand more about inherited disorders. Some of these, such as cystic fibrosis, are linked with just a single gene. One way to develop effective treatments is to target and correct defective genes. **Gene therapy** and **genome editing** are techniques being developed to do this.

1 How might the HGP help in the way we deal with human diseases?

2 Name one condition resulting from a defective single gene.

Is personalised medicine possible?

Most conditions and diseases are associated, not with single genes, but with many genes that interact in complex ways. These genes also interact with the environment. Understanding these genes, and the regions of DNA that vary from person to person, is key to effective medicine.

In future, by understanding a person's **genome**, doctors may be able to:

- recommend better preventative medicine
- identify the targets of drugs more effectively
- tailor healthcare to the individual.

3 How might an understanding of genes lead to personalised medicine?

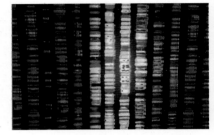

Figure 6.5 Sequencing of the human genome in the early days of the project

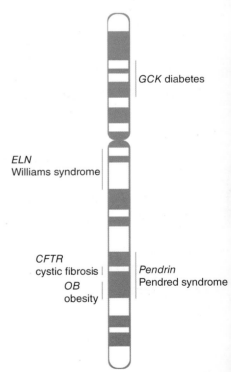

Figure 6.6 A simplified map of chromosome 7. Five genes that are linked with diseases are shown. Note that the chromosome actually has 1800 genes in total

Understanding more about the human genome

Late in the twentieth century, before the work of the HGP, it had been estimated that there were around 100 000 genes in the human genome. We now know that there are many fewer. The HGP and on-going research have revealed that there are 20 000–25 000 genes that code for proteins; this is only around 1.5% of the human genome.

Scientists used to think that the remainder of the DNA – the part that *didn't* code for proteins – was simply 'junk'. We now know that it's teeming with biological activity. Much of this DNA – we're not yet sure how much – is responsible for switching genes on and off. This is called **gene expression**.

Figure 6.7 DNA base sequencing

4 **What percentage of our genome is made up of genes that code for proteins?**

5 **What is the one function of the non-coding regions of our DNA?**

6 **How are genes expressed in a muscle cell, for example?**

7 **The Human Genome Project was estimated to have cost about 2.7 billion US dollars. Evaluate whether it was worth the money.**

DID YOU KNOW?

The 100 000 Genome Project has been set up by the Department of Health. It will analyse complete genomes of people suffering from rare diseases, cancer and infectious disease.

COMMON MISCONCEPTIONS

It's convenient to think that genes just code for production of proteins – that's what we used to think – but far more is involved in gene regulation.

Tracing human migration

Learning objectives:

- explain the importance of understanding the human genome
- discuss the use of the human genome in understanding human migration patterns.

The study and comparison of current human *genomes* is also helping to trace human *migration patterns* across history.

One of the largest studies of its kind – the Genographic Project – was launched in 2005 by the National Geographic Society and computer giant IBM.

Figure 6.8 Volunteers in the project take a swab of DNA from their mouth and send it for analysis

The Genographic Project

Regional centres from around the world have collected DNA samples from local human populations. The project has also sought samples from volunteers from the public.

Unlike the HGP, the Genographic Project has not focused on the *whole* human genome. The first phase of the project focused on analysing data from people's **Y-chromosomes** – inherited down the male line – and from DNA in mitochondria. Mitochondrial DNA is inherited through our mothers.

REMEMBER!

You don't need to know details about the Genographic Project, but it's important to understand the processes and principles that are involved.

1. Suggest why the Genographic Project does not investigate whole human genome.

2. What were the sources of DNA analysed in the first phase of the project?

Representative sampling of the world's populations

Representative sampling of world populations is critical. Little information would be gained if whole populations were excluded from the study.

By 2015, data from 700 000 people in 1000 populations and 130 countries has been sampled.

3. Why is representative sampling important in this type of project?

4. What evidence supports the suggestion that sampling *has* been representative?

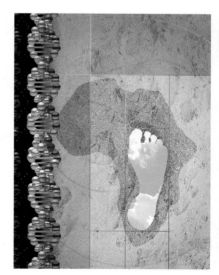

Figure 6.9 The first findings of the project support the 'Out-of-Africa' hypothesis. So, humans *did* originate in Africa

The second phase of the study

Many additional research organisations are now involved in the study. The current phase is also focusing on many thousands of additional regions of DNA across the non-sex (autosomal) chromosomes.

There were ethical concerns with this project. Could the data collected find its way into the hands of medical or pharmaceutical companies and be exploited?

Steps have been taken to ensure regions of DNA were selected for analysis that did not code for proteins and that had no known function. The data, therefore, *cannot* be exploited.

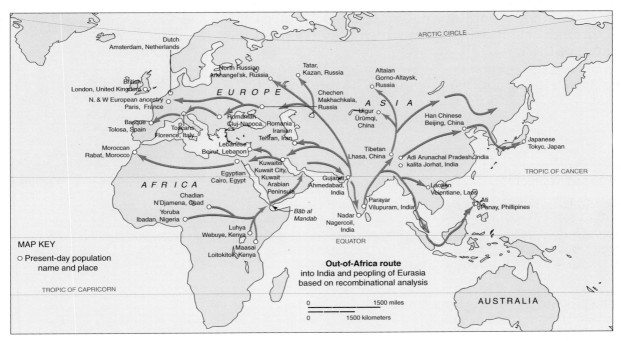

Figure 6.10 The **genomic** evidence to date suggests that early humans migrated out of Africa 60 000 years ago, through Arabia.

5. **What ethical concerns were there about collection of DNA data from populations across the world?**

6. **How were these ethical concerns addressed?**

7. **Give three applications of the mapping of the human genome.**

8. **According to the Genographic project humans began to migrate away from Africa about 60 000 years ago. Suggest a reason for this.**

DID YOU KNOW?

Forensic scientists also look at specific regions of DNA to make comparisons in the technique of genetic fingerprinting. In the UK, DNA 17 analysis was introduced in July 2014. It makes comparisons across 16 regions (also called loci or markers) of our DNA, where there are different numbers of repeating sequences of bases in different people, plus the gender identifier.

The structure of DNA

KEY WORDS

base
complementary
double helix
genetic code
nucleotide
polymer

Learning objectives:

- describe the structure of DNA as repeating nucleotide units
- identify the four bases in DNA
- explain that the bases A and T, and C and G are complementary.

In April 1953, Francis Crick and James Watson, working in the Cavendish Laboratory in Cambridge, produced the breakthrough scientific paper in which they revealed the structure of DNA.

Laboratories in Cambridge are still at the forefront of current work in genomics.

Figure 6.11 On 25 February 1953, Francis Crick walked into the Eagle pub in Cambridge and announced, 'We have found the secret of life'

The DNA double helix

A molecule of DNA has a structure like a ladder that has been twisted. The shape is called a **double helix**.

Figure 6.12 An illustration of DNA. The molecule is like a ladder that has been gently twisted

1 What is the name given to the structure of the DNA molecule?

2 Describe the shape of the DNA molecule.

DNA is a polymer

The two supports of the 'ladder' are alternating sugar molecules and phosphate groups. The 'rungs' of the ladder are made up of chemicals called **bases**. Each base is attached to the sugar molecule of the 'backbone' of the DNA.

There are four different bases in DNA:

- adenine (A)
- thymine (T)
- cytosine (C)
- guanine (G).

The DNA molecule is a **polymer** made up of repeating units called **nucleotides**.

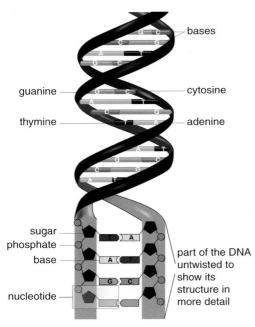

Figure 6.13 Each nucleotide consists of a sugar molecule, phosphate and a base

It is the bases of DNA that make up the **genetic code**. The Human Genome Project revealed that our genome contains approximately 3.3 billion base pairs. Human genes vary in size from a few hundred bases to more than 2 million.

3 Name the four bases that make up the genetic code.

4 How many different types of nucleotide are there?

HIGHER TIER ONLY

The genetic code

Bases always pair up in the same way. The two strands that make up the DNA molecule are said to be **complementary**.

- An A on one side is linked to a T on the other side.
- A C on one side is linked to a G on the other side.

It is the sequence of base pairs in a gene that provides the code for the cell to build a protein.

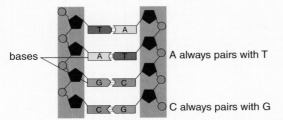

Figure 6.14 Complementary base pairs link the two strands together

5 Create a diagram to show how the bases pair up.

6 How is the structure of a protein determined?

> **REMEMBER!**
>
> You only need to know the *letters* of the four bases.

> **KEY INFORMATION**
>
> Remember that base pairs, and therefore opposite strands, are *complementary*. Remember to spell it correctly, with an 'e'.

> **DID YOU KNOW?**
>
> Other information helped Crick and Watson to work out the structure of DNA. Biochemist Erwin Chargaff discovered that the amount of adenine is usually similar to that of thymine, and the amount cytosine approximates to that of guanine.
>
> The X-ray data of Rosalind Franklin and Maurice Wilkins helped Crick and Watson deduce that DNA had a helical structure.

Proteins

Learning objectives:

- describe how proteins are synthesised according to the DNA template of a gene

- explain that the genetic code of a gene specifies the protein to be made.

We have several hundred thousand proteins in our bodies. Some of these are also found in other organisms. Our hair is the same chemical as feathers, hooves, horns and the shell of a tortoise.

Genes are needed to assemble proteins

Genes provide the code for the assembly of proteins.

Proteins are essential chemicals that are involved in nearly every task in the life of the cell.

Although proteins are very different in structure, they are all made up of the same 20 types of **amino acids**.

It is the number and combination of these amino acids, and the order in which they're arranged, that's important.

A gene provides the code for how a protein is assembled from its constituent amino acids.

The four bases (A, C, T and G) work in threes: there is a three-letter code (or triplet) for each amino acid.

an antibody - its shape corresponds with a matching antigen

collagen - super-coiled triple helix. Tough and found in connective tissue

an enzyme (amylase) - has a complex shape. The attached substrate, in the active site, is shown in yellow

hair protein (keratin) - a fibrous structural protein

muscle protein (actin) - forms filaments in muscles

a hormone (insulin) - its shape matches that of its receptor molecule

Figure 6.15 Proteins have a wide range of molecular shapes. It is its structure that enables a protein to do a particular job

1. **Name three types of protein.**

2. **How many different types of amino acids are found in proteins?**

HIGHER TIER ONLY

Protein synthesis

The sequence of bases in a gene acts as a template for a messenger molecule. The messenger molecule ensures that, as a protein is assembled on a **ribosome**, amino acids are linked together in the required order.

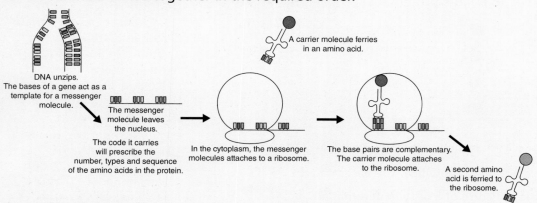

DNA unzips. The bases of a gene act as a template for a messenger molecule.

The messenger molecule leaves the nucleus.

The code it carries will prescribe the number, types and sequence of the amino acids in the protein.

A carrier molecule ferries in an amino acid.

In the cytoplasm, the messenger molecules attaches to a ribosome.

The base pairs are complementary. The carrier molecule attaches to the ribosome.

A second amino acid is ferried to the ribosome.

Figure 6.16a Protein synthesis

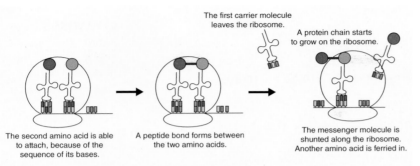

The second amino acid is able to attach, because of the sequence of its bases.

A peptide bond forms between the two amino acids.

The first carrier molecule leaves the ribosome.

A protein chain starts to grow on the ribosome.

The messenger molecule is shunted along the ribosome. Another amino acid is ferried in.

Figure 6.16b Protein synthesis continued

3 **How many base letters code for an amino acid?**

4 **Where are amino acids assembled into proteins?**

Proteins have a unique shape

The amino acid chain does not stay straight for very long. In less than a second, most proteins bend, twist and fold into a particular shape. This is because different amino acids in the chain are attracted to each other, while others repel.

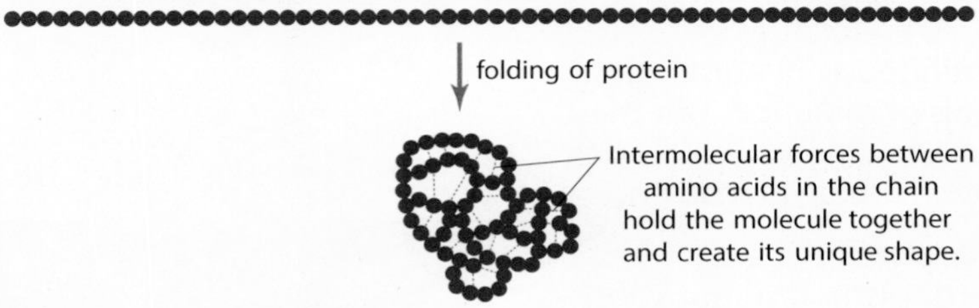

folding of protein

Intermolecular forces between amino acids in the chain hold the molecule together and create its unique shape.

Figure 6.17 The protein folds into a unique shape

The sequence of amino acids in a protein is very important. If the sequence was changed, the protein would be a different shape. The complex shape is essential for the activity of physiological proteins such as enzymes, antibodies and hormones, and structural proteins such as collagen and keratin.

The unique 3D shape of the enzyme produces the active site.

The incoming substrate has a shape that is complementary to the active site.

The substrate locks on to the enzyme.

Figure 6.18 The enzyme and substrate have a complementary shape. Anything that disrupts the intermolecular forces holding the enzyme together will affect its activity

5 **Why does a protein chain fold immediately after synthesis?**

6 **What holds the protein structure together?**

7 **Construct a flow diagram to show how the structure of DNA affects the protein made.**

REMEMBER!

You do not need to know protein synthesis in *detail*, but you need to understand how the genetic code works.

DID YOU KNOW?

One of the smallest proteins in the human body is insulin, the hormone that helps to control our blood sugar. It contains 51 amino acids.

The largest is titin, with around 30 000. Titin helps to keep the muscle proteins that move when a muscle contracts in place.

Mutations

Learning objectives:

- model changes to the base sequences of DNA to illustrate mutations

- describe the negative and, sometimes, positive effects of mutations
- describe how mutations can affect protein function.

KEY WORDS

amino acids
gene mutation

In some animals, a *gene mutation* occasionally causes a dark, or melanic form, to appear.

Today, when dark forms of the peppered moth appear, these are quickly seen by birds and eaten. But 50 years ago, when the air was very polluted, melanic forms were much more successful.

Mutations are changes to our DNA

Mutations, changes to our DNA, occur continuously in our bodies. These changes can be the result of chemicals or radiation, but they also just happen. They are errors that are made when a cell divides, or when the instructions to produce a protein are being copied.

Sometimes, they can affect the way an animal or plant functions, leading to the death of the organism. If a mutation occurs in a body cell, cancer may result (see Chapter 1). If a mutation occurs in cells that produce eggs and sperm, it may lead to a genetic disorder in the offspring.

Sometimes, mutations lead to extra or reduced numbers of whole chromosomes. More commonly, a change in one or two bases of a gene happens.

1 **What is a mutation?**

2 **When can mutations be beneficial?**

Figure 6.19 In industrial areas, peppered moths were camouflaged on trees covered with soot

HIGHER TIER ONLY

Not all mutations have serious effects

Because we have two copies of each gene, it's likely that, if one is faulty, the other will be normal; so the protein can be produced as normal.

In the genetic code, there are instances where several combinations of base triplets code for the *same* **amino acid**. So, a change in a single letter may not lead to a change in the amino acid sequence in the protein.

There are many instances where one amino acid is substituted for another; the protein is changed but is still able to carry out its function.

If a mutation occurs in non-coding DNA, it may affect how a gene is expressed. Occasionally, mutations can be beneficial. Mutations are one of the driving forces of evolution.

3 Name two types of mutation.

4 Why don't all mutations lead to change in a protein?

Mutations can have serious consequences

A few gene mutations can have serious effects. A change in the bases that code for a protein may lead to a different amino acid being assembled into the protein. Some base triplets are designed to terminate a protein. So, some mutations can lead to the synthesis of a shorter protein.

With these changes, the protein will have a different shape. If the protein is an enzyme, it may lose its active site and no longer function.

RESEARCH

Look up the genetic code. Try modelling the effects of changes in the base sequence on the order of amino acids in a protein.

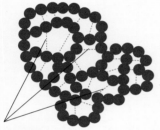

In the normal enzyme, the shape of the active site and substrate are complementary.

The substrate fits into the active site.

Intermolecular forces maintain the shape of the enzyme.

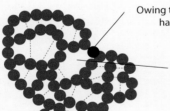

Owing to a mutation, one of the amino acids has been substituted with another – shown in black.

The intermolecular forces between the amino acids are changed.

The enzyme changes shape; the shapes of enzyme and substrate are no longer complementary.

Figure 6.20 A mutation in the active site of an enzyme can lead to loss of its function

5 How can the base sequences of a gene change in a mutation?

6 Create a diagram to show how a change in an amino acid in an enzyme can affect its activity.

DID YOU KNOW?

It is estimated that a mutation occurs for one in every billion (1 in 10^9) nucleotides copied.

Meiosis

Learning objectives:

- explain how meiosis halves the number of chromosomes for gamete production
- explain how fertilisation restores the chromosome number
- understand that the four gametes produced by meiosis are genetically different.

You already know that each cell in the human body has 46 chromosomes. How is this number kept at 46 during sex-cell production and fertilisation?

Meiosis is *a reduction division*

Mitosis is the type of cell division used during growth, or when old or damaged cells need replacing.

But another form of cell division takes place when sex cells are produced in the ovaries and testes in animals, and in the carpels and stamens of plants.

During **meiosis**:

- four **gametes** are produced from one parent cell
- each gamete has half the number of chromosomes of the parent cell (in humans, that's 23 chromosomes instead of 23 pairs).

1 How many gametes are produced from one cell during meiosis?

2 How many chromosomes does a gamete have?

What happens during meiosis?

In meiosis, the DNA of each chromosome is copied, just as in mitosis. But it then divides *twice*, so the chromosome number is *halved*.

So, in mitosis, there is one replication of DNA and one division, but in meiosis, there is one replication of DNA and two divisions.

DID YOU KNOW?

In mammals, Y-chromosomes house the gene that controls gender. Both the X- and Y-chromosomes arose from autosomes around 200 million years ago. During evolution, the Y-chromosome has lost most of its genes.

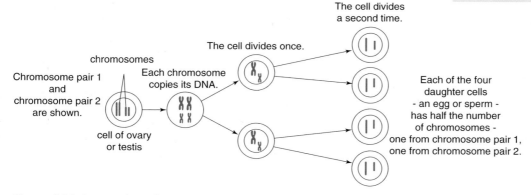

Figure 6.21 An overview of meiosis

When the gametes fuse at fertilisation, the normal number of chromosomes is restored. The zygote then divides by mitosis and the number of cells increases. As the embryo develops cells begin to differentiate.

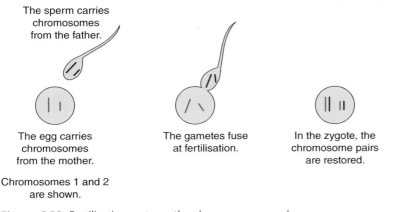

The sperm carries chromosomes from the father.

The egg carries chromosomes from the mother.

The gametes fuse at fertilisation.

In the zygote, the chromosome pairs are restored.

Chromosomes 1 and 2 are shown.

Figure 6.22 Fertilisation restores the chromosome number

REMEMBER!

You should aim to be able to recognise or describe differences and similarities between meiosis and mitosis.

One of each chromosome pair is inherited from our mother; the other from our father. But for each chromosome pair, it's a completely random event as to which of these goes into the egg or sperm.

3 **How many replications of DNA occur in meiosis?**

4 **How many divisions occur in meiosis?**

All gametes are genetically different

Another thing that makes each gamete unique is that there's some exchange of genetic material during meiosis. This contributes to **genetic variation**. When an egg or sperm is produced, it doesn't *just* receive *unchanged* chromosomes from the parent.

What determines our sex?

Each gamete also contains one of the two sex chromosomes – from pair 23. These carry the genes for **sex determination**. Female humans have two X-shaped chromosomes (XX). Males have an X- and a Y-shaped chromosome (XY).

All the eggs produced by the mother contain **X-chromosomes**. Half of the father's sperm contain the X-chromosome; half the **Y-chromosome**.

5 **Explain how meiosis leads to genetic variation.**

6 **In the UK, the proportion of live births was in the proportion of 1.05 males:1.00 females. Suggest why it isn't exactly 1:1.**

		Mother (XX) gametes	
		X	X
Father (XY) gametes	X	XX female	XX female
	Y	XY male	XY male

Figure 6.23 In theory, 50% of live-born children are female; 50% male

Asexual and sexual reproduction

Learning objectives:

- understand that asexual reproduction involves just one parent and produces genetically identical offspring
- understand that sexual reproduction leads to variety in the offspring.

KEY WORDS

asexual reproduction
clone
natural selection
sexual reproduction
variation

Some organisms, such as the malaria parasite, use both asexual and sexual reproduction during their life cycle.

One parent or two?

Asexual reproduction involves just one parent. The offspring are identical to the parent. They are **clones**. New cells are produced by mitosis.

Asexual reproduction can be an advantage:

Figure 6.24 The parasite is transferred to and from humans as the *Anopheles* mosquito feeds on blood

- if the chances of meeting with another individual are rare
- as it produces a large number of identical offspring quickly when conditions are favourable
- as it requires less energy; there's no need to find a mate.

Most organisms reproduce sexually. Two parents are involved. The parents produce sex cells or gametes by meiosis.

	Female gametes	Male gametes
animals	egg cells	sperm
plants	egg cells	pollen grains

Sexual reproduction provides advantages:

- Genetic material comes from both parents, producing **variation**. Offspring are different from each other, and their parents.
- If the environment changes, because of their genetic differences, some offspring are more likely to survive than others. This gives a survival advantage through **natural selection**.

Sexual reproduction of animals and plants can be manipulated by humans. Selection processes can be changed to produce new varieties of plant and breeds of animal for food.

1 Give two advantages of asexual reproduction.

2 Give two advantages sexual reproduction.

Using both forms of reproduction

Flowering plants reproduce sexually by seeds. But many also reproduce asexually. New plants can be produced quickly, without the need for sexual reproduction.

Reproducing asexually gives lots of identical plants quickly, but, if affected by disease, they may *all* die out.

Many fungi, such as moulds, grow asexually most of the time, but sometimes reproduce sexually if different strains meet.

3 Name three organisms that reproduce both sexually and asexually.

4 Why is it an advantage to the gardener when plants reproduce asexually?

Adapting to circumstances

It is thought that organisms reproduce asexually to produce large numbers quickly when the conditions are favourable.

Sexual reproduction is an adaption to increase the chances of survival if circumstances change. With greater variation, at least *some* individuals will survive a change.

The disease malaria is caused by a protist called *Plasmodium*. *Plasmodium* is transferred to and from humans by mosquitoes.

Figure 6.25 Mature strawberry plants produce stems that run along the ground called runners. The new plants on a runner can root and grow. Daffodils produce new bulbs after they have flowered

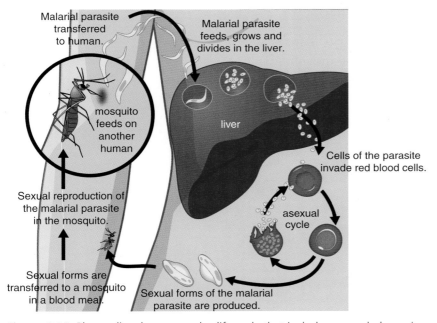

Malarial parasite transferred to human.

Malarial parasite feeds, grows and divides in the liver.

mosquito feeds on another human

liver

Cells of the parasite invade red blood cells.

Sexual reproduction of the malarial parasite in the mosquito.

asexual cycle

Sexual forms are transferred to a mosquito in a blood meal.

Sexual forms of the malarial parasite are produced.

Figure 6.26 *Plasmodium* has a complex life cycle that includes asexual phases in humans, then sexual reproduction in a mosquito

DID YOU KNOW?

Bees, ants and wasps reproduce partheno-genetically. If the female lays a fertilised egg, a female is produced. If she lays an egg that hasn't been fertilised, a male emerges.

In the human, the organism has a plentiful supply of food and warm conditions. Scientists believe that the sexual phase may be triggered as the human immune system, or perhaps antimalarial drugs, start to challenge the parasite.

REMEMBER!

You should aim to able to discuss the advantages and disadvantages of asexual and sexual reproduction in organisms that you are given information about.

5 Where do the asexual and sexual phases of the malarial parasite occur?

6 Why is it an advantage for the parasite to reproduce sexually?

7 Describe the conditions in the liver that favour asexual reproduction.

Genetics

Learning objectives:

- understand and be able to use genetics terms, such as gamete, chromosome, gene, dominant, recessive, genotype, phenotype, homozygous and heterozygous
- know that some human conditions are caused by a recessive allele.

Jenny has cystic fibrosis. She needs physiotherapy to keep her lungs working. Fifty years ago, children with cystic fibrosis rarely lived beyond the age of 5. Today, life expectancy is over 40.

Cystic fibrosis is a gene disorder

Cystic fibrosis is an example of an inherited disorder. Sufferers produce mucus that is thicker and stickier than normal, making it difficult to breathe.

Figure 6.27 Researchers are working with gene and stem-cell therapies to find a possible cure for cystic fibrosis

Cystic fibrosis is linked to a gene on chromosome 7. It codes for a protein called CFTR. The protein controls the movement of water and ions in and out of certain cells, including those of the tissue that lines the lungs.

Each gene can have more than one form, called **alleles**. Cystic fibrosis occurs when alleles of the *CFTR* gene are defective.

A *CFTR* allele is present on each of the chromosomes of chromosome pair 7.

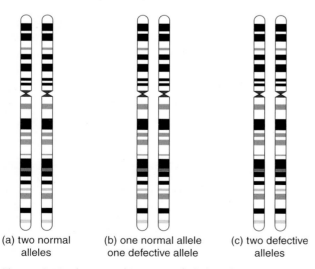

(a) two normal
    alleles

(b) one normal allele
    one defective allele

(c) two defective
    alleles

Figure 6.28 Three combinations of alleles of the *CFTR* gene are possible

If a person has one normal and one defective allele, the presence of the normal allele means that they still produce CFTR protein. They do *not* have cystic fibrosis. But they are a carrier; they could pass on the allele to their children. A person with two defective alleles has cystic fibrosis.

1 **What are different forms of a gene called?**

2 **In Figure 6.28, explain why a person with each combination that is shown has, or does not have, cystic fibrosis.**

An organism's genetics

Many characteristics are controlled by a single gene. Examples are red-green colour blindness in humans and fur colour in mice. Each gene has different forms called alleles. Alleles can be **dominant** or **recessive**. In Figure 6.28b, the person does not have cystic fibrosis. Both alleles are present in the gene pair; the dominant allele is expressed. The recessive characteristic is only shown if there are two recessive alleles.

When describing the genetics of an organism, we give the alleles letters. A dominant allele is given a capital letter; a recessive allele is written in lower case. If we use the letter 'C' for the *CFTR* gene, the combinations of alleles in Figure 6.28 are CC, Cc and cc.

The alleles present for a particular gene make up the organism's **genotype**. How the gene is expressed (the appearance or characteristics of an organism) is the **phenotype**.

When an organism has two alleles of the same type – either dominant or recessive – they are said to be **homozygous** for that characteristic. If the alleles are different, they are described as **heterozygous**.

> **REMEMBER!**
>
> It is important that you know, understand and can use the genetics terms on these pages.

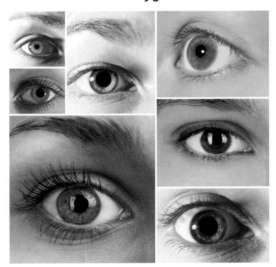

Figure 6.29 Eye colour is controlled by many genes. While we can write down the phenotype, it's not possible to write a genotype

> **DID YOU KNOW?**
>
> Height and skin colour are polygenic – they are controlled by several genes. These characteristics show a continuous gradation across populations.

3 **What is meant by a dominant allele?**

4 **Explain the difference between an organism's genotype and phenotype.**

5 **'A dominant trait is most likely to be found in the population.' Discuss this statement. Do you agree with it? Justify your answer.**

Genetic crosses

KEY WORD
....
Punnett square

Learning objectives:

- use the terms dominant, recessive, genotype, phenotype, homozygous and heterozygous
- know that some human conditions, such as cystic fibrosis, are caused by a recessive allele
- complete or construct a Punnett square to predict the outcome of a genetic cross.

Cystic fibrosis has been recorded since the sixteenth century. One symptom is production of sweat with an increased salt content.

In the 1980s, the gene for cystic fibrosis was mapped to chromosome 7. We can now diagnose cystic fibrosis quickly and predict its inheritance.

Genetic crosses

Let's take an example where a mother with cystic fibrosis has children with a father who does not.

The mother's genotype must be cc, as she has the condition. The father could be CC or Cc. Let's start by assuming he's CC.

All the sperm produced by the father will be C. The eggs produced by the mother will be c. So at fertilisation, *all* the children will be Cc. Their phenotypes will be normal.

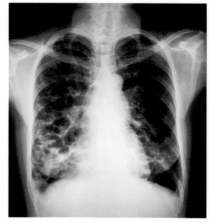

Figure 6.30 This X-ray shows a there is a widening of the airways and a build-up of mucus in the lungs

① **What is the genotype of a person who is homozygous dominant for a gene, B?**

② **What types of gametes will the person produce?**

Genetic diagrams

Let's now assume the father is Cc – he's normal but a carrier. This time, the cross is a little more complicated. It's best to display the cross in a diagram called a **Punnett square** to show the possible outcomes. See Figure 6.31.

Figure 6.32 shows the possibilities when a couple are *both* carriers of the cystic fibrosis allele. We know that as carriers, they both have Cc genotypes. They will produce both C- and c-carrying gametes.

		Mother (cc) gametes	
		c	c
Father (Cc) gametes	C	Cc unaffected (but a carrier)	Cc unaffected (but a carrier)
	c	cc cystic fibrosis	cc cystic fibrosis

Figure 6.31 There is a 50%, or 1 in 2, chance that a child will be born with cystic fibrosis

Mother (cc) gametes		
	C	c
Father (Cc) gametes C	CC unaffected	Cc unaffected (but a carrier)
Father (Cc) gametes c	Cc unaffected (but a carrier)	cc cystic fibrosis

Figure 6.32 The probability of the couple having a child with cystic fibrosis is 1 in 4

The ratio of children without cystic fibrosis to children with it would be 3:1. The ratio gives an *overall* probability, but of course which allele is in the sperm and which allele is in the egg will be completely random.

HIGHER TIER ONLY

3 The ability to taste a chemical called PTC is controlled by a single gene that codes for a taste receptor on the tongue. A man who is heterozygous for gene T has children with a woman who is homozygous recessive for the gene. Draw a Punnett square showing the possible genotypes of the children.

4 For an eye colour gene in parrots, the brown allele is dominant to red.

Draw a Punnett square showing the genotypes and phenotypes of a mating between two heterozygous parrots.

DID YOU KNOW?

Reginald Crundall Punnett was a Professor of Biology, and later Genetics, at the University of Cambridge. With William Bateson, he helped to establish genetics as a science.

REMEMBER!

You need to be able to interpret Punnett square diagrams, and to construct them if working at a higher level. For most of the questions you'll encounter, you'll find a 1:1 ratio, a 3:1 ratio, or – if crossing a homozygous dominant individual with a homozygous recessive – the offspring will all have the dominant phenotype.

Tracking gene disorders

Learning objectives:

- understand the use of a family tree to show the inheritance of a characteristic
- explain economic, social and ethical issues concerned with embryo screening.

KEY WORDS

embryo screening
family tree
in-vitro
  fertilisation

We can predict the possibility of a child being born with some gene disorders. A simple test, using cheek cells or blood, identifies if potential parents are carriers of a disease-causing allele.

Gene disorders caused by dominant alleles

Most, but not all, inherited conditions are caused by recessive alleles. Some are the result of a defective dominant allele.

In these instances, the presence of one allele will cause the characteristic to be seen in the phenotype.

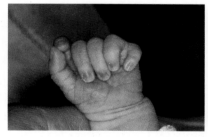

Figure 6.33 A genetic counsellor can advise on the risks of passing on a genetic disorder

1 Write down the three possible genotypes and phenotypes for the polydactyly gene.

2 Draw a Punnett cross between a father who is heterozygous for polydactyly and homozygous recessive mother.

Family-tree diagrams

In genetics, we can use a genetic **family tree** to show how a condition is passed down through a family.

Figure 6.34 Polydactyly – having extra fingers or toes – is caused by a dominant allele

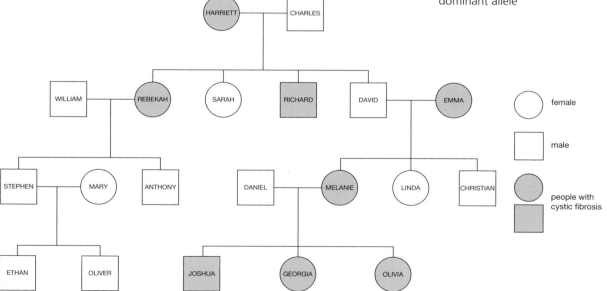

Figure 6.35 An example of the inheritance of cystic fibrosis in a family

Some of the genotypes of the family in Figure 6.35 can be worked out.

We know that Harriett, and the other members of the family affected by cystic fibrosis, *must* be cc. Anyone unaffected will either CC or Cc. Charles *must* be Cc, because both Rebekah and Richard have cystic fibrosis. If Charles were CC, none of his children could possibly have the condition.

3 What can we say about William's genotype?

4 Identify the people in the family tree who are *certain* to be carriers.

Embryo screening

When a couple use *in-vitro* fertilisation (see Chapter 5), a number of embryos is usually produced. A few cells can be removed from an embryo and tested for a defective allele. This is **embryo screening**.

In families where there is a known history of a genetic disorder, embryos can be screened. A decision can then be made whether or not to implant an embryo into the mother. Embryo screening raises ethical, social and economic issues.

Embryos that are not chosen for implantation can be frozen for possible future use, or used for research. Eventually, if they are not used, they are destroyed.

One worry about embryo screening is the potential selection of embryos for social reasons, for instance, to have a baby of a particular sex.

Embryo screening is carefully monitored in the UK. Guidelines on how embryos are used are provided by the Human Fertilisation and Embryology Authority (HFEA).

A defective gene may not have immediate life-threatening consequences. It could suggest an increased risk of a particular disease or decreased life expectancy. So what if the information from screening was obtained by insurance companies? Or a potential employer?

5 Give one possible medical, social and economic consequence of embryo screening.

6 The HFEA was asked to rule on couples who wanted their embryos screened for gene mutations that gave an increased risk of breast cancer. Not all people with these mutations go on to develop breast cancer. What would *your* ruling have been? Give reasons to justify your choice.

DID YOU KNOW?

In the case of many gene disorders, a person may be infertile and so not able to pass the allele on. Most women with cystic fibrosis can become pregnant without difficulty, but only 2–3% of affected men are fertile.

REMEMBER!

Include all the stages when working through a genetic cross. When constructing a Punnett square, write down the genotypes of the parents and the gametes before you work out the cross. Be careful with letters where upper and lower case could be confused.

Gregor Mendel

Learning objectives:

- plan experiments to explore phenomena and test hypotheses
- draw conclusions from given observations
- evaluate data in terms of reproducibility.

KEY WORDS

hypothesis
pure line
reproducibility
valid

The basis of the science of genetics was established by Gregor Mendel, through his work on pea plants in the nineteenth century. Mendel is often called the 'Father of modern genetics'.

Designing the investigations

Mendel chose to use the garden pea in his experiments. Peas were ideal for these reasons:

- A wide range of varieties was available.
- For each trait chosen, differences are sharply defined, with no intermediate forms.
- Fertilisation is easily controlled. A pea plant will fertilise itself, or can be cross-fertilised.
- Peas are easy to cultivate.
- Peas grow and flower, and seed can be collected for sowing the following growing season.

Mendel bred plants for each of the traits for 2 years. By that time, no other characteristics appeared in the offspring. These were **pure lines**, necessary to give **valid** results.

1 **Why did Mendel select the pea for his research?**

2 **Explain how breeding pure lines was necessary to give valid results.**

Carrying out the investigations

Mendel crossed a pure-breeding purple-flowered plant with a white-flowered plant. When he collected and sowed the seeds, all the plants produced purple flowers.

Mendel checked his results. It did not make any difference if the pollen grains came from the purple-flowered plants or the white.

Next, Mendel allowed these purple-flowered plants to self-fertilise. He collected the seed and grew the plants.

His results were 705 purple-flowered plants to 224 white-flowered plants.

The white-flowered plants, hidden in the first generation, had reappeared in the second!

He carried out experiments on plants with all seven traits and always obtained 3:1 ratios in the second generations.

Figure 6.36 Gregor Mendel was a monk in at a monastery in Brünn, now Brno, in the Czech Republic

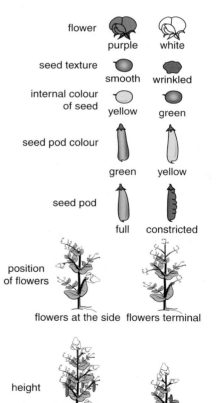

Figure 6.37 Mendel chose varieties with seven different traits, or characteristics

Mendel produced a **hypothesis**. From his observations, he suggested that:

- the inheritance of each characteristic was in 'units', which were passed on to descendants
- each plant had a pair of 'factors' governing a particular characteristic – in a dominant and recessive form
- one of these factors, at random, would go into a gamete.

He used seed-colour trait to test his hypothesis. In the first generation, all the plants produced yellow seeds.

He crossed one of these yellow-seed producing plants with a green-seed producing plant. He predicted a 1:1 ratio of plants in the offspring. He obtained 58 yellow seeds and 52 green seeds.

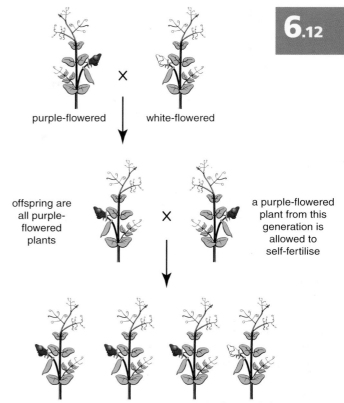

purple-flowered   white-flowered

offspring are all purple-flowered plants

a purple-flowered plant from this generation is allowed to self-fertilise

ratio of 3 purple-flowered plants: 1 white-flowered plant

Figure 6.38 Mendel's experiment with flower colour

3 **Draw genetic diagrams to show Mendel's crosses between purple-flowered and white-flowered peas.**

4 **Did Mendel's results support his hypothesis?**

The reproducibility of Mendel's results

Scientists repeat others' experiments to investigate the **reproducibility** of the original data.

Scientist	Number of yellow seeds	Number of green seeds
Gregor Mendel (1866)	6022	2001
Carl Correns (1900)	1394	453
Erich von Tschermak-Seysenegg (1900)	3580	1190
William Bateson (1909)	11903	3903
Charles Hurst (1902)	1310	445
Robert Lock (1908)	1438	514

Mendel's experiments were repeated by different scientists.

5 **Calculate the ratios of yellow to green seeds for each scientist.**

6 **Comment on the reproducibility of Mendel's results.**

DID YOU KNOW?

We now understand the biochemical basis of purple- and white-flowered peas. The purple-flowered allele codes for an enzyme that is involved in the synthesis of the purple pigment anthocyanin.

MAKING CONNECTIONS

Mendel's experiments are an excellent example of working scientifically, for experimental design, collection of large amounts of data and testing a hypothesis.

KEY CONCEPT

Genetics is simple – or is it?

Learning objectives:

- explain how certain characteristics are controlled by a single gene
- understand many characteristics are the result of multiple genes interacting
- describe the search for genes linked to disease.

KEY WORDS

complex diseases
Mendelian
 inheritance

Huw is one of the 1 in 12 males in the UK who is colour blind. The condition affects just 1 in 200 females.

Colour blindness

Huw suffers from red-green colour blindness. It's the most common form. He confuses colours that have some red and green in them.

A single gene determines whether a person has normal colour vision or is red-green colour blind. A defective allele of this gene means that the cells in Huw's eyes (called cones) that detect red and green don't function properly.

The gene is on the X-chromosome.

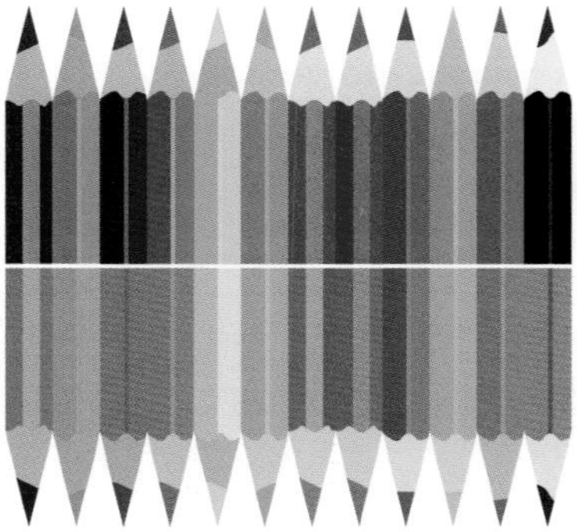

normal colour vision

red-green colour blind

Figure 6.39 The lower image shows how this set of pencil crayons might look to someone with red-green colour blindness

1. **What is red-green colour blindness?**

2. **Where is the gene for red-green colour blindness located?**

3. **How many alleles will a male and female have for this characteristic?**

Genomics and inherited disease

Observations of populations, and more recently, investigations using genomics have confirmed that most of our characteristics are controlled by not just one gene, but usually by many.

The same is true of genetic diseases. Single-gene disorders, such as cystic fibrosis, follow what is known as **Mendelian inheritance**. But these single-gene disorders are rare.

Most diseases are referred to as **complex diseases**. These include heart disease, diabetes and obesity. They result from the interaction of many genes, non-coding regions of the genome and environmental factors.

Complex diseases do cluster in families but they have no clear pattern of inheritance, because multiple genes are involved.

Figure 6.40 Research doesn't just focus on the *human* genome; 84% of the genes known to cause human disease have a zebrafish equivalent

3 Give two examples of a single-gene disorder.

4 Give two examples of a disease linked to multiple genes.

Studying pathogen genomes

Scientists are also analysing the genomes of human pathogens, to help us understand and control infectious disease. We now have complete genomes of many, including MRSA. MRSA is a type of bacterium that is resistant to antibiotics.

One day, it will be possible to analyse the genomes of pathogens and quickly select the best treatment, according to genomics of the strain of bacterium.

Figure 6.41 A nurse with an MRSA patient

5 Describe how MRSA has arisen.

6 How could analysis of the MRSA genome help in the control of an outbreak?

7 Write a scientific argument to say why it is important for our government to invest money into researching pathogen genomes.

DID YOU KNOW?

Recent excavations in London unearthed the graves of medieval plague victims. Scientists analysed DNA of the plague bacterium isolated from victims' teeth. The genomes of the 1349 bacterium, and the one that caused an outbreak in Madagascar in 2014, are *identical*.

KEY INFORMATION

Remember that most of our characteristics are controlled by multiple genes. Those controlled by a single gene are rare.

MATHS SKILLS

Fractions, ratio, proportion and probability

Learning objectives:

- understand and use fractions and percentages
- understand and use ratio and proportion
- understand and use probability when predicting the outcomes of genetic crosses.

KEY WORDS

fraction
percentage
probability
proportion
ratio

Predicting the outcome of genetic crosses quantitatively matters in understanding the inheritance of human characteristics, or of genetic diseases.

Fractions and percentages

In sections 6.8 and 6.10, you looked at the inheritance of cystic fibrosis.

The fraction of affected children =

$$\frac{\textbf{the number of affected children}}{\textbf{the total number of children shown}} = \frac{1}{4}$$

A quarter are born with cystic fibrosis.

When comparing fractions with different denominators it is helpful to convert them to a **percentage**.

The percentage of affected children =

$$\frac{1}{4} \times 100 = 25\%$$

If we know the total number in a population, we can calculate the number in a fraction (or percentage) of that population.

In the UK, 48%, or 48/100 people, in a 2014 survey, had blue eyes. So, in the UK population of 63 500 000 in 2014:

The number of people in the UK with blue eyes =

$$\frac{48}{100} \times 63\,500\,000 = 30\,480\,000$$

1 **In a cross between a heterozygous and a homozygous recessive individual for a recessive disorder, what fraction of the offspring would be expected to have the condition**

2 **Three alleles of the MC1R gene are involved in the inheritance of red hair. One of them, R160W, is found in 9% of the UK population. What number of people in the UK have this allele?**

			mother	
			Cc	
			C	c
father	Cc	C	CC normal	Cc carrier
		c	Cc carrier	cc cystic fibrosis

Figure 6.42 Punnett square showing the cross between two carriers of cystic fibrosis.

DID YOU KNOW?

In the early 1900s, a mathematical law for calculating the frequency of a genotype in a population of organisms was developed by Godfrey H. Hardy.

ADVICE

Remember that geneticists can only work with probabilities and not certainties.

Ratio and proportion

In the Punnett square in Figure 6.42, there are three normal children to every one child with cystic fibrosis: a **ratio** of 3:1.

In other words: For every 4 people, 1 will probably have cystic fibrosis and 3 will not. These are often described as the **proportions** of a given population.

If we have a ratio of 3:1 in a population, we would *expect* that as the number of normal children increases, so will the number of those with cystic fibrosis.

In genetics crosses, expected numbers will follow a certain multiplication factor, known as 'the constant of proportionality', so are said to be directly proportional.

3 The allele for Huntington's disease is dominant to the allele for normal. What ratio of offspring would you expect if a heterozygous couple had a family?

4 In a town with a population of 100 000, 40 people are affected with cystic fibrosis. Assuming direct proportionality, how many affected people would you expect in a population of a city with 1½ million people?

Probability

A couple with a history of cystic fibrosis in the family visit a genetic counsellor to assess the **probability** – or likelihood – of their children being born with cystic fibrosis.

Probabilities can be expressed as a fraction, percentage, ratio or on a scale of 0 to 1, where 1 represents certainty. If an event happens once for every four trials, then the probability of this happening is ¼ or 25% or 0.25. In biology, we often work with the decimal representation and ratio.

The genetic counsellor draws up a Punnett square.

The *probability* of producing a child with cystic fibrosis will *never change* for this couple – It is always ¼.

In reality, probability will not be exact because fertilisation is random.

5 The first child of a couple who are carriers of cystic fibrosis does not have the condition. They think the chances of having an affected child will be increased next time. Explain to them why it is not.

6 Suggest why, when working with plants, the ratio of the outcomes were very close to those expected, while those in a family might not be.

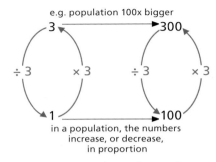

Figure 6.43 Numbers of normal children, and those with cystic fibrosis, increase by a multiplication factor of three.

		mother	
		Cc	
		C	c
father / Cc	C	CC child ①	Cc child ②
	c	Cc child ③	cc child ④

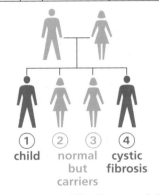

There is a probability that one child with cystic fibrosis will be born in every four children.
Two in every four will be carriers.
That is the case in this family.

Figure 6.44 Predicting the probability of children from two carriers of cystic fibrosis.

Check your progress

You should be able to:

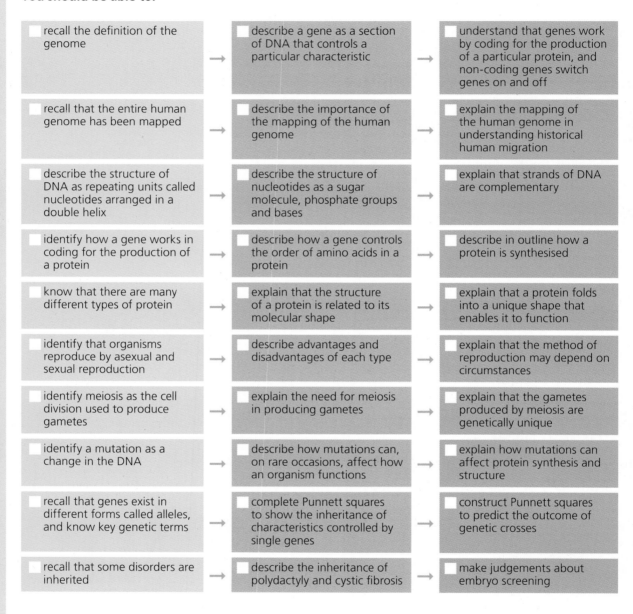

recall the definition of the genome →	describe a gene as a section of DNA that controls a particular characteristic →	understand that genes work by coding for the production of a particular protein, and non-coding genes switch genes on and off
recall that the entire human genome has been mapped →	describe the importance of the mapping of the human genome →	explain the mapping of the human genome in understanding historical human migration
describe the structure of DNA as repeating units called nucleotides arranged in a double helix →	describe the structure of nucleotides as a sugar molecule, phosphate groups and bases →	explain that strands of DNA are complementary
identify how a gene works in coding for the production of a protein →	describe how a gene controls the order of amino acids in a protein →	describe in outline how a protein is synthesised
know that there are many different types of protein →	explain that the structure of a protein is related to its molecular shape →	explain that a protein folds into a unique shape that enables it to function
identify that organisms reproduce by asexual and sexual reproduction →	describe advantages and disadvantages of each type →	explain that the method of reproduction may depend on circumstances
identify meiosis as the cell division used to produce gametes →	explain the need for meiosis in producing gametes →	explain that the gametes produced by meiosis are genetically unique
identify a mutation as a change in the DNA →	describe how mutations can, on rare occasions, affect how an organism functions →	explain how mutations can affect protein synthesis and structure
recall that genes exist in different forms called alleles, and know key genetic terms →	complete Punnett squares to show the inheritance of characteristics controlled by single genes →	construct Punnett squares to predict the outcome of genetic crosses
recall that some disorders are inherited →	describe the inheritance of polydactyly and cystic fibrosis →	make judgements about embryo screening

Worked example

A couple's second child was born with a condition called phenylketonuria (PKU). The condition is caused by a recessive allele.

The family tree shows that there is a history of PKU in the mother's (8) family. The father (7) was unaware of the condition in his family.

The couple would like another child. A genetic counsellor draws a family tree.

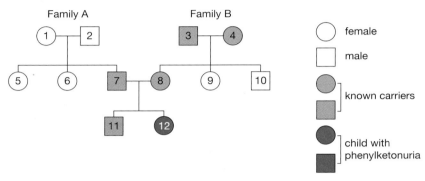

The student has identified the genotype correctly.

1 **The genetic counsellor talks about the father's family's medical history.**

  a **What is the genotype of the father (7)?**

    Pp

  b **What can be deduced about the genotype of the father's parents, 1 and 2? Explain your answer.**

    One is a carrier.

The answer would have been better with more explanation. One parent (or both) must be carriers. For 7 to be a carrier, he must have inherited one recessive allele. So, one parent must have passed on a recessive allele. The other parent would have passed on a dominant allele (otherwise 7 would have PKU). Their other allele *could* have been the recessive allele, which was not passed down to any of the three children.

  c **Draw a genetic diagram to show how 12 came to have PKU.**

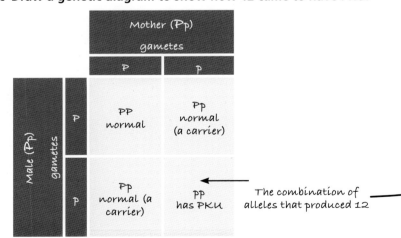

The combination of alleles that produced 12

The student has included the parental genotypes, the gametes produced, the possible genotypes and phenotypes produced, and has highlighted the combination of alleles that must have produced 12.

End of chapter questions

Getting started

1 The genetic term used to describe the entire DNA of an organism is: `1 Mark`

    A phenotype   B genome   C double helix   D allele

2 The process of cell devision that results in the formation of gametes is: `1 Mark`

    A natural selection   B mitosis   C meiosis   D cloning

3 Use the words provided to complete the sentences.

    alleles   bases   chromosomes   different positions

    DNA   each chromosome   identical positions   one chromosome

    Alleles for a particular characteristic are located at _____ on _____ of the chromosome pair. `1 Mark`

    The structure of _____ is like a twisted ladder. The rungs of the ladder are chemicals called _____ . `1 Mark`

4 Explain how scientists can use information from the mapping of the human genome. `2 Marks`

5 People can either roll their tongue into a U-shape, or are unable to roll their tongue.

    a The diagram shows a pair of chromosomes.

        Is the allele for tongue-rolling dominant or recessive? Explain your answer. `2 Marks`

    b Give the other possible genotypes related to tongue rolling. `1 Mark`

    c Give the phenotypes for these genotypes. `1 Mark`

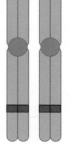

genotype TT
can roll tongue

Going further

6 Draw two lines to match the types of cell division with where they occur.

1 Mark

Cell division

Where it occurs

Asexual reproduction in bacteria

Meiosis

Growth of a human embryo

Mitosis

Sperm production

7 Which of the following organisms show *both* asexual and sexual reproduction?

1 Mark

A malarial parasite **B** human **C** daffodil **D** strawberry **E** mushroom

8 Describe the process of sex determination in human offspring.

2 Marks

You can use a diagram to help with your description.

9 In a species of mouse, black coat colour is dominant to white.

Two black mice mate.

a Complete the Punnett square below to show the genotypes and
phenotypes of the offspring.

2 Marks

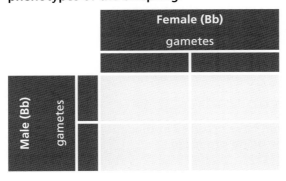

b Calculate the probability of producing a white mouse. Express your
answer as a fraction.

1 Mark

c The chromosome containing the genes for coat colour was found to have
220 million base pairs. Write this number in standard form.

1 Mark

10. Scientists have investigated how the mutation rate of certain organisms is affected by the size of their genome.

Here are some data from various groups of organism.

What conclusion can be drawn from the data?

Suggest how the confidence in your conclusion could be increased.

`2 Marks`

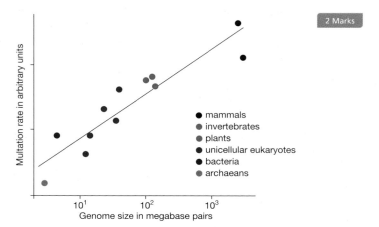

More challenging

11. Explain how a mutation can affect the production of a protein. `2 Marks`

12. All the offspring produced in asexual reproduction are genetically identical. `1 Mark`

Give one disadvantage of this.

13. The base sequence on two opposite strands of DNA is said to be: `1 Mark`

    A an allele   **B** complementary   **C** dominant   **D** a gene   **E** a double helix

14. Mutations can occur in genes. Mutations can cause genetic disorders in humans. Phenylketonuria (PKU) is a genetic disorder caused by a gene mutation. People with PKU produce an inactive enzyme. `6 Marks`

    The normal base sequence and the mutated base sequence which can cause PKU are shown below.

    normal base sequence .... C T C G G C C C T............

    mutated base sequence .... C T T G G C C C T............

    Describe and explain how the changes that have occurred in the mutated base sequence can result in an inactive enzyme being produced.

Most demanding

15. Explain how meiosis leads to the production of gametes that are different genetically. `2 Marks`

16. A bacterial cell divides using binary fission and produces 128 cells in 175 minutes. Calculate the time between each division. `1 Mark`

17 In Estonia, a large percentage of the population have blue eyes. In an Estonian town with a population of 14 000, a survey reveals that 12,600 have blue eyes. Assuming direct proportionality, how many people would you expect to have blue eyes in another Estonian town with a population of 20 000?

`1 Mark`

18 A team of scientists analysed the amount of DNA bases in the cells of 32 species of bacteria.

`6 Marks`

They used their data to plot a graph of the proportion of bases in the bacterial cells that were T (thymine) over the proportion of bases that were A (adenine).

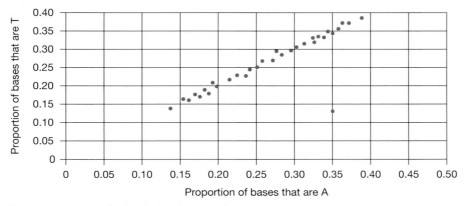

What can you conclude from the graph?

Evaluate the data obtained. In this evaluation, comment on:

- any anomalous results.
- why the proportions of A and T do not add up to 1.
- any additional investigations that could be carried out.

`Total: 40 Marks`

VARIATION AND EVOLUTION

THERE ARE DIFFERENCES BETWEEN AND WITHIN SPECIES.

- Variation between individuals is continuous or discontinuous.
- Variation can be measured and the data represented graphically.
- Hereditary material is passed from one generation to the next through the genetic material, DNA.

VARIATION BETWEEN SPECIES CAN DRIVE NATURAL SELECTION.

- Variation between individuals of the same species means that some organisms compete more successfully than others.
- When an animal or plant is more successful it passes more of its genes to the next generation. Less successful organisms will not pass on as many genes. This is natural selection.
- The science of genetics was established by the work of Austrian monk, Gregor Mendel.
- Genetics allows us to understand the inheritance of certain characteristics in humans and many other organisms.

CHANGES IN THE ENVIRONMENT CAN LEAD TO EXTINCTIONS.

- Some individuals within a species, and some species, are less well adapted than others.
- Changes in the environment can lead to the extinction of those individuals that are less well adapted.
- Gene banks can be used to preserve hereditary material.

IN THIS CHAPTER YOU WILL FIND OUT ABOUT:

WHAT CAUSES VARIATION AND WHAT ARE ITS EFFECTS ON THE INDIVIDUAL?

- Variation has genetic and environmental causes.
- Mutation, sexual reproduction and meiosis are processes that lead to variation.
- Variation results in differences in phenotypes and genotypes in a population.

HOW DO VARIATION, A STRUGGLE FOR EXISTENCE AND NATURAL SELECTION LEAD TO THE EVOLUTION OF NEW SPECIES?

- In any population, there is a struggle for existence.
- Because of variation, some individuals are better suited to the environment, so they reproduce and pass on their genes to the next generation, a process called natural selection.
- Natural selection acts on populations, and if the environment changes a new species may result, by the processes of evolution and speciation.
- In 1858, Charles Darwin and Alfred Russel Wallace proposed separately the theory of evolution.
- The evidence for natural selection and evolution comes from observations of current organisms, the fossil record and biochemistry, including DNA.
- During the 100 years that followed the theory, many scientists pieced together the connection between inheritance, genetics, evolution, genes and the genetic code.
- Humans can also change the genetic make-up of organisms by selective breeding and genetic engineering.
- Fossils are used to create evolutionary trees to help understand the relationships between groups of organisms.

WHAT ARE THE CAUSES OF EXTINCTIONS?

- Extinction arises from changes in the environment, diseases and the introduction of new species that are predators or out-compete existing species.
- The five mass extinctions to date are thought to have been caused by climate change, or a catastrophic event such as a volcanic eruption or a collision with an asteroid.

Variation

Learning objectives:

- recall that differences in the characteristics of individuals in a population is called variation
- understand the genetic and environmental differences leading to variation.

Harriett and Imogen are sisters. Their differences in appearance are the result of genetic and environmental variation.

Genetic and environmental variation

The sisters inherited genes from their parents. They are similar to, but not identical to, their parents. They also resemble each other.

Individuals of every organism that have been produced by sexual reproduction are different. With the exception of identical twins, we all have different genomes. Our genome gives us a unique set of characteristics.

Some of the sisters' features have nothing to do with genetics. Harriett wears spectacles and has braided hair. Differences that arise during development are part of **environmental variation**. Some environmental causes of variation are under our control, but others are not.

Figure 7.1 The two sisters have features in common, but they also have differences

Some human characteristics – the person's phenotype – are the result of the interaction of the genome with the environment.

Some characteristics in humans affected by:

Our genetics	The environment	A combination of both
blood group	hair colour and length	height
eye colour	language	skin colour
natural colour of hair	scars	weight
shape of earlobe	tattoos	sporting achievements

1. Give two characteristics in humans that result from:
   - **genetic variation**
   - **environmental variation.**

2. Give two characteristics in humans that result from a *combination* of genetic and environmental influences.

Variation and survival

Variation is essential to survival. There's usually a large amount of genetic variation within a population. This is essential to survival. Scientists studying populations of organisms can see how variation has led to changes. Variation is crucial to **evolution**. Each species in a stable environment is there because it has adapted to survive in that habitat.

Figure 7.2 Brown bears are adapted to life in the forest; polar bears to life on Arctic ice. Polar bears evolved from brown bears more than 350 000 years ago

3 How is an organism able to live in a specific habitat?

4 Suggest two characteristics that enable polar bears to live on the Arctic ice.

The mechanism of genetic variation

Genetic variation is the product of three processes that lead to differences in genomes: mutations, sexual reproduction and meiosis.

Mutations occur continuously. They happen spontaneously, by accident. Most mutations have no effect, but a few are harmful. Rarely, mutations give rise to new phenotypes where the new characteristics are advantageous. When this occurs, the change in an organism can be quite rapid – over generations, rather than millions of years. Examples include the peppered moth and antibiotic resistance in bacteria.

> **REMEMBER!**
>
> The three processes that lead to genetic variation. Remember also that some mutations are not expressed in an organism, but where they are, most are harmful.

Figure 7.3 A series of mutations in rats and mice have made some populations resistant to warfarin poison

5 List three processes leading to genetic variation.

6 Give an example of where a mutation has been beneficial to an organism.

> **DID YOU KNOW?**
>
> Warfarin disrupts the blood-clotting mechanism. It is used as a drug to help control strokes and heart attacks. When used as a rat poison, the animals bleed to death. As a result of mutations, and the fact that rats reproduce rapidly, many rat populations are now resistant to the poison.

The theory of evolution

Learning objectives:

- recall that all species of living things have evolved from simple life forms
- explain how evolution occurs through natural selection.

KEY WORDS

evolution
natural selection

Around 1¾ million species of eukaryote have been named and described. But the real number of species alive today may be more like 9 million.

The theory of *evolution* states that the species alive today have descended from simpler organisms.

An early theory of evolution

An early theory was put forward by French scientist Jean-Baptiste Lamarck in 1809.

Lamarck believed that organisms survived by adapting to their environment. He suggested that body parts developed during an organism's life were passed to the next generation.

Figure 7.4 Insect species outnumber all others.

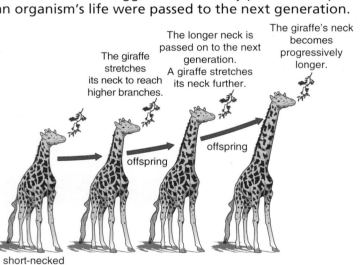

The giraffe stretches its neck to reach higher branches.

The longer neck is passed on to the next generation. A giraffe stretches its neck further.

The giraffe's neck becomes progressively longer.

offspring

offspring

short-necked ancestor

Figure 7.5 Lamarck's view of evolution

Lamarck also thought that one organism simply developed into another, rather than becoming extinct.

1. **Describe how Lamarck thought living organisms develop.**

2. **Why was Lamarck's theory wrong?**

How does evolution happen?

The environment does influence the phenotype, of course. But not in the way Lamarck thought. And these changes are *not* passed to the next generation. A body builder's muscles, for instance, are not passed on to their children.

Differences in phenotype are because individuals of a species show *variation*.

In a population, differences will give some individuals, in that particular environment, advantages compared with others.

In animals, these advantages might make an organism better able to catch food, resist disease or attract a mate.

Organisms best suited to the conditions would survive and breed. If the desirable characteristics were genetic, they would be passed on to their offspring. These inherited characteristics are the result of genes and alleles.

This is called **natural selection**. Nature is, effectively, selecting the organism with the most useful genes. Evolution happens through natural selection.

3 Suggest characteristics that might help plants to survive.

4 What is natural selection?

Environmental change

Environmental change accelerates the rate of evolution. This could be a change in the climate. Or it could be through geographical isolation. Organisms can become isolated by a mountain range, valley or water. Each population will develop differently, dependent upon the demands of the unique habitat.

The animals – or plants – with the characteristics that are most suited to the new environment will survive to produce offspring.

Over time, organisms will develop that are quite different from the original population. Different species will evolve in the different locations. These can no longer breed with the original population to produce fertile offspring.

> **DID YOU KNOW?**
>
> Lamarck's thoughts about evolution were wrong. But it is now clear that our genome *is* influenced by the environment. Addition of a chemical group – a methyl group – to our DNA is key to its regulation.

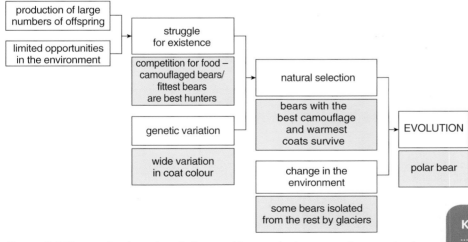

Figure 7.6 The mechanism of evolution, and how polar bears may have evolved from brown bears that were isolated by ice

> **KEY INFORMATION**
>
> Evolution explains the diversity of life on Earth and how organisms evolved from their ancestors.

5 What changes in an organism's environment might lead to different selective pressures?

6 At what point has a new species been produced?

The origin of species by natural selection

KEY WORDS

evolution
natural selection

Learning objectives:

- explain the evidence that led Darwin to propose the theory of evolution by natural selection
- describe the process of natural selection.

Charles Darwin, along with Alfred Russel Wallace, proposed the theory of evolution by natural selection.

Evidence for Darwin's theory

From 1831 to 1836, Charles Darwin travelled around the world on board the H.M.S. *Beagle*.

A short period of the voyage, in 1835, was spent on the Galápagos Islands. Those 19 days were to shape the ideas for his theory.

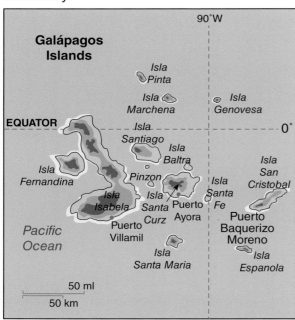

Figure 7.7 The Galápagos Islands are 600 miles off the coast of South America. They arose from the sea by volcanic action

Darwin realised that animals and plants that lived there must have originated from mainland South America.

Some species on the islands were similar to those on the mainland, but *not* the same. Others were very different. The giant tortoises were all different from island to island. Darwin also noted three species of mockingbird, each living on a different island.

DID YOU KNOW?

Darwin applied Thomas Malthus' work on human populations to other living organisms.

He realised that species were not fixed: they can change.

Darwin recorded his observations and collected specimens of the animals and plants from the islands.

1. **What part of the *Beagle's* voyage was critical to Darwin's theory?**

2. **What observations were key to Darwin's theory of evolution?**

Artificial selection

On his return to England, Darwin spent time breeding domestic pigeons. Darwin was convinced that the different breeds must be descended from the rock dove, or 'pigeon' (on the right in Figure 7.9).

These changes had been produced in a short space of time. Could a similar process be at work in nature?

3. **Name the process humans use to produce new breeds of animal.**

4. **Which animal did Darwin think was the ancestor of all pigeon breeds?**

A struggle for existence

Darwin knew that organisms produced more offspring than they needed to maintain their numbers. But numbers in populations of organisms were reasonably stable. The death rate must therefore be high.

There was, Darwin said, a 'struggle for existence'. Because of variation in a species, some individuals survived better than others. Darwin called this **natural selection**.

When populations of organisms become separated and isolated, natural selection operates differently on the populations. This is what must have happened to animals and plants arriving on the Galápagos Islands.

Key to Darwin's thinking about **evolution** was that species change. This was contrary to the thinking of the time, which was that all the species on Earth were created. The domestication of animals and plants by humans was clear evidence that species can be changed.

5. **What did Darwin think produced a 'struggle for existence'?**

6. **What processes led to the development of new species on the Galápagos Islands?**

7. **Compare and contrast Darwin's theory of evolution with that of Lamarck.**

giant Pinta tortoise

marine iguana

flightless cormorant

Galápagos penguin

Figure 7.8 Animals of the Galápagos are very unusual

Figure 7.9 Darwin's pigeons – on the left – are at the Natural History Museum

KEY INFORMATION

Ideas that led Darwin to formulate his ideas on evolution came from three sources:

- his observations, notes and specimens
- his realisation that selective breeding could change species
- knowledge of geology and fossils.

Fossil evidence

Learning objectives:

- understand how, and the situations in which, fossils are formed
- understand how fossils are used as evidence for evolution of species from simpler life forms.

KEY WORDS

fossil
radiometric dating

Darwin's work in geology was important. He formulated a theory on how volcanic islands had formed.

Fossils are the remains of organisms that lived millions, or hundreds of thousands, of years ago.

How are fossils formed?

If an organism dies in water, or is washed into water, conditions may be right for it to be fossilised.

Most **fossils** are formed as a dead animal or plant becomes buried in mud, silt or sand. Oxygen must be excluded. Otherwise, microorganisms would feed on the body and it would quickly decay.

During the process of fossilisation, the hard parts of an animal (bones, shell or teeth) or hard woody parts of a plant become replaced by minerals.

Figure 7.10 Fossils were well known in Darwin's time. *Ichthyosaurus* – featured here in a commemorative stamp – was found in 1811 by fossil hunter Mary Anning

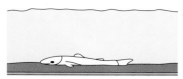

An organism dies.
It sinks to the bottom of the water.

The organism becomes covered in sediment.
The soft parts of the body decay.
The sediment begins to turn to rock.

More sediments settle. Other organisms die. If conditions are right, they begin to fossilise.
The sediments are compressed as further layers are added.
An exchange of minerals occurs between the skeleton and the water
The skeleton is turned to rock - it has become a fossil.

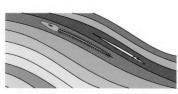

The rock layers become lifted up.
They are eroded by wind and rain.

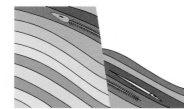

Faults in rocks and erosion expose the fossils.

Figure 7.11 How organisms become fossilised. Fossils may later become exposed by erosion and earth movements

1. **Describe how mineralised fossils are formed.**

2. **Which parts of an animal or plant are usually fossilised?**

Other remains of organisms

As well as mineralised remains, fossils can be formed in other ways.

Sometimes traces of organisms are left, such as footprints, burrows or spaces left by plant roots. Occasionally, whole organisms are found. Whole insects or plant parts have been found trapped in amber, a hardened form of tree sap. Occasionally, trunks of trees have been found after being buried in mud containing volcanic ash.

The conditions must be just right for fossil formation. It is actually a very rare occurrence.

3 What fossil traces of organisms have been found?

4 Where have whole fossilised organisms been found?

Evidence from the fossil record

As layer upon layer of rock is added, each must be younger than the one beneath it. We can tell the relative ages of the rock layers. Nowadays, with **radiometric dating**, it is also possible to find the absolute age of rocks.

The fossil record shows us how organisms that lived in the past differ from those around today. It also shows how long a species existed for, by when it enters and leaves the fossil record. It shows us extinction, and the slow and successive appearance of new species.

7.4

COMMON MISCONCEPTION

Remember that fossils are not usually the remains of the organisms themselves. The remains have usually been replaced by minerals.

DID YOU KNOW?

Palaeontologists have been collecting fossils along the 'Jurassic Coast' in Dorset, for 200 years. Tilting of the rock layers has meant that rocks range from 250 million years old – the Triassic Period – at Lyme Regis in the west, to 65 million years – the end of the Cretaceous Period – and younger at Studland in the east.

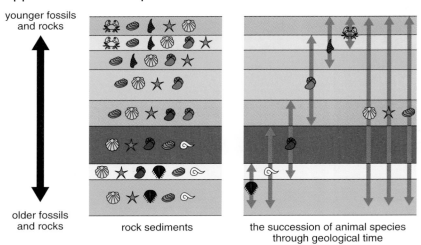

younger fossils and rocks

older fossils and rocks

rock sediments

the succession of animal species through geological time

Figure 7.12 Palaeontologists record the fossils found in rock sediments. Rocks are dated and a picture is built up of the organisms that we know were present at the time

5 How are the relative ages of fossil species worked out?

6 How can scientists find the actual age of fossils?

How much have organisms changed?

Learning objectives:

- understand why the fossil record is incomplete
- use the fossil record to understand how much, or how little, organisms have changed as life developed on Earth.

KEY WORDS

fossil record
missing link
palaeontologist

About 250 000 fossil species have been discovered, but one estimate suggests that this represents less than 1 in 20 000 of the species that ever lived. There must, of course, also be large numbers that remain undiscovered.

Fossils *are missing from the record*

The **fossil record** is incomplete. Conditions must be just right for fossil formation. The chances of this are small. In addition, geological activity over many years will have destroyed most fossils. And the older a fossil is, the more likely it is to have been lost.

Many early life forms were microorganisms or they were soft bodied. So fewer traces of them exist. Sometimes, however, conditions have enabled soft-bodied animals to be found.

Figure 7.13 **Palaeontologists**, working with scientists using medical techniques, created these three-dimensional models of an early type of green alga.

And some of the Archaea and Bacteria have left traces of the unique chemicals that they have produced. There is now clear evidence of life going back 3.5 billion years.

Nevertheless, the fossil record is very incomplete. And Darwin appreciated that if life evolved, then there must be intermediate forms, or **'missing links'**, between different organisms, as they evolved.

Figure 7.14 Darwin was delighted when an *Archaeopteryx* was found in Bavaria in 1861. It was reptile-like but had feathers. It was a critical piece of evidence showing that birds have evolved from reptiles

1 **Why do we have few fossils of early life forms?**

2 **Which early organisms have left traces?**

Fossil horses

Often, we can learn from the fossil record how much, or how little, life on Earth has changed.

When you look at the horse, for example, there is an extensive fossil record of horse-like animals. And new ones are still being found. We can trace the evolution of the modern horse, *Equus*, from an animal the size of a small dog, called *Hyracotherium*, which had toes. These toes have been lost during evolution and hooves developed.

3 **From which animal are horses thought to have evolved?**

4 **Describe the trends in the evolution of the horse.**

Some life has changed very little

In contrast, other organisms, such as certain bacteria, have changed little in billions of years. It can also be easy to distinguish fossil archaeans from fossil bacteria because of traces of the unique chemicals in their cell membranes.

5 **Give an example of a type of organism that has changed very little in billions of years.**

6 **Explain how we know this.**

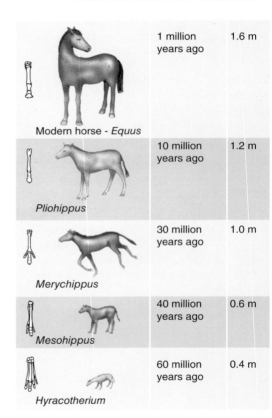

Modern horse - *Equus*	1 million years ago	1.6 m
Pliohippus	10 million years ago	1.2 m
Merychippus	30 million years ago	1.0 m
Mesohippus	40 million years ago	0.6 m
Hyracotherium	60 million years ago	0.4 m

Figure 7.15 The evolution of the modern horse. We now know that this evolution is not a simple straight line – their evolutionary tree branches and has offshoots – with some intermediate stages actually being smaller

Darwin and Wallace

KEY WORDS

mimicry
warning
  colouration

Learning objectives:

- recall how Darwin and Wallace proposed, independently, the theory of evolution
- describe how Alfred Wallace gathered evidence for evolution, including warning colouration and mimicry.

Naturalist Alfred Russel Wallace had independently developed very similar ideas to those of Darwin about evolution.

Publication of the theory

Wallace was already considering ideas about evolution when he began to travel in 1848. While on an expedition to Malaysia he wrote to Darwin, with his ideas.

Darwin actually felt that Wallace should have priority in publication of the theory. But he was persuaded by his friends to write a short paper on his work.

On the Tendency of Species to form Varieties; and on the Perpetuation of Varieties and Species by Natural Means of Selection. By CHARLES DARWIN, Esq., F.R.S., F.L.S., & F.G.S., and ALFRED WALLACE, Esq. Communicated by Sir CHARLES LYELL, F.R.S., F.L.S., and J. D. HOOKER, Esq., M.D., V.P.R.S., F.L.S., &c.

[Read July 1st, 1858.]

London, June 30th, 1858.

MY DEAR SIR,—The accompanying papers, which we have the honour of communicating to the Linnean Society, and which all relate to the same subject, viz. the Laws which affect the Production of Varieties, Races, and Species, contain the results of the investigations of two indefatigable naturalists, Mr. Charles Darwin and Mr. Alfred Wallace.

Figure 7.16 Short papers by Darwin and Wallace were read together to the Linnean Society in London 1858

Darwin immediately set about writing a book on his theory. *'On the Origin of Species by Means of Natural Selection, or the preservation of favoured races in the struggle for life'* was published in November 1859.

1. **Why did Darwin not publish his findings on returning from his world voyage?**

2. **Who were the co-authors of the theory of evolution?**

Colours and warning colouration

Wallace had researched **warning colouration**, the purpose of which is to deter predators. Warning colours are usually bright and conspicuous and in certain patterns.

COMMON MISCONCEPTION

It is important to remember that Darwin and Wallace are co-authors of the theory.

Darwin often gets more credit. But he did present large amounts of evidence to support his original hypotheses.

Predators learn to avoid animals with warning colouration and patterns. They learn that the animals taste bad, or will cause injury.

Wallace realised that warning colouration must be favoured by natural selection.

Figure 7.17 Yellow or orange and black stripes are common warning colours. Wasps sting and the cinnabar moth caterpillar produces bitter-tasting alkaloids

3 Explain how warning colouration is an example of natural selection.

4 Give an example of two animals that show warning colouration.

Mimicry in animals

Plants trick animals using **mimicry**. Sometimes animals, too, have evolved to mimic other organisms. Unrelated species will have similar colours and patterns.

Predators will avoid a mimic because it looks like a distasteful organism. For example, the hoverfly has warning colours similar to a wasp. Mimicry is seen as one of the best examples of the efficiency of natural selection.

Some butterflies can be found in different forms that each mimic an unpleasant-tasting species.

A different type of mimicry is where two species that have evolved similar colours and patterns are both distasteful, not just one.

5 Explain how mimicry is an example of natural selection and evolution.

6 In an example of mimicry, one species is distasteful, and the other is not. What might happen if the non-distasteful species increases in number?

non-mimic form

three different species of distasteful butterfly

three forms of the African swallowtail butterfly that mimic each distasteful species on the left

Figure 7.18 The African swallowtail butterfly mimics unpleasant-tasting butterflies

DID YOU KNOW?

Evolutionary biologists have found that colour patterns in a species are controlled by just a handful of genes. Genes controlling colour patterns in different species seem to be located in the same region of their genomes.

A new species

Learning objectives:

- understand that when natural selection operates differently on populations, a new species is produced
- understand that during evolution, new species are formed when populations become so different that they can no longer successfully interbreed.

KEY WORDS

hybrid
interbreeding
speciation

Changes that occur in populations of organisms, over time, lead to the formation of a new species. But when do separated populations of organisms become a new species?

Isolation and natural selection lead to a new species

Imagine two populations of a species that have been separated for a long time. Environmental conditions will be different in the two locations.

Differences will emerge in the characteristics – the phenotypes – of the two populations. Eventually, two new species are produced. The organisms can be regarded as a new species when they are no longer able to successfully breed with each other, or interbreed to produce fertile offspring.

Some animals and plants do sometimes interbreed – either naturally, or having been encouraged to by humans. But **hybrids**, if they do happen, are infertile. The key to our definition of a species is that *successful* **interbreeding** – that produces fertile offspring – is not possible between two different species.

Figure 7.19 A mule is a cross between a female horse and a male donkey

1. **Write down the definition of a species.**

2. **A horse has 64 chromosomes, and a donkey has 62. A mule has 63 chromosomes. Suggest why mules are infertile.**

Observing new species

All the evidence suggests that different species evolved from common ancestors. The process of **speciation** is a slow process, occurring over millennia.

But we can see historical evidence of speciation and changes occurring in current populations that may lead to new species.

Figure 7.20 Scientists can study evolution in the fruit fly, *Drosophila*, which has a short life cycle

3 If the theory of evolution is correct, why can we not see new species emerging?

4 Suggest what characteristics of an organism are necessary for us to investigate speciation.

The Galápagos finches

We can see good evidence of speciation in the different species of finch on the Galápagos Islands. Birds from mainland South America would have been ground-dwelling seed eaters. They would have found different conditions when they arrived on the different islands, including variations in climate and food supplies.

Competition among the birds for these food supplies would have produced a struggle for existence. Slight variations in beak shape enabled some birds to exploit slightly different food supplies, for example, small seeds, nuts, cacti and insects. Birds also selected different habitats in which to live, so as to survive.

There are now 13 species of finch living on the islands.

A tree finch.

A cactus finch feeding on a prickly pear cactus.

The woodpecker finch finds short twigs, breaks them to the right lenght, then uses them to spear and extract insects from trees.

Figure 7.21 Three species of Galápagos finch, all decended from the same ground-dwelling mainland finch

Evidence of natural selection and evolution?

Learning objectives:

- understand how scientific theories develop over time
- plan experiments to test hypotheses.

KEY WORDS

lichen
melanism

The peppered moth is white, 'peppered' with black spots. In 1848, in Manchester, the first mutant black, or melanic, peppered moth was recorded. By 1898, 95% of the moths were melanic.

The hypothesis

Lepidopterist James W. Tutt suggested an explanation. Polluted city air was laden with sulfur dioxide and soot, killing **lichens** and blackening the bark of trees. Dark moths were well camouflaged on tree trunks and less likely to be eaten by birds. In the countryside, the opposite would have been true: pale-coloured moths were camouflaged, but melanic forms were quickly seen and eaten.

In the heavily polluted Britain of the 1950s, the melanic moths appeared to have a selective advantage over the pale form.

Figure 7.22 In Victorian times urban areas like Manchester became heavily polluted

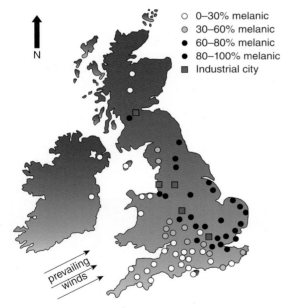

○ 0–30% melanic
◐ 30–60% melanic
● 60–80% melanic
● 80–100% melanic
■ Industrial city

Figure 7.23 In the 1950s, the melanic form was spreading, in particular downwind from industrial areas

DID YOU KNOW?

A lepidopterist is someone who specialises in the study of moths and butterflies.

1. **What caused the dark form of the moth to be produced?**

2. **What was Tutt's hypothesis?**

Testing Tutt's hypothesis

Amateur lepidopterist Bernard Kettlewell carried out a series of experiments.

His first experiment was in heavily industrialised woodland near Birmingham, with around 90% melanic moths.

Kettlewell released a large number of peppered moths, marking the underside of the wings with a spot of paint. He then caught 770 peppered moths. Recaptured moths from the first part of the experiment were identified by the spot of paint. Others were wild.

	Number of melanic moths	Number of pale moths
No. of moths released initially	447	137
No. of marked moths recaptured	149	20

In 1955, Kettlewell carried out a similar study in unpolluted woodland in Dorset. Here, none of the local moths were melanic.

	Number of melanic moths	Number of pale moths
No. of moths released initially	473	496
No. of marked moths recaptured	30	62

Figure 7.24 Kettlewell's photographs of peppered moths, on a lichen-covered tree and a green leaf

Kettlewell demonstrated that the distribution of peppered moths was correlated with the level of pollution.

3 Suggest how Kettlewell might have improved the way he controlled variables.

4 Discuss possible difficulties that Kettlewell might have had with carrying out this experiment.

Support of Kettlewell's study

Since then, anti-pollution measures of the 1950s and 1960s have taken effect. The dark form has decreased in abundance, just as predicted by the hypothesis. **Melanism** has also been observed in other species, such as ladybirds.

Although there has been some criticism of Kettlewell's work in the past 15 years, Professor Mike Majerus concluded in one of the most recent studies that the moth is 'the proof of evolution'.

5 Why did air quality improve in the UK in the 1950s and 1960s?

6 What other evidence supported Kettlewell's findings?

COMMON MISCONCEPTION

Do not think that pollution affects the moths. Melanic moths arise anyway, by random mutations, and they survive if they have a selective advantage.

DID YOU KNOW?

Scientists have used molecular genetics to show that one mutation of a single gene leads to moths becoming melanic.

KEY CONCEPT

Evolution: fitting the pieces of the jigsaw

Learning objectives:

- understand the work of Mendel, Darwin and Wallace
- appreciate that the contributions of many scientists led to gene theory being developed.

The theory of evolution was not well received. Most people believed that God had created everything.

Evolution and genetics: Mendel rediscovered

While Mendel was carrying out his work on inheritance, Darwin was developing the theory of evolution. But no one yet knew how organisms passed on characteristics to their offspring.

Mendel's paper of 1866 was not widely read. Darwin certainly never read it. It is said, however, that Mendel read Darwin's work with interest.

In the early twentieth century, Mendel's work was rediscovered. William Bateson recognised the importance of Mendel's work and had it translated into English.

1. **Why was the importance of Mendel's work not recognised until after his death?**

2. **How were human characteristics thought to be inherited, in the nineteenth century?**

Figure 7.25 A cartoon of Darwin from the 1870s. People could not accept that humans shared their ancestry with apes

Population genetics

The early geneticists studied the inheritance of certain characteristics that gave either/or phenotypes in the offspring. They did not understand how genetics worked to give subtle variations in a population. 'Mendelism' seemed incompatible with natural selection and evolution.

Then, in the 1920s, population genetics emerged. It recognised that **genetic variation** within populations, not just individuals, drove evolution.

Professor John B.S. Haldane, Sir Ronald A. Fisher and Sewell Wright, working independently, developed **mathematical models** of natural selection and mutation. They showed that Mendelian genetics *was* consistent with evolution.

Figure 7.26 'Jack' Haldane (centre). His father, J.S. Haldane (right) is famous for his work on haemoglobin

3 Why wasn't evolution thought to be compatible with Mendelism?

4 Who resolved the problem, and how?

Evolution and genes

In the twentieth century, the work of many scientists led to the **gene theory** being developed.

In the late nineteenth century, it was seen that during meiosis and fertilisation, chromosomes separated then reunited, just as Mendel had described his 'units' of inheritance.

The finding that it was genes, located on chromosomes, that controlled this inheritance, was proved by American scientist, Thomas Hunter Morgan, in the 1920s.

Figure 7.27 Morgan worked in the 'Fly room' at Columbia University for 17 years. He showed that eye colour in fruit flies corresponded with transmission of the X-chromosome

In 1941, George W. Beadle and Edward L. Tatum, showed that genes work by controlling enzyme synthesis.

The structure of DNA was discovered in 1953 (see Chapter 6). Just one important stage was missing. How do genes control protein, and therefore, enzyme synthesis? It was Francis Crick's team that deciphered the genetic code. This was the final piece of the jigsaw.

The science of genomics has, and will, continue to shed light on the processes of evolution.

5 Explain how the work of Thomas Hunter Morgan influenced scientists' understanding of the genetic code.

6 Why do you think it took so long for scientists to discover how the genetic code works?

DID YOU KNOW?

J.B.S. Haldane also carried out experiments on human physiology – many on himself. He published studies on the peppered moth and, later, genes and enzymes.

REMEMBER!

Except for Darwin and Mendel, you do not need to know about the scientists who developed gene theory. But you do need to appreciate that it involved the contributions of many scientists.

Antimicrobial resistance

Learning objectives:

- recall that bacteria develop that are resistant to antibiotics, which is evidence of evolution
- understand the mechanism by which antibiotic resistance develops
- understand the effects of the development of antibiotic resistance on the treatment of disease.

Antibiotics saved countless lives during the Second World War and after. In 1947, however, within 4 years of the first antibiotic – a type of penicillin – being introduced, the resistance of some bacteria to the antibiotic was observed.

Developing resistance

Bacteria and other microorganisms are becoming resistant to the drugs that have been designed to kill them.

The mutation of genes in pathogenic bacteria produces new strains of the bacteria. Some of these strains may be resistant to an antibiotic used on a patient. So the bacteria will not be killed.

As they reproduce, the genes that give the bacteria resistance spread throughout the population. Antibiotic resistance is an example of natural selection. Bacteria can evolve quickly because of their rapid reproductive rate.

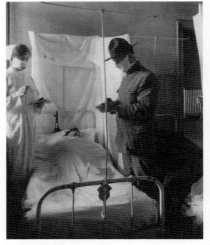

Figure 7.28 An American soldier is given a dose of penicillin. Penicillin significantly reduced the deaths of soldiers from infection

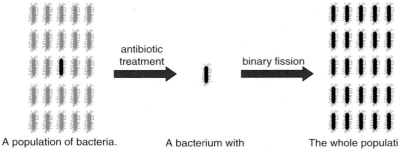

A population of bacteria. A mutation occurs. One or more bacteria in the population are resistant to the antibiotic.

A bacterium with the resistant gene(s) survives.

The whole population of the bacterium is now resistant to the antibiotic.

Figure 7.29 A simplified diagram of how resistance to antibiotics spreads throughout a population of bacteria. In reality, more than one mutation is required

Certain bacteria, such as *Staphylococcus aureus*, are building up resistance to antibiotics very quickly. *Staphylococcus* that are resistant to the antibiotic methicillin are called MRSA.

1. Describe how antibiotic resistance develops.

2. Explain how this resistance is evidence for Darwin's theory of evolution.

Antimicrobial resistance

An antibiotic will not work against a resistant bacterium. An alternative antibiotic must be used. Gradually, bacteria become resistant to more and more antibiotics.

For serious infections, certain antibiotics are kept in reserve, the so-called 'last-resort' antibiotics.

Many scientists consider antibiotic resistance – or, more correctly, **antimicrobial resistance** (AMR), because resistance is also developing against antivirals, antiprotozoals and antifungals – to be the worst public health problem facing us today.

Dame Sally Davies, the UK's Chief Medical Officer, says Britain faces 'returning to a nineteenth century world where the smallest infection or operation could kill' us. Major advances in organ transplants and cancer treatment would be wiped out.

3 How do we currently treat bacteria that are resistant to common antibiotics?

4 Why is it suggested that we are returning to a medical world of the nineteenth century?

Growing resistance

AMR would occur anyway, but it is happening at a quicker rate than it should because of the over-prescription of antibiotics.

Antibiotics are often prescribed, inappropriately, for minor viral infections such as colds or respiratory infections. Doctors are also often persuaded by patients to prescribe them.

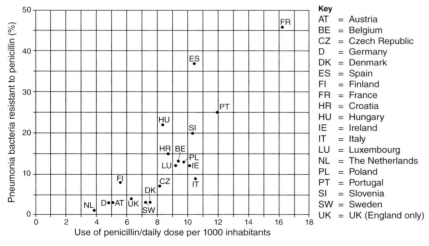

Key
AT = Austria
BE = Belgium
CZ = Czech Republic
D = Germany
DK = Denmark
ES = Spain
FI = Finland
FR = France
HR = Croatia
HU = Hungary
IE = Ireland
IT = Italy
LU = Luxembourg
NL = The Netherlands
PL = Poland
PT = Portugal
SI = Slovenia
SW = Sweden
UK = UK (England only)

Figure 7.30 Antibiotic use and penicillin-resistant bacteria across Europe. In some countries antibiotics are available 'over the counter' without a prescription

5 What is the cause of growing resistance to antibiotics?

6 Describe trends in resistance of pneumonia bacteria to penicillin in the European countries.

KEY INFORMATION

Remember that antibiotic resistance would occur anyway. It is just our misuse of antibiotics that is speeding the process up.

DID YOU KNOW?

MRSA is the term used for strains of the bacterium that are resistant to the antibiotic methicillin and, nowadays, to other commonly used antibiotics.

7.10

Combatting antimicrobial resistance

KEY WORD
....
AMR

Learning objectives:

- describe how to reduce the rate of development of antibiotic resistance
- understand the requirement for, and the impact of, new antibiotics.

In 2015, a new method for growing bacteria enabled scientists to isolate 25 new antibiotics. One of them – teixobactin – is described as a 'game-changer'. Researchers could not find any mutations of *Staphylococcus aureus* or *Mycobacterium tuberculosis* that were resistant.

Slowing down antimicrobial resistance (AMR)

Many people, when prescribed antibiotics, do not complete the course. But completing the prescribed course is essential – all bacteria must be killed. Any bacteria left by not completing the course will be more resistant to antibiotics than the least resistant ones that were killed first.

Patients must never share antibiotics or skip doses, or save some for later on. Doctors must limit the prescription of antibiotics and not prescribe antibiotics inappropriately, such as for non-serious or viral infections.

The widespread use of antibiotics, for example in agriculture, should also be restricted.

1. **Why must a course of antibiotics always be completed?**

2. **What advice is given to doctors to reduce the problem?**

Figure 7.31 People can pledge their support to cut unnecessary use of antibiotics with the *Antibiotic Guardian*. November 18 is European Antibiotic Awareness Day

ADVICE
....
Remember the ways of reducing AMR.

Modelling the future

The British government commissioned three reports on **AMR**, published in 2014 and 2015.

In one of the reports, scientists had modelled the projected numbers of deaths from AMR in the future. These numbers were based on current estimated death rates and the rate of developing resistance.

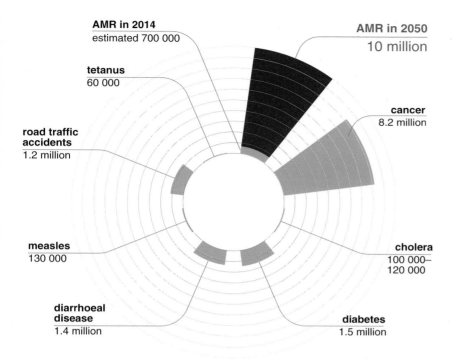

Figure 7.32 The global death rates from various diseases and road traffic accidents, and current and projected death rates from AMR

3 Compare current death rates from different diseases and road traffic accidents with current and projected death rates from AMR.

4 What assumptions are made when modelling the number of deaths from AMR in 2050?

5 The report also suggested that AMR would cost the world up to $100 trillion by 2050. Suggest why.

Can we solve the problem?

AMR is a global concern, and reducing the problem cannot be achieved by action in the UK alone. Governments across the world are encouraging pharmaceutical companies to develop new antibiotics.

The problem for drugs companies is partly economic. The testing, development and trialling of new antibiotics is slow and costly. And, any new antibiotic is likely to be used sparingly, or as a last resort. So pharmaceutical companies will not sell much of their new antibiotic. They will lose money.

Another problem is that, as new resistant strains emerge, an antibiotic may not be effective for long.

But there is also cause for optimism. Advances in genetics and genomics will change the way that infection by resistant pathogens is detected, reported and treated.

6 Discuss the scientific and economic development of new antibiotics.

7 Should governments finance pharmaceutical companies to develop new antibiotics?

DID YOU KNOW?

Most current antibiotics work by affecting cell wall production, or protein or DNA synthesis in bacteria.

Scientists have tricked tuberculosis bacteria into becoming suicidal. They have blocked synthesis of an enzyme, which leads to the lethal build-up of a sugar.

Selective breeding

Learning objectives:

- describe the process of selective breeding
- recall how selective breeding enables humans to choose desirable characteristics in animals
- explain how selective breeding can lead to inbreeding.

Humans have been domesticating animals and plants for thousands of years. DNA evidence suggests that dogs were the first animals to be domesticated, perhaps as early as 33 000 years ago.

Selective breeding

Among the first animals to be domesticated were goats, then sheep. They would have been kept and bred for their meat, milk and hides or skins, which were used for clothes and shelter.

Animals in a population show genetic variation. Humans would have selected those with the characteristics required – the ones that produced the most meat or milk, for instance – and allowed them to breed.

From the offspring of those animals, the humans would then have selected those animals producing the largest yields and bred those. This was repeated over many generations.

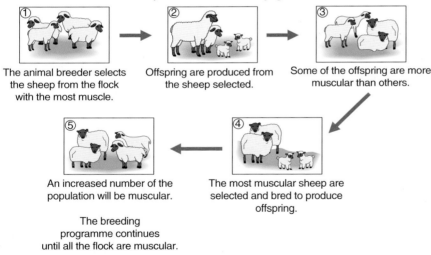

① The animal breeder selects the sheep from the flock with the most muscle.

② Offspring are produced from the sheep selected.

③ Some of the offspring are more muscular than others.

④ The most muscular sheep are selected and bred to produce offspring.

⑤ An increased number of the population will be muscular.

The breeding programme continues until all the flock are muscular.

Figure 7.33 Sheep are bred for their lean meat (muscle) and quick production of meat.

This is called **selective breeding**. Darwin referred to it as artificial selection, in contrast to natural selection.

The outcomes of different selective breeding programmes are different **breeds**. The breed produced will depend on the characteristics required.

Related characteristics are often selected together, such as quality *and* volume of milk produced in cattle.

Figure 7.34 The Belgian blue (top) has been bred for its meat, which is muscle. The Jersey (bottom) has been bred to produce rich, creamy milk

1. **Suggest what other characteristics sheep may be bred for.**

2. **Describe the technique of selective breeding.**

Domesticating dogs

Breeding dogs for hunting, or for their appearance and temperament, has involved selective breeding. The dog breeder selects dogs with characteristics of the breed.

Popular breeds of large dogs look different, but they are all the same species. They can still interbreed.

A person buying a dog looks for its 'pedigree'. This is a record of the animal's parentage, to show how true a dog is to the breed.

3. **List some characteristics that dogs are bred for.**

4. **Explain why, even though they look very different, all dogs belong to the same species.**

Problems with inbreeding

Pedigree dogs are bred with individuals of the same breed. This maintains their characteristics. They will have a limited gene pool, or range of genes in the population. They will become inbred. Usually, the incidence of rare disease alleles will increase in inbred animals.

Studies show that **inbreeding** puts dogs at risk of birth defects and genetically inherited health problems, and makes them prone to disease. Unnaturally small or large dogs are particularly prone to problems.

> **REMEMBER!**
>
> It is important to be able to understand and describe the principles of selective breeding.

> **DID YOU KNOW?**
>
> Analysis of mitochondrial DNA suggests that modern dogs are related to the first dogs – which would have been kept for hunting – which originated in Europe between 19 000 and 32 000 years ago.

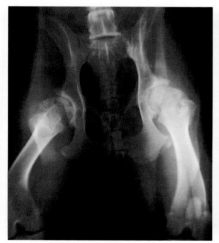

Hip dysplasia, where the hip joint is not correctly formed, and size are correlated in large dogs.

Many Dalmations are predisposed to deafness.

In bulldogs the large head and narrow hips mean that sometimes puppies have to be born by Caesarean section.

Figure 7.35 Some problems with inbreeding in certain dog breeds

5. **Explain why breeding dogs with other dogs of the same breed causes problems.**

6. **List some of the problems with this process.**

7. **Evaluate whether the benefits of breeding pedigree dogs outweigh the health risks to the dog.**

Producing new plant varieties

Learning objectives:

- describe the process of selective breeding
- recall how selective breeding enables humans to choose desirable characteristics in plants.

KEY WORDS

environmental change
genetic variation
mixed population

Lack of *genetic variation* in potatoes cultivated in Ireland in the 1850s was a factor that contributed to the potato famine. Many people died of starvation. Most of the potatoes grown were one variety: the Lumper.

Producing new plant varieties

In modern agriculture, new crop varieties are bred for disease resistance. Increased yields and improved quality of food crops are also the main goals of plant breeders. Other beneficial characteristics are that crops:

- grow and mature quickly
- have a distinctive taste, aroma or colour, for example, in strawberries
- have long shelf-life, store well or can be frozen.

Selective breeding enables crop growers to plant large areas with identical plants, giving maximum yields. But this does mean that many crops are genetically uniform. The whole crop could be lost if there is **environmental change**.

1 Suggest three characteristics that crop plants could be bred for.

2 What could happen to a crop made up of just one variety if the environment changed?

Traditional plant breeding

In traditional plant breeding, a plant with a desirable trait is crossed with another plant having the same or another desirable trait. The plants selected would be from a **mixed population**, or existing varieties. Pollen is transferred from the flowers of one plant to the other. The plant is prevented from self-fertilising or being cross-pollinated by another plant.

The process is repeated over several generations until the plants breed true for the required characteristics.

Plants in horticulture are developed, for instance, for the size and colour of their flowers and scent.

Clematis is a wild flower that grows in hedgerows and climbs trees. In the nineteenth century, plant breeders in Britain, France, Belgium and Germany raced to produce varieties with the largest, most colourful flowers.

Figure 7.36 In a plant-breeding programme, plants of one variety or species are crossed with another

REMEMBER!

You should aim to be able to explain the benefits of and risks of selective breeding.

Figure 7.37 Wild clematis (top left) and two modern clematis varieties. Plant breeders still use wild clematis because it is tolerant of clematis wilt disease

3 How does the plant breeder control which pollen is transferred to which female flower?

4 Explain why it is still useful to use wild versions of a cultivated plant in breeding programmes.

Producing a new variety

It can be very difficult to combine several desirable characteristics in a plant, for example, yield, disease resistance and perhaps other characteristics. And it is expensive.

A traditional plant-breeding programme will commonly take 12–15 years. It will involve the selection and crossing of suitable individual plants, various selection processes and a series of trials.

> **DID YOU KNOW?**
>
> Many plant scientists are now using marker-assisted techniques in plant breeding. We can identify genetic markers in a plant's DNA linked to key important characteristics such as nutritional qualities, disease resistance, yield, etc.

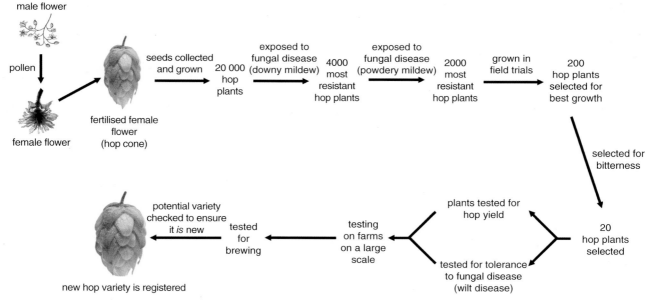

Figure 7.38 A hop-breeding programme uses traditional plant-breeding techniques

5 Explain why a plant-breeding programme can take up to 15 years.

6 What three characteristics are hops bred for?

Biology

Genetic engineering

Learning objectives:

- give examples of how plant crops have been genetically engineered to improve products and describe how fungus cells are engineered to produce human insulin.

- describe the process of genetic engineering

KEY WORDS

genetic engineering
GM crops
vector

Genetic engineering involves taking specific genes from one organism and introducing them into the genome of another. Scientists can now, more or less, transfer genes from any organism, including plants, animals, bacteria or viruses.

Producing human insulin

Patients with Type 1 diabetes need regular injections of the hormone insulin. Since the early 1920s, insulin was extracted from the pancreas of pigs or cattle. But these types of insulin differ *slightly* from human insulin in the amino acids they contain. They had some side effects.

With **genetic engineering** it became possible to genetically engineer the bacterium, *Escherichia coli*, and the fungus, yeast, to produce 'human' insulin. This is *identical* to the insulin produced by the human body.

Yeast produces a more complete version of the insulin molecule. Less processing is required, so this method is often preferred.

Figure 7.39 Human insulin production in India. This photograph shows the purification process

1. What is genetic engineering?
2. Name two organisms that can be genetically engineered to produce insulin.

Genetically engineered plants

Genetic engineering has transformed crop production. Genes from many organisms, often not even plants, are cut out of their chromosomes and inserted into the cells of crop plants. Such crop plants and other organisms are called genetically modified, **GM crops** or GM organisms (GMOs).

Plants have been engineered to be resistant to disease, and to increase yields, such as producing bigger, better fruit. Several types of crop plant have been produced that are resistant to diseases caused by viruses.

In the wet summer of 2012, potato plants became exposed to the potato blight fungus. In 2014 British scientists produced a GM potato that is resistant to potato blight. Genes from two wild relatives of the potato were inserted into the Desiree potato variety.

300 AQA GCSE Biology: Student Book

Figure 7.40 Pesticides are sprayed over crops to protect them from diseases. Disease resistant GM plants don't need the pesticides.

3 **Give two reasons for the genetic modification of plant crops.**

4 **What types of organism cause disease in plants?**

HIGHER TIER ONLY

The genetic engineering technique

Enzymes are used to remove the required gene, or genes, from the organism that carries the gene(s).

The gene is transferred, using a **vector**, to the organism that is to be modified. The vector is often a plasmid. The gene is inserted and sealed into the plasmid DNA using another enzyme. Bacteria have plasmids, and so do some eukaryotes, such as yeast. Viruses may also be used as vectors, including tobacco mosaic virus. Viruses that have had other genes modified, so that they are not infective, have been used in vaccine production.

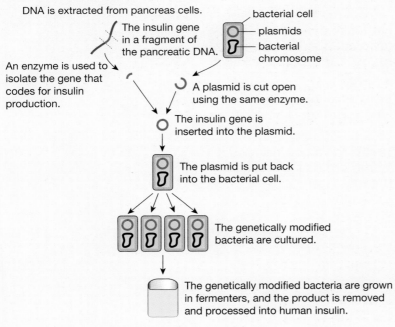

DNA is extracted from pancreas cells.

The insulin gene in a fragment of the pancreatic DNA.

bacterial cell

plasmids

bacterial chromosome

An enzyme is used to isolate the gene that codes for insulin production.

A plasmid is cut open using the same enzyme.

The insulin gene is inserted into the plasmid.

The plasmid is put back into the bacterial cell.

The genetically modified bacteria are cultured.

The genetically modified bacteria are grown in fermenters, and the product is removed and processed into human insulin.

Figure 7.41 Insulin production using *Escherichia coli*

5 **How is the required gene removed from the donor organism?**

6 **How is the gene transferred to the organism that is to be genetically modified?**

DID YOU KNOW?

Genes for human insulin were cloned and transferred to *E. coli* in 1978. The first drug produced using genetic engineering was human insulin.

KEY INFORMATION

Remember: if genes are transferred to plants, this needs to be at an early stage of their development. Older organisms have too many cells that would need to be modified.

Genetically modified crops: the science

KEY WORDS

food security
GM crops
larva
order

Learning objectives:

- explain the benefits of, and concerns about, genetic modification
- explain the ethical concerns of genetic engineering.

Biotechnologists in Britain aim to increase wheat yields to 20 tonnes per hectare by 2035.

Feeding the world

Many people think that we have the ability, and a moral obligation, to achieve global **food security** by growing GM organisms. **GM crops** have increased yields and will grow in poor soil and harsh environments. Scientists have also fortified plant crops with extra nutrients.

1 How might GM crops help to achieve global food security?

2 What characteristics are being introduced into GM crops?

Herbicide- and insecticide-resistant crops

Weeds reduce crop-plant yields and encourage fungal disease. GM crops have had genes inserted that make them resistant to a particular herbicide. As the crop grows, the field is sprayed with herbicide. The crop is unaffected, but the weeds are killed.

Another key area is the development of GM crops that resist insect attack. *Bacillus thuringiensis* (Bt) is a soil bacterium that produces a natural insecticide. The gene for this has been inserted into crops.

Bt insecticide is a protein that kills the caterpillar, or **larva**, that eats the crop plant. It only works on some **orders** of insect, such as butterflies and moths, which are the most serious pests.

Figure 7.42 After a few bites of the leaves of this Bt peanut, the caterpillar died

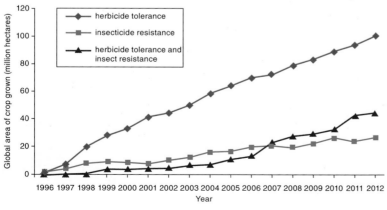

Figure 7.43 Global production of herbicide-tolerant and insect-resistant plants

3 Why are herbicide-tolerant genes inserted into crops?

4 The reproductive rate in insects is very fast. Suggest why some populations are becoming resistant to the Bt gene.

Ecological effects

GM crops growing in the wild have been reported in many countries. 'Escaped' populations usually die out. GM crops may also cross-pollinate a wild or cultivated relative of the crop. GM oilseed rape is genetically capable of cross-pollinating with eight wild relatives.

GM crops have been reported to be swapping genes with other GM crops. A study of 288 oilseed rape plants in America showed that two had 'man-made' genes for multiple pesticides.

Of concern is the *Bt* gene's effect on non-target organisms. Only insects that eat the crop should be affected. However, pollen of some GM crops is carried in the wind. The pollen could be toxic to other insects that are essential pollinators of crops and other plants.

The monarch butterfly

A monarch butterfly larva, feeding on milkweed

Figure 7.44 One study in 1999 showed that the growth of monarch butterfly caterpillars was affected when its foodplants were dusted with Bt pollen

Studies suggest, however, that the density of pollen in the wild would rarely come close to the levels needed to harm the caterpillars. The monarch butterfly is in decline, and research is ongoing to find out why.

5 Explain the possible spread of GM plants.

6 What are the concerns about Bt plants?

7 Do you consider GM products to be safe? Justify your answer.

ADVICE

Be careful what you read about genetic modification. Newspapers report scientific information second hand. Look for the original source of the information.

DID YOU KNOW?

Researchers aim to start growing a vitamin-enhanced 'super-banana' in Uganda. The bananas are genetically engineered to have increased levels of β-carotene, which is converted to vitamin A by the body. A deficiency in vitamin A can be fatal.

Is genetic modification safe?

Learning objectives:

- explain the concerns that people have about genetic modification
- explain the possible safety issues of genetic engineering in agriculture and medicine.

GM food crops and production of hormones by GM have had beneficial effects. But are GM products safe?

Safety concerns of GMOs

Insulin can be produced by yeast or by *Escherichia coli* bacteria. *E. coli* lives in our large intestines. Yeast, in the wild, lives on the surface of fruit. If they escaped from biotechnology production facilities might they be dangerous?

Genetic modification of viruses that are used to transfer genes causes real concern. Viruses can only reproduce in the cells of other organisms, so their genetic manipulation is worrying. The solution is to inactivate genes related to infection.

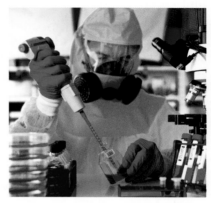

Figure 7.45 The Ebola outbreak in 2014. Development of a vaccine has involved inserting genes from the ebola virus into a respiratory virus. The benefits need to be weighed against risks

1. **Why are people concerned about GMOs escaping from fermenters?**

2. **How are viruses used to transfer genes made safe?**

Antibiotic-resistance genes

When organisms or cells are modified, not all cells take up the foreign gene. The cells that have not, need to be screened out. When the cells are modified, a **genetic marker gene** is inserted into the organism's genome as well as the required gene. This marker gene is usually one giving antibiotic resistance. When the culture of cells is treated with antibiotic, any cells that have not taken up the required gene successfully will be killed. Resistance is created to the antibiotic kanamycin, which is rarely used today.

3. **Why is a genetic marker gene used when genetically modifying organisms?**

4. **Why has the gene for resistance to the antibiotic kanamycin been chosen as the genetic marker?**

Assessing the safety of GM products

The EU's approach to GM crop development is the most stringent in the world.

One significant concern is whether a GM food might cause allergic reactions. No GM products to date have been found to produce new allergies, or are any more or less allergenic than their non-GM versions. GM foods, modified appropriately, could become the solution to, rather than the cause of, food allergies.

Most scientists have found no differences that are statistically significant between other health impacts of GM crops and non-GM versions.

Recent studies have suggested that some GM products may cause cancer. But these studies are controversial.

GM plants are commonly made resistant to a herbicide called glyphosate, or Roundup®. Some harmful effects could be caused by the herbicide, and not the genetic modification itself.

Glyphosate is described as harmless to humans and other animals, and is said to break down quickly in the environment. But now some scientists have identified correlations between glyphosate use and an increase in conditions ranging from autism to cancer.

Figure 7.46 A GM field of maize sprayed with glyphosate. Use of this herbicide has increased in the USA since the introduction of GM crops

Figure 7.47 Environmental activists protesting against Monsanto's GMO programs and demanding that foods be labelled

5. **Discuss the evidence for GM crops being harmful.**

6. **Suggest why crop treatments, and not the crops themselves, might be harmful.**

REMEMBER!

It may be helpful to be prepared to analyse and discuss negative and positive aspects of genetic modification from information that you are given.

DID YOU KNOW?

US scientists have engineered a bacterium that, to live, needs an amino acid that does not occur in nature. If any of the organisms escape the culture vessel, they cannot survive.

Ethically wrong, or essential?

Learning objectives:

- explain the benefits of, and concerns about, genetic modification
- explain the ethical issues of genetic engineering in agriculture and medicine.

KEY WORDS

gene therapy
genome editing
vector

Glowing GM fish were created during attempts to produce a fish that would glow in polluted water. Unfortunately, these GM fish glowed all the time.

The ethics of genetic modification

Was the sale of the Glofish® ethical? Or its development in the first place?

Should we be inserting genes from one organism into another? Very often these species are not even closely related. And should we be inserting human genes into other organisms?

Some religious teachings suggest that it is wrong to change natural organisms that God has created. What right do we have to change the genome of another organism? Is it acceptable to do this for the benefit of the human race? Is genetic engineering *ever* justified?

Figure 7.48 Glofish® didn't work as a pollution indicator, but quickly went on sale in pet shops and then on the Internet.

1 Give one ethical reason in favour of genetic modification.

2 Give one ethical reason against genetic modification.

Producing and marketing GM foods

Farmers cannot collect and sow the seed from GM crops because they will not breed true. They have to buy more for the next growing season. The seed companies are perceived as exploiting poor farmers.

Food producers are required to label GM foods, making it clear when a food is genetically modified. Customers can decide whether or not to buy them.

GM food must be clearly labelled. And food manufacturers must ensure that no other food is contaminated with GM ingredients. Unscrupulous food manufacturers might substitute cheaper GM ingredients for more expensive conventional ones.

3 Explain why farmers should not collect seed from a GM crop to sow the following year.

4 Why should GM foods be labelled?

5 How could GM contaminants be detected in food?

Gene therapies

Introducing genes from other organisms into humans would be unethical. But modern medicine has been investigating **gene therapy** in humans to overcome inherited disorders.

Most gene therapy is in its experimental stages. It would largely centre on inserting a normal version of an allele into cells that carry a defective version of that allele.

A deactivated virus is normally used as a **vector** to transfer the gene into the cell. But a major difficulty is how to deliver the replacement gene effectively.

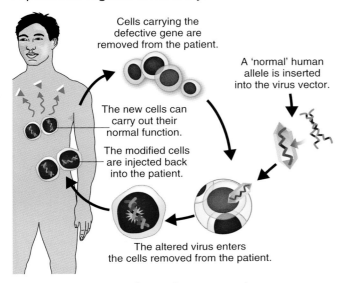

Cells carrying the defective gene are removed from the patient.

A 'normal' human allele is inserted into the virus vector.

The new cells can carry out their normal function.

The modified cells are injected back into the patient.

The altered virus enters the cells removed from the patient.

Figure 7.49 One type of gene therapy procedure

In the early days of gene therapy trials there were many failures. Some patients died from a massive immune response or developed leukaemia, probably because of the viral vector.

Currently there are thousands of gene therapy trials. In the meantime, another gene technology is showing great potential for improving human health. **Genome editing** involves replacing or removing sections of the DNA of the genome using 'molecular scissors'. Although not new, recent developments have improved the precision and efficiency of the DNA editing process.

Any successful gene therapy would not prevent the person from passing on an inherited condition to their children. It is the patient's body cells that are being modified. Any modification of reproductive cells, or germline gene therapy, would be both technically and ethically difficult. This type of therapy involving reproductive cells is currently not permitted by law.

6 Why were early attempts at gene therapy unsuccessful?

7 Explain one of the limitations of gene therapy.

8 What ethical argument could you give against developing gene therapy?

> **ADVICE**
>
> **You should aim to be prepared to analyse information and present ethical arguments for and against genetic modification.**

Figure 7.50 Many trials have been focused on correcting a severe immune system abnormality called SCID, or 'bubble-boy syndrome'

> **DID YOU KNOW?**
>
> One of the most publicised applications of gene therapy is for cystic fibrosis. The gene therapy formulation has been based on a plasmid rather than a virus, sprayed into the lungs of patients.

Cloning

Learning objectives:

- describe how cuttings and tissue culture are used to produce new plants
- describe the use of embryo transplants and cloning in animals.

KEY WORDS

adult cell cloning
cuttings
embryo transplant
therapeutic cloning
tissue culture

Ground-breaking research in China showed that the errors in DNA that lead to the blood disorder beta-thalassaemia could be successfully corrected in human embryos.

Cloning GM cells

When plant or animal cells are genetically modified, it is essential that they all carry the newly introduced gene, or genes. Cells are first screened, often with an antibiotic, so that cells without the new gene are killed.

The cells are then cloned. GM plant cells are cloned in small groups (**tissue culture**). The plant cells grow into small plants or plantlets on culture media containing nutrients and plant hormones.

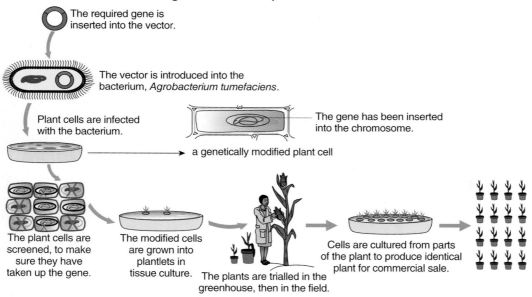

The required gene is inserted into the vector.

The vector is introduced into the bacterium, *Agrobacterium tumefaciens*.

Plant cells are infected with the bacterium.

The gene has been inserted into the chromosome.

a genetically modified plant cell

The plant cells are screened, to make sure they have taken up the gene.

The modified cells are grown into plantlets in tissue culture.

The plants are trialled in the greenhouse, then in the field.

Cells are cultured from parts of the plant to produce identical plant for commercial sale.

Figure 7.51 One method used to genetically modify and then clone plants

Tissue culture is also used to produce many other plants, such as rare plants that are in danger of extinction.

Tissue culture is not the only way of producing new, identical plants. Gardeners produce many plants using **cuttings**. A leaf and stem are cut and dipped in hormone powder to encourage rooting. The cutting then grows into a new plant.

1. What technique is used to grow GM cells into plantlets?

2. What technique is used by gardeners to produce identical plants?

Embryo transplants

Cloning is also important in animals. It enables the animal breeder to produce many animals that have identical characteristics. This can be important commercially, perhaps related to producing high milk volumes or quality beef. Cloning in animals is done by **embryo transplants.**

A developing embryo is removed from a pregnant female early in the pregnancy, so the cells have not yet become specialised. The cells are separated, grown for a while in culture and then transplanted into host mothers.

3 **What is the purpose of embryo transplants?**

4 **Describe the technique of embryo transplantation.**

Adult cell cloning

You have read about **therapeutic cloning** (Chapter 1). **Adult cell cloning** is an identical process, but the embryo produced is actually implanted into a female.

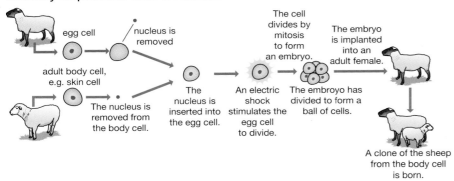

Figure 7.52 Dolly the sheep was the first mammal to be produced by adult cell cloning, in 1996

To produce Dolly, a specialised cell was made to behave as if it was a recently fertilised egg. It showed that cell specialisation is not irreversible. The donor cell was an udder cell of a 6-year-old white sheep.

Dolly lived to the age of 6½. She produced four healthy lambs in 1998 and 1999 in the normal way.

Since Dolly, scientists have cloned other animals using the same type of cloning technique, and inserted human genes into some of them, to produce human proteins such as blood clotting factor.

The research has also made possible the technique of therapeutic cloning, although adult cell cloning in humans is currently not permitted.

5 **Describe how adult cell cloning is carried out.**

6 **Describe the differences between adult cell cloning and therapeutic cloning.**

REMEMBER!

You need to know about tissue culture, cuttings, embryo transplants and adult cell cloning.

DID YOU KNOW?

Scientists genetically modify plant cells in one of two ways: using plasmids inserted into the soil bacterium *Agrobacterium tumefaciens,* or using a gene gun, which shoots DNA into plant cells.

The tree of life

Learning Objectives:

- describe how living things have been classified into groups using a system devised by Linnaeus
- describe how new models of classification have developed.

In 2015, scientists discovered a new species of snake in a remote region of Western Australia. When a new species is discovered, scientists try to classify it.

Classifying organisms

All species are classified using the system developed by Swedish scientist Carl Linnaeus. Linnaeus divided all organisms into large groups called kingdoms. These were sub-divided into smaller groups, and these groups into smaller groups still: kingdom, phylum, class, order, family, genus, species. The smallest group is the species.

Linnaeus named species using the **binomial system**. Each organism has two names – a generic name and species name.

1. Who devised the classification system that we use today?

2. What system do we use to name an individual type of organism?

Characteristics become more specific at each consecutive level.

The number of organisms in each level decreases.

Organisms have more and more characteristics in common.

Kingdom	Animalia	(animal)
Phylum	Chordata	(with a 'notochord' – in vertebrates, this develops into the spinal column)
Class	Mammalia	(mammal)
Order	Primates	(primate)
Family	Hominidae	(modern humans and extinct immediate ancestors of humans)
Genus	*Homo*	(modern humans: 'wise man')
Species	*sapiens*	

Figure 7.53 The classification of humans. Our scientific name is *Homo sapiens*. There are at least ten animals in our genus, but all are extinct except for humans

Evolutionary trees

An **evolutionary tree** shows how scientists think organisms evolved as they diverged from common ancestors.

There have been many versions of evolutionary trees, and new models continue to be developed. These developments have occurred because:

- Scientific examination has improved as microscopes have developed and our understanding of the physical characteristics (external and internal structures) of organisms has grown.
- Our understanding of biochemistry and biochemical processes has improved.

3. How have new models of classification developed?

4. Name the three domains that have been proposed. It may help to refer back to section 1.5 in Chapter 1.

Modern evolutionary trees

The information for modern evolutionary trees comes from organisms that are alive today and from the fossil record. But now we also look at base sequences of portions of DNA, or the structures of other chemicals such as proteins.

Scientists look at areas of DNA that show large variations between species. As organisms evolve, differences in their DNA accumulate. So the degree of genetic difference – the number of changes in the base sequences of a group of related organisms – can give us an estimate of the relative times at which these branches occurred.

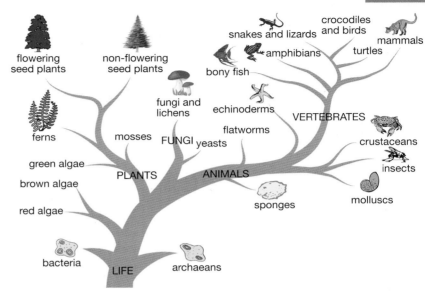

Figure 7.54 One version of an evolutionary tree. Some scientists have now added an extra branch to produce three domains (see Chapter 1)

-GTTGCGATTCTGACCTTAGA-	Sequence 1
-GTAGCCATACTGACCTTAGA-	Sequence 2
-GTAGCGATACCGACCTTACA-	Sequence 3
-GTTGCGATTCCGACCTTACA-	Sequence 4

The distance value between sequences 1 and 2 = $\dfrac{\text{number of changes}}{\text{length}} = \dfrac{3}{20} = 0.15$

The distance value between sequences 3 and 4 = $\dfrac{\text{number of changes}}{\text{length}} = \dfrac{2}{20} = 0.10$

Figure 7.55 How similar are DNA sequences in different organisms? Calculations can show scientists that Sequences 3 and 4 are more closely related than sequences 1 and 2

Computer programs can generate comparisons between the sequences of equivalent lengths of DNA. New evolutionary trees can be built.

DNA evolutionary trees can help to clear up uncertainties in trees that are based on anatomical features. But we cannot be certain of these evolutionary relationships. And obtaining DNA from extinct species is difficult.

5 Explain how modern evolutionary trees are constructed.

6 Using the information in Figure 7.55, calculate the distance values between the other DNA sequences.

DID YOU KNOW?

The 'Barcode of Life' is an international initiative devoted to developing a 'DNA barcode' database of living species, which can be used in identification.

REMEMBER!

You type a scientific name in *italics*, or if it's handwritten, underline it. The genus name has a capital letter but the species name does not. Avoid thinking of some organisms on an evolutionary tree as 'higher' than others. We cannot rank organisms.

Extinction ... or survival?

Learning objectives:

- list the causes of extinction
- explain how new predators, competitors and diseases can lead to extinctions.

The Svalbard Global Seed Vault – the 'Doomsday Vault' – is buried deep in the Norwegian permafrost. It was built to safeguard seed varieties from across the world from loss of genetic diversity or famine.

Mass extinctions

The fossil record shows that there have been at least five mass **extinctions** over geological time.

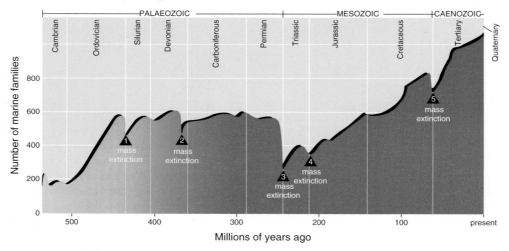

Figure 7.56 The five mass extinctions. The current rate of extinctions is comparable

Changes to the Earth's environment over geological time must have been a factor. The warming or cooling of the Earth, changes in seasons or ocean currents, and the changing position of the continents would have contributed.

Around 66 million years ago, however, it is thought that a massive asteroid impact killed three-quarters of life on Earth, including the dinosaurs. The dust thrown up would have blocked out sunlight. Plants would have declined, followed by herbivores, and then the carnivores that were dependent on the herbivores. Volcanic eruptions and sea-level change could also have contributed.

The Earth's environment is constantly changing. When the environment changes, organisms either adapt to these changes, or they become extinct. A species is considered to be extinct when there are no remaining individuals of that species still living.

1 Approximately when were the five mass extinctions?

2 Why can we only estimate the number of extinct species?

DID YOU KNOW?

According to one estimate in the scientific journal, *Nature*, humans have a century or perhaps two left. Can the sixth mass extinction be stopped?

Invasive or alien species

Another important cause of extinctions is the introduction of a new species into a location. These may:

- be new predators
- compete with native organisms that are present for food
- introduce new diseases.

On the Galápagos Islands the animals had few natural predators. But then mammals were introduced, including rats, dogs and feral cats. Cats and dogs attacked birds and destroyed the nests of birds, tortoises and turtles. Projects are under way on the islands, or have already been carried out, to eradicate these predators.

In Britain, harlequin ladybirds are a new species. They eat other ladybirds and outcompete them for food. Scientists are worried that our native ladybirds might become extinct.

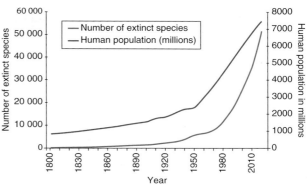

Figure 7.57 Humans are now the main cause of environmental change. This graph shows growth of the human population over time and the estimated numbers of extinctions

3 Why are scientists worried that some animals on the Galápagos Islands might become extinct?

4 Suggest one way of preventing these extinctions.

The verge of the sixth extinction?

Some scientists believe we are entering a sixth mass extinction, and that humans are the major contributing factor, through:

- transformation of the landscape
- overexploitation of species
- pollution
- introduction of alien species.

One serious concern in recent years has been the decline of bee populations. It is estimated that bees pollinate over 70% of the 100 crop species that provide 90% of global food supplies. If bees become extinct, the human race will follow soon after.

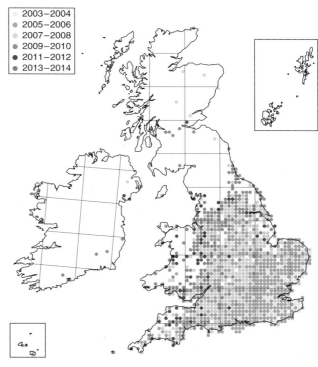

Figure 7.58 The spread of the harlequin ladybird, 2004–14

5 What may contribute to the sixth extinction?

6 How might bees contribute to human extinction?

7 'Humans are responsible for the extinction of species in modern times.' Do you agree with this statement? Justify your answer.

ADVICE

Remember that while humans have contributed to extinctions, we alone have the ability, although not always the inclination, to prevent many further extinctions.

MATHS SKILLS

Using charts and graphs to display data

Learning objectives:

- understand when and how to use bar charts
- understand how to show sub-groups on bar charts
- understand how to plot histograms.

KEY WORDS

demographic
discrete
grouped bar chart
stacked bar chart

The International Union for Conservation of Nature and Natural Resources (IUCN) monitors species in danger of extinction. It must make this huge amount of data understandable to the public.

Bar charts

Scientists use graphs and charts to display, summarise and analyse data. They can be used to simplify the presentation of complex data and highlight patterns and trends.

Bar charts are used to display data collected for distinct groups. If the data points in the groups can be counted then this data is known as **discrete** data.

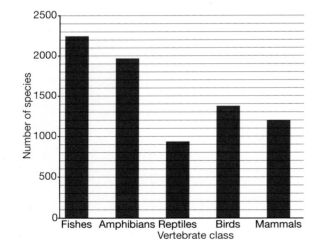

Vertebrate class	Number of species threatened with extinction
Fishes	2248
Amphibians	1961
Reptiles	931
Birds	1373
Mammals	1200

Figure 7.59 Data of the number of species in each class threatened with extinction are transferred from a table to a bar chart.

When drawing a bar chart:

- the *height* of the bars is proportional to the measured number or frequency
- the *width* of the bars should be consistent
- the bars do not touch each other; they represent distinct groups
- the bars can be placed in *any* convenient order. This shouldn't be haphazard.

1 Draw a bar chart for the groups of organisms, shown below, found as fossils in a particular rock type.

Group of organism	Fossils found in %
Algae	14
Arthropods	33
Molluscs	3
Sponges	22
Worms	3

2 What percentage of fossil organisms were unclassified?

DID YOU KNOW?

In his book of 1871, *The Descent of Man*, Darwin discussed the diversity of skull size and shape in different people.

More complex bar charts

Other types of bar chart could be used to provide further information about these endangered species.

From Figure 7.58, we don't know the number of species in each vertebrate class, so we don't have any indication of what proportion of each vertebrate class is endangered.

We can add this information to produce a dual bar chart. (also known as a **grouped bar chart**).

A similar type of bar chart is the **stacked bar chart**. We often see these in the media. Here, different sub-groups are stacked on top of each other to produce a single bar for each group included.
Note that bar charts can be constructed horizontally as well as vertically.

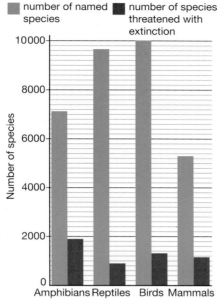

Figure 7.60 A dual bar chart for four of the vertebrate classes.

Histograms

A histogram is a statistical diagram, similar to a bar chart. In a histogram, the area of each bar is proportional to the frequency of the class interval. The data are continuous instead of being discrete. They're commonly used to show the variation of a characteristic, such as height. Remember:

* the variable on the x-axis is continuous, so there are no gaps between bars
* the size of the categories is represented by the area of the bars, although for many of the histograms you will plot the ranges on the x-axis have been chosen to be equal.

Histograms can be used to show the range of variation across groups of organisms. One characteristic determined for both living humans and extinct human ancestors is the cephalic index (CI).

$$\text{Cephalic index} = \frac{\text{Maximum head width}}{\text{Maximum head length}} \times 100$$

The CI varies between human populations of different **demographic** groups and over time.

Cephalic Index (C.I.)	Boys in %	Girls in %
$72.0 \leq CI < 75.0$	7.3	4.8
$75.0 \leq CI < 80.0$	28.4	32.8
$80.0 \leq CI < 85.0$	30.0	42.6
$85.0 \leq CI < 90.0$	34.3	19.6

③ Construct a bar chart to illustrate the proportion of endangered species of fish:

Number threatened with extinction	2248
Number of species	33100

> **REMEMBER!**
>
> Ensure that you choose sensible scales when drawing bar charts and histograms and draw them accurately.

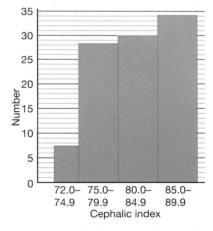

Figure 7.61 The histogram shows the range of the CI for the boys in a group of school children. Note that the bar for lowest category is narrower.

④ Plot a histogram for the range of CIs of girls in the group.

⑤ Plot a histogram for the range of heights in your biology group.

Check your progress

You should be able to:

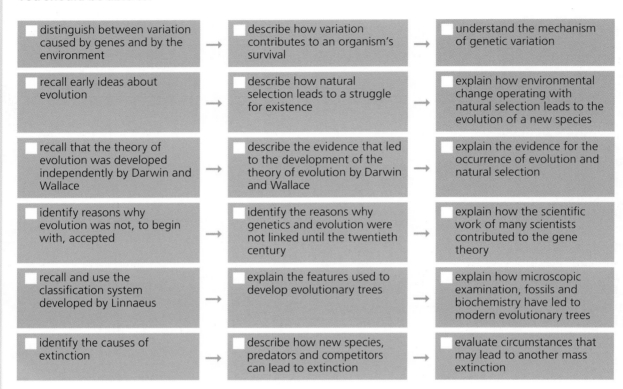

distinguish between variation caused by genes and by the environment → describe how variation contributes to an organism's survival → understand the mechanism of genetic variation

recall early ideas about evolution → describe how natural selection leads to a struggle for existence → explain how environmental change operating with natural selection leads to the evolution of a new species

recall that the theory of evolution was developed independently by Darwin and Wallace → describe the evidence that led to the development of the theory of evolution by Darwin and Wallace → explain the evidence for the occurrence of evolution and natural selection

identify reasons why evolution was not, to begin with, accepted → identify the reasons why genetics and evolution were not linked until the twentieth century → explain how the scientific work of many scientists contributed to the gene theory

recall and use the classification system developed by Linnaeus → explain the features used to develop evolutionary trees → explain how microscopic examination, fossils and biochemistry have led to modern evolutionary trees

identify the causes of extinction → describe how new species, predators and competitors can lead to extinction → evaluate circumstances that may lead to another mass extinction

Worked example

Scientists in Germany have investigated the resistance of rats to two poisons – warfarin and bromadiolone.

They studied populations of rats in four towns where resistance to the poisons had been reported, and one town – Ludwigshafen – where no resistance had been reported to date.

Some of their results are shown below.

Town	Not resistant to either poison (%)	Resistant to warfarin alone (%)	Resistant to both poisons (%)
Dorsten	44	56	0
Drensteinfurt	90	5	5
Ludwigshafen	100	0	0
Olfen	21	21	58
Stadtlohn	5	8	87

1 **Explain why the data collected are evidence for natural selection.**

In a population, some rats will be resistant to the rat poisons, and survive their use. These will pass on their resistant genes. These will spread through the population as the rats breed, until most of the rats become resistant.

2 **Suggest reasons for the differences in the degree of resistance to the poisons.**

Resistance is greatest in Stadtlohn and Olfen. There is no resistance to the poisons in Ludwigshafen.

This could be because of the locations of the towns. If the towns are close together, populations can interbreed so that the resistant gene(s) spread. Ludwigshafen is further away.

This is a concise answer, but one that is not fully detailed. The student should have mentioned that some rats in a population will be resistant to the rat poison because of mutations. It is also important to say that these mutations will spread rapidly through the population because of the rats' rapid reproductive rate.

In this answer the student has overlooked the fact that mutations arise spontaneously in populations, as well as being spread by sexual reproduction.

These could also arise at a greater rate in Stadtlohn and Olfen, perhaps because of the numbers of rats in the different towns, which we do not know from the data given.

The data suggest that resistance arises first to warfarin, followed by bromadiolone. You do not need to know the reason for this, but warfarin was a 'first generation' poison, while bromadiolone was developed because rats became resistant to warfarin.

End of chapter questions

Getting started

1. Draw lines to match the scientists with the scientific work they carried out. `1 Mark`

   Scientist

   Charles Darwin

   Gregor Mendel

   Alfred Wallace

   Scientific work

   Carried out genetics experiments on peas

   Wrote *On the Origin of Species*

   Described warning colouration

   Recorded numbers of peppered moths

2. Describe how a fossil is formed. `2 Marks`

3. Why is our fossil record incomplete? `1 Mark`

4. What is a hybrid? `1 Mark`

5. Describe what is meant by genetic variation. `1 Mark`

6. Use the information in this photograph to discuss two of the person's features that have genetic causes, and two features that have environmental causes. `2 Marks`

7. Draw a bar chart for the groups of organisms shown below, found as fossils in a particular type. `1 Mark`

Group of Organism	Fossils found %
Algae	11
Arthropods	30
Molluscs	4
Sponges	23
Worms	2

8. Calculate what percentage of fossil organisms were unclassified. `1 Mark`

Going further

9 How can a species new to an area, country or island lead to the extinction of other species?

3 Marks

10 What is meant by the term 'natural selection?'

1 Mark

11 The occurrence of a dark form of the two-spot ladybird was monitored by scientists between the 1960s and the 1980s.

a Some of the scientists' results are shown below.

Year	Frequency of the dark form (%)
1960	47
1965	37
1970	27
1975	19
1980	12
1985	10

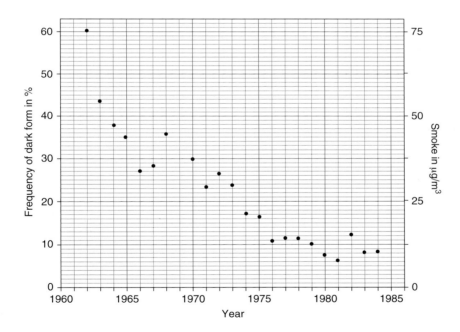

The scientists also recorded the yearly amounts of smoke in the air.
The readings for smoke concentration have already been plotted on the graph.

Plot the frequency of the dark form of the ladybird from 1960 to 1985 on the graph above.

Connect the points with a smooth curve.

4 Marks

b Suggest a reason for the change in frequency of the dark form of the ladybird. Predict what may have happened to the ladybird population after 1984.

2 Marks

More challenging

12 Describe how a human gene that codes for insulin production can be transferred to a bacterium.

`2 Marks`

13 Give one ethical objection to genetic engineering.

`1 Mark`

14 Some scientists believe we may be on the verge of entering a sixth mass extinction. Suggest a human action that could be a potential contributing factor.

`1 Mark`

15 A plant breeder is trying to produce a new variety of wheat that will withstand drought.

Describe how the plant breeder would do this.

`6 Marks`

Most demanding

16 Scientists have genetically engineered some crops to be resistant to herbicides.

Explain how herbicide resistance is beneficial to genetically modified (GM) crop production.

`2 Marks`

17 A scientist is investigating variation in bone mineral density measurements between human populations of different demographic groups over time.

The table below shows the data from a sample of people taken from a demographic.

Draw a histogram to show the data below.

bone mineral density = BMC / W [g/cm^2]	Number of people
81–100	7
101–120	18
121–140	31
141–160	35
161–170	27
171–180	11

`2 Marks`

18 **MRSA is a bacterial infection.**

The graph shows the number of cases of hospital patients with MRSA infections from 1993 to 2005.

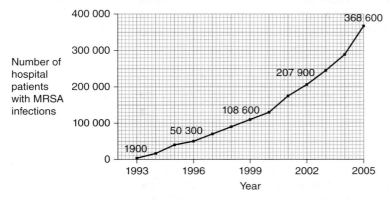

Explain the trend in the graph, even though the patients were treated with antibiotics.

6 Marks

Total: 40 Marks

ECOLOGY IN ACTION

IDEAS YOU HAVE MET BEFORE:

ORGANISMS IN AN ECOSYSTEM DEPEND ON EACH OTHER FOR SURVIVAL.

- Food chains show how energy is transferred between living organisms.
- Food webs are a series of inter-connected food chains.
- A change in the population of one organism affects other organisms in the food chain.
- Insects pollinate crops to provide food for human populations.
- Some organisms survive by living in or on another species.

ORGANISMS AFFECT AND ARE AFFECTED BY THEIR ENVIRONMENT.

- Organisms can affect populations of other plants and animals in their habitat.
- Some animals feed on dead animal and plant matter.
- If one organism is removed from a food web, the whole food web can be affected.
- Adaptations allow organisms to survive in their habitat.

HUMAN ACTIVITIES PRODUCE CARBON DIOXIDE, WHICH IMPACTS ON THE CLIMATE.

- Increased carbon dioxide emissions have caused global temperatures to increase.
- Humans produce toxic substances that can accumulate in food chains.
- Humans frequently destroy ecosystems to use the land for other purposes.

ALL ORGANISMS HAVE TO UNDERTAKE CERTAIN LIFE PROCESSES.

- Plants use carbon dioxide from the atmosphere to photosynthesise and produce simple sugars, such as glucose.
- Plants and animals need to take in water from their habitat; water is needed for chemical reactions.
- Dead plant and animal material is broken down by microbes.
- Plants need mineral nutrients to grow and stay healthy.

IN THIS CHAPTER YOU WILL FIND OUT ABOUT:

WHAT FACTORS AFFECT LIVING ORGANISMS IN A HABITAT?

- How energy is lost from each trophic level in a food chain.
- Why there is a maximum size to food chains.
- How pyramids of biomass show the mass of organisms in a food chain.
- How different groups of decomposers break down dead animal and plant matter.
- How distribution and numbers of species are sampled.

HOW DO PLANTS AND ANIMALS WITHIN A COMMUNITY INTERACT?

- How a change in an abiotic (non-living) factor affects a community.
- How changes in biotic (living) factors affect a community.
- How organisms are adapted to survive in extreme habitats.
- Why animals compete for resources in a habitat.

HOW DO HUMAN ACTIVITIES AFFECT BIODIVERSITY?

- What biodiversity is and why it is important to maintain a good level of biodiversity.
- How human population growth has impacted on the use of land and how this affects biodiversity.
- How waste from human activities has affected the atmosphere and biodiversity.
- How biodiversity can be maintained.

HOW ARE MATERIALS IN A COMMUNITY CYCLED?

- How carbon is cycled from organisms to the atmosphere.
- How water is cycled to provide a constant source of fresh water.
- How microorganisms help in cycling materials through an ecosystem.
- How we use the process of decay to produce compost and biogas.

KEY CONCEPT

Learning about ecosystems

Learning objectives:

- describe what an ecosystem is
- explain the importance of high biodiversity
- explain what is meant by a self-supporting ecosystem.

KEY WORDS

biodiversity
community
ecosystem
habitat
interdependence
population
self-supporting
  ecosystem

Ecology looks at how organisms survive, how they relate to other organisms and their physical environment, and the features that make them successful in their habitat.

Ecosystems and biodiversity

Living organisms are affected by their environment. For example, if plants cannot absorb enough water for their needs, they wilt and may die. But plants also affect their environment. Plant roots hold soil particles together and stop the wind from blowing the soil away.

An **ecosystem** is defined as the interaction of a community (of living organisms) with the non-living parts of their environment. Ecosystems can be:

- natural, for example, oceans, lakes, puddles and the rainforest
- artificial, for example, fish farms and planted forests.

In every ecosystem there are different living organisms. **Biodiversity** is the range of different plant and animal species living in an ecosystem. The living organisms in an ecosystem are described as producers, consumers and decomposers. Decomposers break down dead organisms.

Figure 8.1 A marine ecosystem

1. What is an ecosystem?

2. Describe the biodiversity of the marine ecosystem in Figure 8.1.

Exploring ecosystems

A **population** is the total number of one species in an ecosystem. A community is all the plants and animals living in an ecosystem. The place where a living organism lives in the ecosystem is its **habitat**.

REMEMBER!

A community is all the different plants and animals in an ecosystem and a population is the total number of one species.

High biodiversity is important in an ecosystem because:

- it allows a wide variation of food sources reducing the dependence of a species on a particular food source
- it provides us with food, medicines, the atmosphere and water.

Taking more of a species than we need from the environment (for example, by overhunting, deforestation and overfishing), and not replacing them, means that populations are not sustainable. This puts species in danger of extinction.

3 **Explain the difference between**

  a **an ecosystem and a habitat**

  b **a community and a population.**

4 **Describe the ecosystems shown in Figures 8.1 and 8.2.**

Ecosystem interactions

Some natural ecosystems have high biodiversity, such as the rainforest or shallow tropical coral reefs. They provide a wide variety of food throughout the year and shelter, above and below the ground or water. Other natural ecosystems, such as the Arctic tundra or deep-sea thermal vents, have low diversity, with species adapted to survive those extreme environments.

Artificial ecosystems have low biodiversity: the variety and number of plants is limited, so only a small number of animal species can survive.

All ecosystems are **self-supporting**: all the requirements for living organisms to grow and survive are present. They do need an external energy source, which is usually the Sun.

All animals depend on plants for oxygen and food. Plants depend on animals for carbon dioxide, pollination and seed dispersal. This is called **interdependence**.

5 **How is an ecosystem self-supporting?**

6 **Explain how living organisms are interdependent.**

7 **Construct an argument putting forward reasons for the claim that we should protect areas of high biodiversity such as rainforests.**

Figure 8.2 How are these ecosystems similar and different?

DID YOU KNOW?

Deserts are advancing and taking over fertile land. In Mali (West Africa) the Sahara advanced about 350 km in approximately 20 years.

Changing abiotic factors

Learning objectives:

* identify factors that affect ecosystems
* explain changes in the distribution of species in an ecosystem
* describe stable and unstable populations.

Living organisms are affected by many different factors as they grow and try to survive.

What are abiotic factors?

Many factors affect where an organism lives. **Abiotic factors** are physical (that is, non-living) conditions that affect the **distribution** of an organism. These factors include:

* temperature
* light intensity
* oxygen levels for animals that live in water
* carbon dioxide levels for plants
* moisture levels
* soil pH and mineral content for plants
* wind intensity and direction.

Biotic factors are caused by living organisms affecting other populations in their ecosystem. Biotic factors include:

* food availability
* new pathogens
* new predators
* **competition** between species.

Abiotic and biotic factors change over time. For example, temperature varies daily, monthly, seasonally and over many years.

1 **Choose one abiotic and one biotic factor. Explain how a lack of each factor would affect a plant and an animal.**

2 **Apart from blocking light, how else might trees affect the grass that grows beneath them?**

Looking at changes

The numbers and types of organisms can gradually change across a habitat. These changes are easy to see on the sea shore, where there are distinct zones of organisms due to changing tides (see Figure 8.4).

Figure 8.3 What abiotic factors affect each of these organisms?

REMEMBER!

Abiotic factors are non-living. Biotic factors are living organisms. Changes in these factors cause changes in the distribution of organisms.

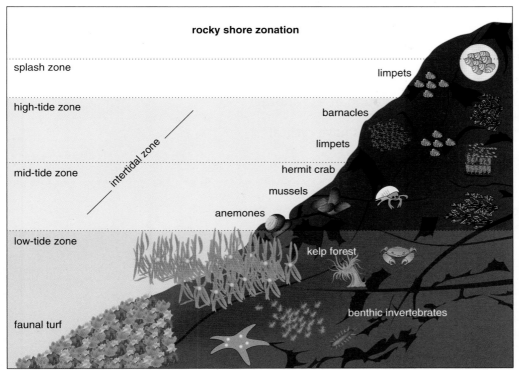

rocky shore zonation

splash zone

high-tide zone

interidal zone

mid-tide zone

low-tide zone

faunal turf

periwinkles

limpets

barnacles

limpets

hermit crab

mussels

anemones

spiral wrack

bladder wrack

saw wrack

kelp forest

benthic invertebrates

Figure 8.4 Zones on the sea shore

3 Name the abiotic factors that change on the sea shore. How can these factors affect distribution of seaweeds?

4 Suggest why limpets are found higher up the shore than anemones?

Looking at how distribution of species can change

Factors that affect the distribution of organisms do not work alone. Two or more factors can interact to form very different environments within a habitat.

Animals such as rabbits and sheep graze on plants. The amount of grazing affects the numbers of plant species found.

Little grazing allows a few plants to out-compete others. As grazing increases, more plant species grow because dominant plants are controlled by the animals, allowing weaker species to grow. Only specially adapted plants can resist the effect of intensive grazing and survive.

A stable community is where the biotic and abiotic factors are in balance so that population sizes remain fairly constant.

5 What is a stable community?

6 Describe and explain the impact of grazing on plant species as shown in figure 8.5.

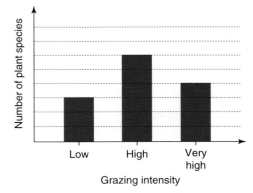

Number of plant species

Low   High   Very high

Grazing intensity

Figure 8.5 How does grazing affect plants?

DID YOU KNOW?

Plants are rare in deserts but after rain the distribution changes. Large numbers grow, flower and make seeds quickly while the water is available.

DID YOU KNOW?

Tropical rainforests and ancient oak woodlands are stable communities.

Investigating predator–prey relationships

Learning objectives:

- describe how changes in one population affect another
- explain interdependent relationships
- explain how predator–prey populations have cyclical changes.

KEY WORDS

interdependence
parasitism
cycle
out of phase

All species in a community depend on each other. This is *interdependence*.

A change in population size

Primary consumers (or herbivores) eat producers (plants). Primary consumers are eaten by secondary consumers (carnivores). Some animals are primary and secondary consumers, (they eat plants and animals). They are omnivores, for example, humans.

Consumers that hunt and eat other animals are predators. The animals that they eat are prey. If a predator kills all the prey it will die. In stable communities, the numbers of predators and prey stay in balance.

1 **Explain why humans are primary and secondary consumers.**

2 **Describe what happens to the predator population when:**

   a prey numbers are high

   b prey numbers are low.

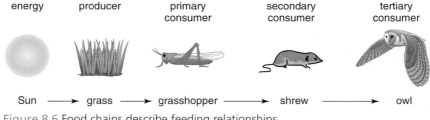

| energy | producer | primary consumer | secondary consumer | tertiary consumer |

Sun ⟶ grass ⟶ grasshopper ⟶ shrew ⟶ owl

Figure 8.6 Food chains describe feeding relationships

Looking at more relationships

Examples of mutually beneficial relationships are:

- Cleaner fish eat dead skin and parasites off large fish. They get food and avoid being eaten by larger fish.
- Burdock fruits have hooks. They attach to the fur of passing animals for dispersal.

In **parasitism** one organism benefits but the other is harmed by the relationship. If parasites kill their host, they die too.

Figure 8.7 Zooxanthellae live in the golden jellyfish tissues. They have a safe place to live and give the jellyfish some of the energy they need from photosynthesis.

Examples of parasitism are:

- Tapeworms attach themselves to the intestinal wall of their human host and absorb nutrients. The person suffers from malnutrition.
- Head lice live on human scalps. They suck blood for food and make the scalp itch.

3 **Explain the relationship between tapeworms and humans.**

4 **Explain what parasitism is.**

Explaining population cycles

In predator–prey relationships the size of each population is dependent on the other. The Canadian lynx was the main predator of snowshoe hares in Canada. Snowshoe hares make up most of its diet. Data from a study between 1845 and 1937 were used to produce the graph showing their relationship (Figure 8.8). The snowshoe hare population rises and falls in a 10-year **cycle**. Because the lynx population is dependent on the snowshoe hare it also rises and falls, but it is **out of phase** (or lags behind) that of the hare population cycle by about 2 years. There is a clear pattern between the populations of the two animals, as shown in the graph.

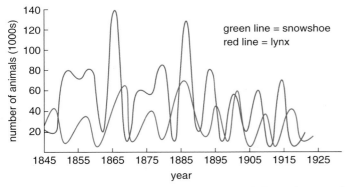

Figure 8.8 The size of the predator population follows the size of the prey population

5 **Explain how a large and a small predator population will each affect the number of prey.**

6 **Explain why the predator cycles and the prey cycles are out of phase with each other.**

> **REMEMBER!**
>
> The population cycle of the predator is out of phase with the cycle of the prey.

> **DID YOU KNOW?**
>
> Scientists study predator–prey cycles to decide if controlling predator populations can help to protect endangered species.

Looking at trophic levels

Learning objectives:

- explain trophic levels
- explain and construct pyramids of biomass
- explain the difficulties in constructing pyramids.

Producers make their own food. Consumers eat plants or other animals. They have a complex relationship.

Describing food chains

Trophic levels describe feeding positions in food chains. Energy is transferred from one trophic level to the next, along the food chain.

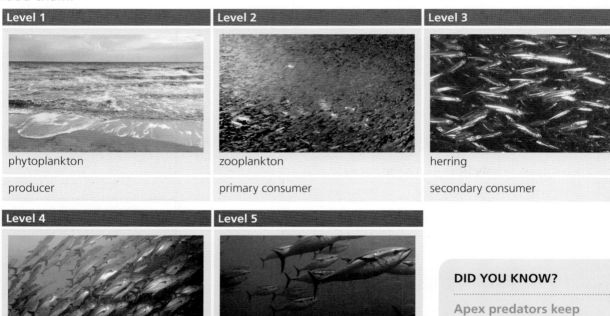

Level 1	Level 2	Level 3
phytoplankton	zooplankton	herring
producer	primary consumer	secondary consumer

Level 4	Level 5
mackerel	tuna
tertiary consumer	quaternary consumer

Figure 8.9 Trophic levels in a food chain

DID YOU KNOW?

Apex predators keep ecosystems stable. This is called trophic dynamics.

Trophic levels in food chains are represented by numbers:

- Level 1: Plants and algae are producers.
- Level 2: Herbivores eat plants/algae.
- Level 3: Carnivores eat herbivores.
- Level 4: Carnivores eat other carnivores. (In some foods chains, as in Figure 8.9, there may be a further level of carnivores.) Carnivores in the top level, with no predators, are **apex predators**.

Decomposers break down dead plant and animal matter. They **secrete** enzymes onto food to digest it. The digested food molecules then diffuse into the microorganism. This is extracellular digestion.

1 What are trophic levels?

2 What trophic level is

    a mackerel

    b zooplankton

    c a producer?

Using pyramids

The mass of organisms at different trophic levels in an area can be used to construct a pyramid of biomass. Level 1 is always at the bottom of the pyramid, Level 2 is above this, and so on (see Figure 8.10).

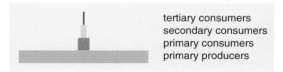

tertiary consumers
secondary consumers
primary consumers
primary producers

Figure 8.10 A simple pyramid of biomass

Pyramids of **biomass** use the dry mass of the organisms because 'wet' mass can vary. The biomass at each level in food chains is always less than the previous level.

3 Why is dry biomass used for pyramids of biomass?

4 Draw a pyramid of biomass for this food chain (parasitic wasps lay eggs in ladybird pupae):

    rose → aphid → ladybird → parasitic wasp

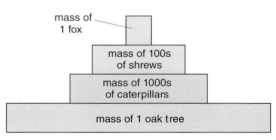

mass of 1 fox

mass of 100s of shrews

mass of 1000s of caterpillars

mass of 1 oak tree

Figure 8.11 A woodland pyramid of biomass

Difficult pyramids

- Pyramids of biomass show the amount of energy in a trophic level more accurately than pyramids of number. But they do have some disadvantages, which are:
- Organisms need to be collected and killed to measure dry mass.
- It is difficult to catch and weigh the organisms.
- Biomass varies. A tree in summer has more biomass than it does in winter.
- Some organisms are omnivores and feed at more than one trophic level.

To avoid killing living organisms, estimates of biomass are often used.

KEY SKILL

Practise drawing pyramids of biomass for different food chains.

5 Use graph paper to draw pyramids of biomass (to scale) for this food chain:

    pondweed (2500 g) → insects (50 g) → frogs (25 g) → trout (10 g)

6 Why would scientists find it difficult to measure the biomass in this food chain?

    phytoplankton → krill → whale → human

Transferring biomass

Learning objectives:

- identify how biomass is lost
- calculate the efficiency of biomass transfers
- explain the impact of biomass loss on the numbers of organisms.

All organisms need energy for their life processes. Biomass and energy get less at each trophic level in a food chain.

Increasing biomass

Plants and algae make food to increase their **biomass,** using energy from the Sun. The biomass is an energy source. Animals use only a small amount of the biomass that they eat to increase their own biomass.

Biomass is lost because:

- some plant material, which cannot be digested, leaves the body as faeces
- some animal material cannot be digested, e.g. bone, horn, hooves, claws and teeth
- biomass eaten by animals is also used in respiration to release energy, and leaves the animal as carbon dioxide and water.

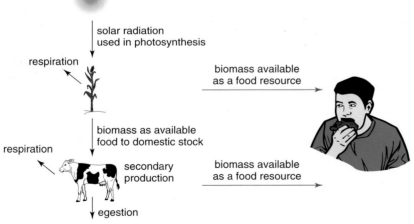

Figure 8.12 How is biomass lost?

1️⃣ **How is energy from the Sun transferred into biomass?**

2️⃣ **How is biomass lost in the food chain?**

Biomass loss in animals

Producers can only transfer 1% of the light energy reaching them to produce biomass.

Herbivores need to eat lots of plant material to convert sufficient amounts into new biomass. They produce huge amounts of faeces that contains all the indigestible plant material. Carnivores eat less often than herbivores because meat is easier to digest. They produce less faeces but it still contains indigestible biomass.

Animals use large amounts of glucose and lose carbon dioxide and water in respiration. The more an animal moves, the more respiration occurs and the more biomass is used.

Mammals and birds transfer lots of energy to their surroundings because they use it to maintain a constant body temperature.

The protein an animal eats, but does not need, is broken down into urea and lost in the urine.

respiration and movement 40 kJ and growth 10 kJ

100 kJ eaten

faeces and urine 35 kJ and methane 15 kJ

Figure 8.13 Biomass transfer in a calf

3 Why do herbivores produce more faeces than carnivores produce?

4 Explain why only some biomass is passed from one **trophic level** to the next.

Looking at numbers

Energy transfer between organisms is about 10% at each trophic level. Look at the diagram. For every 100 kJ that the calf eats, 90 kJ is transferred to either faeces and urine, methane or respiration and movement. Most of this is eventually transferred to heat in the environment.

Just 10 kJ is built into body tissue (biomass) in the calf. This is a biomass transfer **efficiency** of 10%:

$$\text{Efficiency} = \frac{\text{energy used for growth (output)}}{\text{energy supplied (input)}} \times 100\%$$

Food chains rarely have more than five trophic levels because biomass is lost at each trophic level. The amount remaining after Level 5 is too small for an organism to survive.

5 Explain why an elephant needs to eat more food than an alligator needs to eat, to increase its biomass by 1 kg.

6 Look at Figure 8.14.

   a Calculate the percentage biomass used for growth.

   b Explain why lambs that are kept inside grow more quickly than those that are reared outside.

> **REMEMBER!**
>
> Arrows in food chains show energy transfer from one trophic level to the next. Energy is lost at each step.

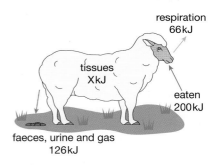

respiration 66 kJ

tissues X kJ

eaten 200 kJ

faeces, urine and gas 126 kJ

Figure 8.14 Biomass transfer in a sheep

> **DID YOU KNOW?**
>
> Commercial food production is more efficient than biomass transfer in nature because there are fewer stages in the food chain. More food is made because less energy is lost.

Competing for resources

Learning objectives:

- describe how competition impacts on populations
- explain why animals in the same habitat are in competition
- explain interspecific and intraspecific competition.

If plants and animals are to survive, they need certain resources from their habitat. How do living things get what they need?

Why do organisms compete?

Organisms only survive if they have sufficient resources for their needs. They compete for the available resources in their habitat. Plants in a community may compete for:

- light
- space
- water
- mineral ions.

Animals often compete for:

- food and water
- mates
- territory.

Figure 8.15 What are these organisms competing for?

Animals and plants that get more of the resources are more successful than those that get less. Successful organisms are more likely to survive and reproduce so the size of their population is more likely to increase. For example, dolphins feed on several foods and are more likely to survive than dugongs that feed on just seagrass. Animals will travel to where food is available.

1 How does competition affect the distribution and number of organisms?

2 Describe how organisms are competing in Figure 8.15.

Competition

When species compete, if they are not perfectly matched one will eventually become more successful than the other. A less successful species may:

- do nothing and become extinct
- stay in its habitat but adopt new survival strategies
- move to another area looking for resources.

Humans are very successful organisms. We compete with animals and plants all over the world.

3 **What happens when two organisms compete?**

4 **Explain why organisms in the same habitat are in competition.**

Types of competition

There are two types of competition:

- **Interspecific competition** is competition between different species.
- **Intraspecific competition** is competition within one species. This may result in territorial behavior.

Intraspecific competition is often more significant than interspecific competition. For example, competition between grey squirrels is likely to affect the population of grey squirrels more than competition with red squirrels. Animals try to avoid competition with other species if they can.

When different species compete with one another for the same resources, it will affect the size and distribution of their populations.

Figure 8.16 How does a farmer compete with other organisms on their farmland?

COMMON MISCONCEPTION

Learn the difference between interspecific and intraspecific competition.

Figure 8.17 What types of competition are these?

5 **Describe examples of interspecific and intraspecific competition.**

6 **Why is intraspecific competition often greater than interspecific competition?**

DID YOU KNOW?

Competition for mates is intense, with many males putting a lot of effort into attracting females, by fighting or by having 'displays' such as the male peacock displaying its feathers.

REQUIRED PRACTICAL

Measure the population size of a common species in a habitat

Learning objectives:

- use scientific ideas to develop a hypothesis
- plan experiments to test a hypothesis
- explain the apparatus and techniques used to sample a population
- explain how a representative sample was taken
- develop a reasoned explanation for some data.

KEY WORDS

hypothesis
meristem
quadrat

It is impossible to count every plant or animal in a habitat, but it can be estimated by taking samples of the organisms from the habitat. The larger the sample, the more accurate your estimate of the population size is likely to be This allows population sizes to be compared between different areas.

Plants can be sampled more easily than animals because they cannot move around.

Developing a hypothesis

It is summer time. Asha and Jalan are going to investigate the population size of daisies in trampled and un-trampled areas of the school field. The centre of the school field is used to play cricket, tennis and train for athletic events. The students run, jump and lie on the grass. Not many students go on the parts of the field away from these areas.

Asha and Jalan need to make a **hypothesis** before they do the investigation. They have made some observations on the field and now know that:

- trampling on soil compacts it
- trampling on plants can destroy the **meristems** as well as crush delicate leaves and flowers
- daisy leaves are not very delicate
- in well-trampled nearly bare ground, plants will have to tolerate large variations in temperature where there is no grass to protect them
- daisy plants have very long fibrous roots and thick leaf cuticles
- plants compete with each other for limited resources.

1 **How do you think compaction may affect**

    a   **the soil?**

    b   **the daisies growing in the soil?**

These pages are designed ❗ to help you think about aspects of the investigation rather than to guide you through it step by step.

Figure 8.18 A daisy population

Figure 8.19 Leaf, root and flower of a daisy

2 If meristems, leaves and flowers can be crushed by trampling how may this affect the daisies?

3 How can periods of very high temperatures affect the daisies?

4 Suggest a hypothesis for the fieldwork investigation that Asha and Jalan are going to do.

Planning an investigation

The teacher has given Asha and Jalan this equipment. They know that it is important to take random samples, make repeat observations and count whole daisy plants.

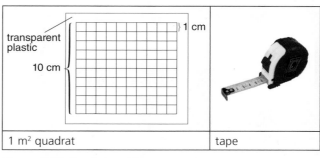

5 Suggest how Asha and Jalan can use this equipment to collect representative data to test the hypothesis from Question 4.

Figure 8.20 Quadrat and tape

6 Why is it important to count whole daisy plants and not just daisy flowers? How many daisies are in the **quadrat**?

7 How can Asha and Jalan calculate the mean number of daisy plants in each area?

8 Another student has asked Asha and Jalan for a copy of their results. Suggest why they asked for it.

Developing explanations

Asha and Jalan did their investigation. Look at their data.

Quadrat position on tape (m)	Un-trampled area	Trampled area
0	3	6
5	4	6
10	3	7
15	2	6
20	2	4
25	4	7
30	3	6

Figure 8.21 Quadrat placed over daisies

ADVICE

Sampling many small sections of an area gives a representative sample of the whole area.

Scientists always use evidence to explain the data they collect. The evidence can be from knowledge or observations. They also relate their data to their original hypothesis.

9 Suggest what Asha and Jalan can conclude from their data

10 Explain the data using the information that Asha and Jalan knew before they started the investigation.

11 Was the hypothesis in Question 4 proved or disproved?

Adapting for survival in animals

Learning objectives:

- recall why animals have adaptations
- explain some adaptations
- use surface area to volume ratio to explain some adaptations.

Abiotic factors vary depending upon where an ecosystem is located. But every ecosystem has living organisms.

Why are adaptations important?

Animals have special features called adaptations that allow them to survive and be successful in their habitat. Animals with more successful adaptations are better able to compete for food, mates and other limited resources.

Figure 8.22 What adaptations do these animals have?

Camouflage is used by predators (so the prey cannot see them) and prey (so they cannot be seen by predators). Many arctic animals have white coats in winter (to blend in against snow) and grey-brown coats in summer (to blend in against plants). Polar bears always have white coats because they have no predators on land.

1 Why are adaptations important?

2 Describe three different adaptations.

Behavioural adaptations

Some ecosystems have extreme conditions. They may have little available water. Animals have to stop body temperatures getting too high or too low. They can do this by changing their **behaviour**.

In hot temperatures animals:

- are most active during the cooler mornings, and they move to shady areas to keep cool
- remain in burrows during hot days and cold nights, where temperatures are more stable.

In cold temperatures some animals:

- hibernate over winter when food is scarce. Their metabolic rate is lowered to reduce the need for food
- migrate to warmer countries where there is more food.

Most reptiles struggle to survive in very cold or hot temperatures, because they do not have internal mechanisms to control their body temperature. They need to bask in the sun to increase body temperature or stay in the shade to keep cooler.

3 **Explain how reptiles in the desert control their body temperature.**

4 **Suggest some behavioural adaptations of meerkats.**

Figure 8.23 Meerkats live in the Kalahari desert

More adaptations

Some adaptations are **structural** (part of the body) and others are **functional** (how the body operates).

In hot temperatures, many animals:

- keep insulating fat and body hair to a minimum to prevent overheating. Camels only have fat in their humps
- do not sweat (e.g. camels) to prevent water loss, and need little to drink to cope with water scarcity
- have a large surface area to volume ratio (sometimes called SA:V) to maximise heat loss through the skin
- have special adaptations to increase their SA:V, for example, elephants have large ears to increase their surface area.

> **REMEMBER!**
>
> Animals in cold climates often have smaller surface area to volume ratio, and those in hot climates have large surface area to volume ratio.

Figure 8.24 What adaptations do camels have?

In cold temperatures, many animals have:

- thick fur and blubber for insulation
- a small surface area to volume ratio to reduce heat loss
- small ears to reduce surface area to volume ratio, to reduce heat loss.

> **DID YOU KNOW?**
>
> Predators have adaptations for hunting and prey have adaptations for escaping.

5 **Explain how surface area to volume ratio can be an adaptation to temperature.**

6 **Explain the difference between structural and functional adaptations, using your own example to illustrate your answer.**

Adapting for survival in plants

Learning objectives:

- identify adaptations of plants and bacteria
- explain the importance of adaptations
- explain a range of plant adaptations.

KEY WORDS

epiphyte
extremophile

Like animals, plants also have many different adaptations that allow them to colonise ecosystems all over the world.

Looking at more adaptations

Plants are adapted in many ways, for example, deciduous trees lose leaves in winter (to protect plants from frost) and flowers die (due to there being fewer pollinators).

Plants in hot climates are adapted by:

- having a reduced surface area (reduces water loss)
- storing water in their tissues (conserves water).

Plants in cold climates are adapted by:

- growing close to the ground (avoids damage by wind)
- having small leaves (conserves water).

Figure 8.25 Suggest the climates where these plants live

Extremophiles are organisms that can survive in very extreme environments, including:

- very high or low temperatures
- high salt concentrations in water
- high pressures.

Extremophiles survive conditions that would kill most species. Some bacteria live in hot springs or around deep-sea vents. Others live in extreme cold.

1 **Explain adaptations shown by arctic plants.**

2 **What are extremophiles?**

Explaining some adaptations

Organisms with a larger surface area to volume ratio lose more heat and water. Plants are adapted to drier environments by:

- having a reduced surface area
- storing water
- having extensive root systems that are wide (to absorb water quickly when it becomes available) or deep (to find underground water stores).

Cactus plants live in deserts that have:

- hot and dry days
- cold nights
- little rainfall.

Adaptations of cacti include:

- spine-like leaves to reduce surface area, transpiration and damage by primary consumers
- fleshy stem
- extensive shallow roots
- thick waterproof cuticle
- round shape (reduces surface area to volume ratio).

Bacteria living in deep-sea vents are extremophiles. They survive in temperatures between 40°C and 80°C because their enzymes do not denature.

3 **Explain how extremophiles are adapted for survival.**

4 **Use ideas about surface area to volume ratio to explain why pine trees have needle leaves.**

More adaptations

Epiphytes grow in rainforests. They are adapted to live above ground level by:

- growing on other plants allowing leaves to absorb sunlight through the trees
- having roots that absorb rain and moisture from humidity and minerals from leaf litter
- having upturned leaves to store water.

5 **Explain adaptations to temperature using surface area to volume ratio.**

6 **Explain why there are lots of cacti in the desert but no roses.**

7 **Studies have shown that tropical and desert plants have different densities of stomata. Suggest and explain why this might be the case.**

> **ADVICE**
>
> Remember that plants open their stomata for photosynthesis and so lose water by evaporation.

Figure 8.26 Marram grass, which lives on sand dunes, has tightly curled leaves to reduce their surface area. How does this give marram an advantage?

Figure 8.27 How are epiphytes adapted?

> **DID YOU KNOW?**
>
> Some plant roots are adapted to grow in sea water.

Cycling materials

KEY WORD

....

decomposers

Learning objectives:

- recall that many materials are recycled in nature
- explain the stages in the water and carbon cycles
- explain the importance of recycling materials.

All materials in the living world are recycled to provide the building blocks for future organisms.

Natural recycling

Water, carbon, oxygen and nitrogen are some of the materials that are recycled.

The water cycle provides fresh water for plants and animals on land. It describes how water moves on, above or just below the ground. Water molecules move between rivers, oceans and the atmosphere by precipitation, evaporation, transpiration and condensation.

Carbon is found in compounds that make up living organisms. Plants get carbon from air during photosynthesis. Animals get carbon from plants. The carbon in dead plants and animals is released by **decomposers** during the decay process.

Figure 8.28 Decomposers, such as the fungi shown here, clean up the environment

1 Give an example of a decomposer that breaks down carbon in dead animals and plants.

2 What is the role of decomposers?

Explaining the water cycle

Water is continuously recycled by:

- precipitation: water droplets in clouds get bigger and heavier. they fall as rain, snow or sleet.
- evaporation: water evaporates as it is heated by the Sun's energy; water vapour is carried upwards in convection currents.
- transpiration: water vapour is released into the air through stomata in leaves.
- condensation: water vapour rises, cools and condenses back into water droplets that form clouds.

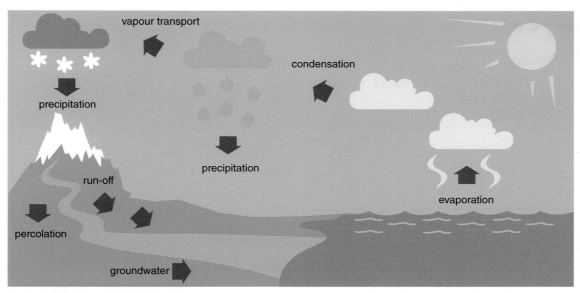

Figure 8.29 The water cycle

The water cycle is important because it circulates water that:

- maintains habitats
- maintains internal fluids and transport systems
- is needed for chemical reactions
- is a reactant in photosynthesis.

3 **Explain the stages in the water cycle.**

4 **In what stage of the water cycle can water be:**

    a **a gas?**

    b **a solid?**

Explaining decay

Worms, woodlice and maggots break down waste and dead material into smaller pieces for decomposers to digest.

Decomposers are microorganisms (bacteria and fungi). They break down the smaller pieces of dead material. Decomposers release waste carbon dioxide, water, thermal energy and nutrients (that plants use).

Decomposers decay waste in compost heaps and sewage works. Factors that speed up decay are:

- plenty of microbes
- warmth
- plenty of oxygen
- some moisture.

Life processes depend on enzymes. Factors that increase enzyme action speed up decay.

5 **Explain the role of microorganisms in cycling materials.**

6 **Why is the water cycle important?**

Figure 8.30 Woodlice live on rotting wood and help to break it down for decomposers

DID YOU KNOW?

The Body Farm is a research site in the USA where scientists study human decay to support forensic analysis.

ADVICE

Learn the stages of the water cycle and the decay process.

Cycling carbon

KEY WORD

carbon sink

Learning objectives:

- recall that plants take in carbon as carbon dioxide
- explain how carbon is recycled
- interpret a diagram of the carbon cycle.

Materials are cycled through ecosystems continually. As one process removes them, another releases them into the ecosystem again.

Looking at carbon

The amount of carbon on Earth is fixed. Most carbon is found combined with other elements, for example in fossil fuels and carbonate rocks. Carbon is also found dissolved in water in rivers, lakes and oceans. A small amount is in the air in carbon dioxide.

Producers use carbon dioxide to photosynthesise. The carbon is used to make carbohydrates, proteins, fats and DNA that form new biomass, which is eaten by consumers.

All organisms respire to release energy for cellular processes. They release waste carbon dioxide back into the environment, which is then used by plants.

1. **How do organisms cycle carbon dioxide?**

2. **Why is most carbon dioxide on Earth not available for photosynthesis?**

Figure 8.31 Burning fossil fuels releases carbon, as carbon dioxide, into the air

Explaining the carbon cycle

The atmosphere contains about 0.04% carbon dioxide, which is enough for every plant to produce biomass (food), by photosynthesis. This process transfers energy from the Sun into chemical energy.

When animals eat plants, they absorb carbon from them. Carbon passes along food chains, even when organisms die and decay. Energy is transferred along the food chain and to the environment at each trophic level.

Carbon dioxide is returned to the atmosphere by:

- Plants, animals and decomposers respiring:
    glucose + oxygen → carbon dioxide + water
- burning (combustion) of fossil fuels and wood:
    fossil fuel/wood + oxygen → carbon dioxide + water

The continual cycling of carbon is shown in Figure 8.32.

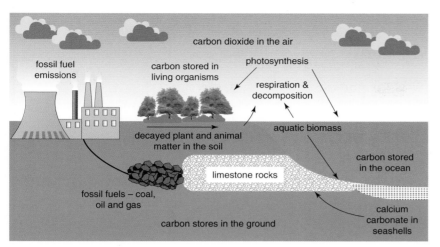

Figure 8.32 The carbon cycle

Burning fossil fuels and wood releases more carbon into the atmosphere. Scientists agree that levels of carbon dioxide in the atmosphere will continue to rise if current rates of fossil fuel burning continue. Increased levels of carbon dioxide in the atmosphere cause global warming.

3 **Name two processes that release carbon dioxide.**

4 **Use Figure 8.32 to describe the carbon cycle.**

> **REMEMBER!**
>
> The carbon cycle returns carbon from organisms to the air for plants to use in photosynthesis.

Carbon stores

Shells of marine organisms contain carbonates. Corals and microscopic algae cover themselves with calcium carbonate. Shells of dead organisms fall to the sea floor. Over millions of years they are compressed to form limestone. Carbon dioxide reacts with rain to form carbonic acid. This acid rain weathers limestone and releases carbon dioxide.

Figure 8.33 What causes weathering of limestone statues?

Carbon dioxide can be absorbed by oceans and held in a **carbon sink**. During volcanic eruptions and forest fires, massive amounts of carbon dioxide are released into the atmosphere.

5 **Why is the carbon cycle important to all living organisms?**

6 **Explain the processes involved in the carbon cycle.**

> **DID YOU KNOW?**
>
> Between 1000 and 100 000 million metric tonnes of carbon pass through the carbon cycle every year.

Investigating decay

Learning objectives:

- recall the factors needed for decay
- describe how different factors affect decay
- explain extracellular digestion.

KEY WORDS

biogas
compost
extracellular
 digestion
generators
secrete

Waste materials produced by living organisms and dead organisms are recycled to provide nutrients and resources for future generations.

The decay process

When dead organisms decay they break down into simpler chemicals; for example, minerals and carbon dioxide. These materials are recycled and used by living plants for growth.

The decay process is faster when:

- there are more microorganisms
- more oxygen is available
- temperature closer to the optimum
- there is some moisture.

Gardeners use the decay process to make **compost** from plant material. This takes months but is speeded up by placing the compost bin in the Sun, mixing the contents to let oxygen in and adding moisture. Dead and decaying plant material is broken up into small pieces by earthworms, maggots and insects. This increases the material's surface area to make it easier for other decomposers to digest it. Compost is a natural fertiliser.

Figure 8.34 Why are compost bins not sealed?

1. What factors are needed for decay?
2. How is compost made?

> **REMEMBER!**
>
> Learn about earthworms, maggots and insects and their roles in decay.

Using decay

The energy in plant biomass can be released by anaerobic bacteria. They produce **biogas**. Biogas is produced naturally in marshes, septic tanks and sewers.

Biogas provides a cheap fuel source for cooking, heating, electricity and fuel for vehicles. **Generators** (large tanks) are used for biogas production. The generators hold rotting organic material (dung, farm and garden waste) and bacteria.

Figure 8.35 What do biogas generators contain?

The biogas contains:

- methane (50–75%)
- carbon dioxide (15–45%)
- water vapour
- small amounts of other gases.

The waste slurry is removed continuously from the generator and used as fertiliser.

3 Explain how biogas is produced.

4 Suggest how biogas production is affected by temperature.

Enzymes and decay

Enzymes control respiration and reproduction. Enzymes in bacteria have an optimum temperature of 37°C and in fungi it is 25°C. Enzymes work best at these temperatures so decay is faster. As microorganisms respire, heat is released, the detritus gets hotter and decay is faster.

Aerobic bacteria cause most decay in compost heaps. They respire using oxygen to release energy for growth and reproduction. The more oxygen present, the more decay happens.

Bacteria and fungi live in or on dead material. They **secrete** digestive enzymes onto the food to digest it into soluble substances that can be absorbed. This is **extracellular digestion** because it happens outside cells.

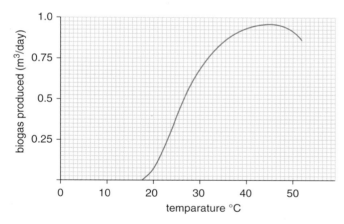

Figure 8.36 Temperature and biogas production

5 Explain the role of bacteria in decay.

6 Explain how increasing the temperature and the amount of oxygen affects the rate of decay in compost heaps.

7 Why are decomposers essential for life? What would happen if they did not exist?

REQUIRED PRACTICAL

Investigate the effect of temperature on the rate of decay of fresh milk by measuring pH change

KEY WORDS

lipase
universal
 indicator

Learning objectives:

- describe how safety is managed, apparatus is used and accurate measurements are made
- make and record observations and make accurate measurements
- evaluate methods and suggest possible improvements and further investigations.

These pages are designed ❗ to help you think about aspects of the investigation rather than to guide you through it step by step.

The natural process of decay in milk is too slow to get results in class. This investigation uses lipase enzyme to speed up the process. This investigation models the decomposition process.

Lipase enzymes break down protein and fat in milk. Digestion of fat by lipase enzymes produces fatty acids (and glycerol) that neutralise alkalis.

Making accurate measurements and working safely

In this investigation milk and sodium carbonate solution are added to water in a test tube. Then **lipase** (an enzyme) is added. Carrying out this method includes accurately measuring small volumes of chemicals and enzymes (less than 10 cm³). To do this a 10 cm³ measuring cylinder or a 10 cm³ calibrated dropping pipette could be used.

It also involves heating solutions to different temperatures and keeping them at this temperature during the investigation. This could be done by putting hot water into a beaker and standing the test tubes inside the beaker or by using water baths at different constant temperatures

It also needs the pH change to be measured. This could be done by using an indicator; **universal indicator** solution is pink in solutions around pH 10. The fatty acids produced when lipase digests fats lower the pH of the solution, changing universal from pink to colourless. Alternatively a pH probe connected to a data logger could be used.

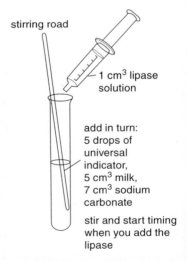

stirring road

1 cm³ lipase solution

add in turn:
5 drops of universal indicator,
5 cm³ milk,
7 cm³ sodium carbonate

stir and start timing when you add the lipase

Figure 8.37 Investigation method and equipment

1 Which piece of apparatus measures small volumes most accurately?

2 Explain which apparatus to change the temperature of the solutions will give the most accurate measurements.

3 Which technique to measure pH change will allow you to make more accurate measurements?

4 The chemicals used in the investigation are hazardous. Give two safety precautions that should be taken when doing the investigation.

Making and recording observations

Zak was going to carry out this investigation. Before he started he needed to design a table for his results. Tables need to fit in all the data collected, be fully labelled and include units. He had decided on five values for his independent variable. Zak has decided to complete three repeat tests for each value and to calculate the mean value. He completed a few trial runs of the procedure before he started and found that the indicator changed colour in about 4 minutes.

5 Why did Zak complete some trial runs?

6 Which is the independent variable in the investigation? What units is it measured in?

7 Which is the dependent variable in the investigation? What units is it measured in?

8 Construct a table that Zak can use that will fit in all the data he is going to collect.

Evaluating the investigation

Zak hypothesised that the warmer the solutions the lower and faster the pH will fall. Zak then completed his investigation using beakers of hot water to change the temperature, a 10 cm³ calibrated dropping pipette to measure the solutions and a pH probe. His results are shown below.

Zak reflected on his technique and the apparatus that he used.

9 Explain how the evidence supports or refutes Zak's hypothesis.

10 How could Zak change his apparatus to make his findings more valid and accurate?

11 Why has Zak used a pH probe?

> **ADVICE**
>
> During an investigation, you may find that you need to make changes to improve your method. These changes should be explained in your evaluation.

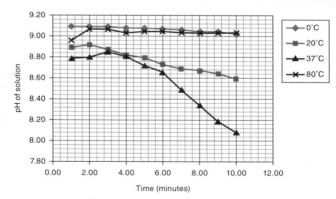

Figure 8.38 Results for pH of solution plotted against Time in minutes

Changing the environment

Learning objectives:

- recall causes of environmental change
- describe the impact of environmental change
- explain the impact of an environmental change.

KEY WORDS

dredging
environmental
  change
global warming

Human life is dependent on a rich biodiversity for survival. Our actions are decreasing biodiversity, and are doing so at an alarming rate.

HIGHER TIER ONLY

Thinking about environmental changes

Environmental changes can be:

- natural, caused by
  › seasonal changes
  › geographic location
- caused by humans.

Environmental changes affect the distribution and behaviour of organisms found there. Changes may be short lived or long lasting.

These changes may cause migration, enable the survival of some well-adapted organisms or lead to the death of organisms.

Environmental changes include temperature changes, or the availability of water and atmospheric gases.

1   **What causes environmental change?**

2   **Suggest how organisms react to environmental change.**

Environmental changes and impact

Examples of changes include:

- seasonal changes
  › low temperatures and food shortages cause some animals to migrate, hibernate or have a dormant stage
  › hot seasons may cause drought; rainy seasons may cause flooding
  › hot temperatures reduce the available oxygen in water
- Geographic changes
  › land bridges sinking and stopping animals from moving between continents, for example, hairy mammoths
  › with **global warming**, mountainous regions are becoming warmer; mountain species are having to compete with lowland species whose habitats are extending upwards

Figure 8.39 What caused these environmental changes?

> with global warming, the seas are becoming more acidic (and warmer), which means that the shells of calcareous organisms (such as bivalve molluscs) are dissolving and becoming thinner
> erosion by rivers and tides may gradually destroy habitats.

Examples of environmental change caused by humans include:

- burning fossil fuels, which causes global warming. Higher sea temperatures are associated with lower dissolved oxygen levels.
- intensive farming, causing desert regions.
- **dredging** sea beds for building has endangered some marine ecosystems.

3 Describe the impact of some natural environmental changes.

4 Describe the impact of some environmental changes caused by humans.

The impact of rising sea temperatures

Coral reefs have a rich biodiversity. They cover 1% of the Earth's surface but are home to 25% of all marine fish species. They form from coral polyp colonies that have a hard exoskeleton of calcium carbonate. Coral reefs develop near the shores of tropical oceans, where sea temperatures are 24 °C to 26 °C. They can only grow in water that is less than 25 m deep. Corals feed on algae.

Global warming is increasing sea temperatures. This causes bleaching of corals, which means that algae living in them cannot survive. Warmer seas are also endangering some marine species which are unable to adapt to the increased temperatures. Around 20% of coral reefs have been destroyed in just 50 years.

5 Evaluate the impact of rising sea temperatures on biodiversity.

> **REMEMBER!**
>
> Evaluate the impact of changes. Explain the intended changes, unintended changes and short- and long-term impacts.

> **DID YOU KNOW?**
>
> Volcanic eruptions can cause land masses to appear (for example, the Hawaiian Islands). The lava makes mineral-rich soils for successful plant growth.

Figure 8.40 Dredging sand for building has changed the marine ecosystem

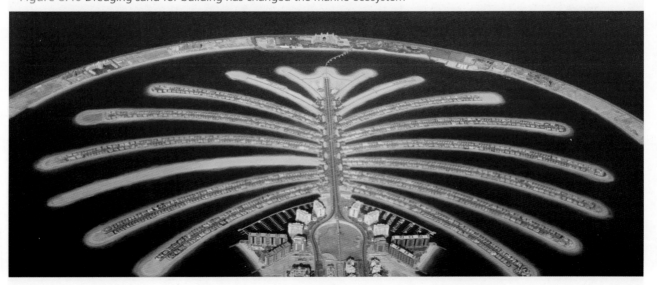

Learning about land use

Learning objectives:

- identify why land use has changed
- describe the effects of changing land use
- evaluate a change in land use.

KEY WORDS

eutrophication
run-off

Scientists agree that human activity has changed environments and that we need to protect the Earth's biodiversity.

Human population growth

The world's population has increased rapidly, from 1 billion (1000 million) in 1880 to about 7 billion in 2012. People use increasing amounts of the Earth's resources, resulting in a decrease in the land available for other organisms.

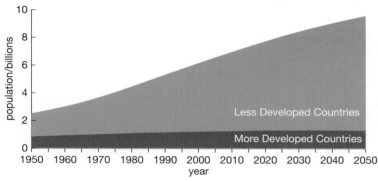

Figure 8.41 Why do less developed countries have a greater population increase?

Humans are using more land for:

- farming
- building
- quarrying
- dumping waste.

Sudden increases in the populations of animals and plants are usually controlled by predators, food availability, disease or toxic waste products. Humans have no predators, grow large amounts of food and have the ability to cure or prevent disease.

Figure 8.42 Intensive farming using fertilisers produces large crops

1 How do humans change the use of land?

2 Why do farmers use fertilisers?

Using the land

As our population increases, biodiversity decreases:

- More land is needed for homes, shops, factories and roads. Building sites destroy habitats. Roads divide habitats making it harder for organisms to find food and mates.
- New quarries are mined to provide stone, slate and metal ores for building materials. Habitats are destroyed.
- More farmland is needed and fertilisers are used. Many farms grow one crop over huge areas. This affects food availability for insect pollinators. There are fewer available nesting sites for birds.
- More waste is sent to landfill, and more sewage and industrial waste are produced. This can pollute the land.

Figure 8.43 An abandoned quarry

Human population is growing exponentially. This means that the increase in any year is greater than in the previous year because the birth rate is greater than the death rate.

3 Suggest the impact if human populations keep rising.

4 Describe the impact of two land-use changes.

Polluting water

Figure 8.44 Why are there no fish in this water?

If water contains a lot of fertilisers (caused by **run-off** from farmers' fields) or sewage, the nitrates and phosphates in the water increase and so then algal growth increases. Algae cover the water surface and prevent light from reaching water plants. The plants and algae die. Bacteria respire as they break down dead plants and use up oxygen in the water. The other living organisms in the water die. This is called **eutrophication**.

5 Explain the impact of fertiliser run-off.

6 Evaluate a change in land use.

7 Why do we need to maintain local and global biodiversity?

REMEMBER!

Humans reduce land availability for other organisms by building, farming, quarrying and dumping waste.

DID YOU KNOW?

When landfill sites are closed, the waste is left to degrade and stabilise. The land is re-used for parks, golf courses and even airports.

Changing the landscape

Learning objectives:

- identify the reasons for deforestation
- describe the impact of peat bog destruction and deforestation
- evaluate the destruction of peat bogs and forests.

Many species in tropical forests and *peatlands* are struggling to survive as their habitat is destroyed.

Why are landscapes changing?

Huge areas of tropical forest are being destroyed. This **deforestation** is happening to:

- provide land for cattle and rice fields
- grow crops, for example, oil palm and sugar cane to make biofuels. Biofuel crops are sometimes grown at the expense of food crops.

Cleared forests are often used to grow a monoculture (one crop) over huge areas.

Figure 8.45 Peat cut to be used for fuel

Peat bogs form over thousands of years in marshy areas. Decomposers cannot completely break down plant material in acidic conditions with little oxygen, so peat forms. Peat stores carbon.

Peat is used as a fuel and as cheap compost by gardeners. Compost improves soil quality to increase food production.

1 Why are forests cut down?

2 What is peat and how is it used?

The impact of changing landscapes

Forests are often destroyed by burning. Mass destruction of trees has:

- increased the release of carbon dioxide into the atmosphere (due to burning and the respiration of microorganisms that are decaying the remaining plant material)
- reduced the rate that carbon dioxide is removed from the atmosphere (by photosynthesis)
- reduced biodiversity; some of the lost plants and animals may have been useful in the future
- increased methane in the atmosphere because cleared land is used to grow rice in swamp-like fields.

> **REMEMBER!**
> ....
> Trees and plants in peatlands all use carbon dioxide to photosynthesise. Carbon is then stored in these plants.

Figure 8.46 How will forest fires affect the atmosphere?

Figure 8.47 Orangutans are losing their food sources

Insufficient trees are being replaced. Peat is being destroyed faster than it is being made. Peat and trees are both important carbon 'stores' that are being lost. The loss of peat bogs reduces the variety of different plants, animals and microorganisms that live there. Monocultures also reduce biodiversity.

3 Describe and explain the impact of deforestation.

4 How can woodland habitats be preserved?

> **DID YOU KNOW?**
> ....
> About 13 million hectares of forest have been cleared or lost through natural disasters. By 2030, there may only be 10% of our forests left.

Balancing act

There is a massive conflict between:

- the need for deforestation to increase land available for food production
- the use of peat as cheap compost to increase food production
- the need to conserve forests and peatlands as habitats for biodiversity
- the need to reduce carbon dioxide emissions from using peat as a fuel and from burning forests.

5 Evaluate the destruction of peatlands.

Thinking about global warming

KEY WORD

global warming

Learning objectives:

- recall what global warming is
- describe the causes of global warming
- explain how global warming impacts on biodiversity.

The future of the human species on Earth relies on us maintaining a good level of biodiversity, yet we are threatening it by our actions.

What is global warming?

The average global temperature of the Earth and its atmosphere is increasing. This is **global warming**. It is caused by increasing atmospheric levels of:

- carbon dioxide
- methane.

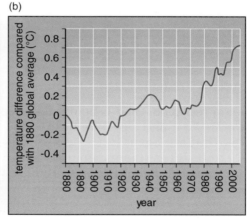

Figure 8.48 (a) CO_2 in the atmosphere 1000–2012; (b) temperature difference from 1880 global average temperature, 1880–2012. What do these graphs tell you?

These gases occur naturally in the atmosphere, but levels have increased over the last 150 years because:

- power plants burn fossil fuels
- petrol is used as fuel in vehicles
- rice crops and cattle farming are increasing
- deforestation and destruction of peatlands are increasing.

As the human population increases more pollution is produced from fossil fuels, particularly by relatively small populations in developed countries.

1 What is global warming?

2 What causes global warming?

Impact on ecosystems

Scientists think that global warming is changing the climate. The average world temperature rise is small (about 0.8°C since 1880), but some species such as coral reefs are sensitive to this.

3 How do human actions affect global warming?

4 Explain how global warming might affect biodiversity.

Case studies

The white lemuroid possum is the first mammal in Australia to have become almost extinct due to global warming. There are just four known adults left. The white possum's habitat spans cooler areas of high-altitude rainforest. Possums are vulnerable to increases in environmental temperature because they cannot maintain their body temperature.

Little terns are vulnerable to high tides and storms. These are happening more often because of global warming. Little terns migrate to the UK each spring and make their colonies just above the high tide line. Their nests are vulnerable to flooding by stormy seas.

Coastal mangrove forests grow in equatorial regions. Increasing numbers of storms and typhoons are undermining the fine sediment that the mangroves grow in. Seedlings cannot root and essential nutrients for the mangrove ecosystems are washed away.

Figure 8.49 Penguins are losing their habitats. Where else can they survive?

REMEMBER!

Learn three ways that global warming affects biodiversity.

DID YOU KNOW?

If sea levels rose by 1 m, half of the world's important coastal wetlands would be threatened.

Figure 8.50 How does global warming affect little terns?

5 Explain how global warming affects biodiversity.

Google search: 'effect of global warming on biodiversity'   357

Looking at waste management

Learning objectives:

- describe how waste production is linked to human population growth
- describe the impact of waste on ecosystems
- explain how waste impacts on biodiversity.

As our population keeps increasing, so our impact on the environment and biodiversity is also increasing.

Pollution

Human population growth and living standards are increasing, particularly in developed countries. We use more resources. As living standards increase, the demand for agriculture, manufacturing and industry increase, producing more waste.

Domestic and industrial waste must be handled correctly to avoid causing more pollution.

Waste substances include:

- **sewage**
- smoke and toxic gases
- herbicides, pesticides and fertilisers
- lead
- paper and cardboard
- plastic products.

Waste can kill plants and animals and reduce biodiversity.

Figure 8.51 Waste tips reduce biodiversity. They are home to decomposers and some birds

1 Why is the amount of waste we produce increasing?

2 Name three types of waste produced by people.

Waste and ecosystems

Causes of pollution include:

- Fertilisers. These can enter waterways, causing eutrophication.
- Toxic chemicals from household and industrial waste. If taken to landfill sites, they can spread into soil and enter waterways. Pesticides and herbicides are also washed into waterways. Toxins build up in food chains, kill organisms and affect feeding relationships.

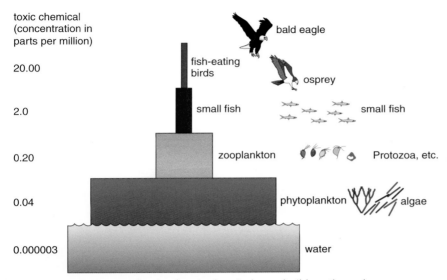

toxic chemical (concentration in parts per million)	
20.00	bald eagle, fish-eating birds, osprey
2.0	small fish, small fish
0.20	zooplankton, Protozoa, etc.
0.04	phytoplankton, algae
0.000003	water

Figure 8.52 Dangerous levels of toxic chemicals can build up through trophic levels

- Sewage. If untreated, chemicals and parasites can enter waterways. Microorganisms that decompose sewage use dissolved oxygen, causing aquatic organisms to die.
- Smoke and gases. Soot in smoke covers and damages trees; organisms may find breathing difficult.

3 How do human actions affect water quality?

4 Suggest how soot damages trees.

Acid rain

Acidic gases are produced when fossil fuels burn. They dissolve in water vapour to make acid (about pH 4). Acid rain:

- damages leaves and roots of plants
- washes mineral ions out of soil, causing mineral deficiencies in plants
- washes aluminium ions from soil into lakes, which affects gills in fish and they cannot survive
- acidifies waterways, so aquatic organisms cannot survive
- can travel in air; acid rain produced in the UK has affected trees and fjords in Norway.

Pollution reduces the available space for other organisms; some species cannot survive. We need to balance our development while sustaining the environment for future generations. Many countries (including the UK) have legislation that controls the use of toxic chemicals, use of landfill sites and treatment of sewage.

5 Explain how dangerous levels of toxic chemicals can build up through trophic levels.

Figure 8.53 How does acid rain affect trees?

Investigating pollution

Learning objectives:

- identify pollution levels using indicator species
- explain how indicator species measure pollution
- compare different methods of measuring pollution.

Scientists monitor environmental change to understand how it affects living organisms.

What is an indicator species?

Living organisms are sensitive to different abiotic conditions. If conditions change – for example, through pollution – the distribution of organisms can also change. Some organisms are used to measure environmental change. They are called **indicator species**.

Lichens grow on trees, roofs and rocks. They are rarely found growing in cities because they cannot survive in polluted air. The sampling of numbers and types of lichens informs scientists about pollution levels.

Lichens are sensitive to sulfur dioxide concentrations in the air.

Very few types and numbers of lichens grow near power stations. The further away they are, the greater the number and different species that are found.

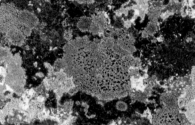

Figure 8.54 The cleaner the air, the leafier the lichens that grow. (a) Leafy lichen; (b) crusty lichen.

1 How does pollution affect living organisms?

2 What is an indicator species?

More indicators

Pollution also affects the distribution and numbers of animals. Aquatic invertebrates are used as indicators of pollution and dissolved oxygen in water, as shown in the following table.

- Sludge worms can live in polluted water. This is because they are adapted to cope with the low oxygen levels in polluted water.
- Mayfly larvae can live in slightly polluted water. There is sufficient oxygen for their needs.
- Alderfly larvae cannot live in polluted water. They cannot survive when oxygen levels are low.

Figure 8.55 Power stations produce sulfur dioxide

Animal	Sensitivity to pollution
stonefly larva	sensitive
water snipe fly larva	sensitive
alderfly larva	sensitive
mayfly larva	semi-sensitive
freshwater mussel	semi-sensitive
damselfly larva	semi-sensitive
bloodworm	tolerates pollution
rat-tailed maggot	tolerates pollution
sludgeworm	tolerates pollution

Pollution affects the distribution of aquatic invertebrates

Water containing lots of different species is a healthy environment.

Pollution levels are also measured directly using:

- probes attached to computers to measure precise conditions, for example, pH, temperature, oxygen and carbon dioxide
- special tests to indicate levels of different chemicals, for example, nitrates.

3 **A water sample from a river contained sludgeworms but no stonefly larvae. Is the water polluted? Explain your answer.**

4 **Suggest two methods to measure the pH of a stream.**

Using indicator species

The animals in three different aquatic habitats near a town were sampled and identified. Unfortunately, the names of the habitats came off the sample jars. The samples were taken from a polluted pond, a stagnant pool near a water outlet and a fast-running stream.

Sample	Animals found
A	stonefly larva, mayfly larva, damselfly larva, alderfly larva, water snipe fly larva
B	rat-tailed maggot, sludgeworm, bloodworm
C	sludgeworm, damselfly larva, freshwater mussel

5 **Which sample came from which habitat?**

6 **Which sample contained the least dissolved oxygen? Explain your answer.**

7 **Suggest one advantage and one disadvantage of using chemical tests instead of indicator species to test for pollution.**

DID YOU KNOW?

Coral reefs are water quality indicators because they only tolerate narrow ranges of temperature, salinity and water clarity.

REMEMBER!

Pollution alters the numbers and distribution of plant and animal species.

Maintaining biodiversity

Learning objectives:

- describe some conservation measures
- describe the impact of breeding programmes
- explain how habitats are regenerated.

KEY WORDS

conservation
monocultures
regeneration
sustainable

Every species depends on others for food, shelter, pollination, etc. Because of interdependence, if one species is removed the whole community is affected.

Protecting ecosystems

Programmes have been developed to reduce negative effects on ecosystems and biodiversity, which are caused by our actions. These measures include:

- Introducing breeding programmes for endangered species.
- The protection and **regeneration** (restoring) of rare habitats.
- Re-introducing field margins and hedgerows on farmland where **monocultures** are grown. Hedgerows are a habitat for many wild species.

Figure 8.56 Why are hedgerows being replanted?

- Reducing deforestation and carbon dioxide emissions. **Sustainable** strategies include replanting trees.
- Recycling resources instead of dumping waste in landfill. Many materials are recycled in the UK.
- Cloning plant species. Cloning can be done quickly and economically using stem cells found in the meristems. This protects plant species from extinction.

Figure 8.57 Why must we recycle waste?

1 **How are ecosystems and biodiversity protected?**

2 **How do hedgerows increase biodiversity?**

Conservation

Conservation programmes are introduced:

- because we have a moral responsibility to protect endangered species
- so more plant species may be identified for medicines
- to minimise damage to food chains and webs
- to protect future food supplies.

Captive breeding programmes are planned to ensure genetic diversity is maintained, such as for the Arabian oryx and for giant pandas. Successful programmes allow species to be reintroduced to the wild. The wild Arabian oryx became extinct in the wild in the 1970s, but a successful breeding programme means the animal has now been reclassified as 'vulnerable'.

Many endangered species, for example rhinos, and tigers, are hunted and poached, despite legal protection. Seed banks store seeds carefully to protect plant species for the future.

3 **Why are conservation programmes introduced?**

4 **How do breeding programmes help endangered species?**

Protecting rare habitats

There are many difficulties and issues associated with organising conservation programmes. Some of these include:

- ensuring long-term funding
- having qualified scientists who understand the issues
- animals and plants do not recognise boundaries
- many organisations and governments may be involved, working locally, nationally and internationally
- lack of 'policing' of protected areas.

Mangrove forests are rich ecosystems that prevent coastal erosion and reduce carbon emissions. They are declining rapidly due to land development, and their use as a fuel and building material. In Abu Dhabi, however, mangroves are increasing due to massive planting programmes over the last two decades. The local environment agency works with land developers and the public to maintain healthy, litter-free sustainable forests. The forests protect seagrass beds, which are the sole food of the endangered manatee.

5 **What difficulties are involved in managing conservation programmes?**

6 **Explain how mangrove forests are protected and regenerated.**

7 **Explain why it is important to protect biodiversity.**

Figure 8.58 Arabian oryx populations are growing

> **REMEMBER!**
>
> Learn the different programmes for protecting ecosystems and biodiversity.

Figure 8.59 Many species depend on sustainable mangrove forests

> **DID YOU KNOW?**
>
> Siberian tigers are endangered because of deforestation, hunting and poaching.

Learning about food security

KEY WORD

food security

Learning objectives:

- identify factors affecting food security
- describe how different factors affect food security
- interpret data to evaluate food security.

About 842 million people in the world do not have enough food to eat.

What is food security?

Food security is when all people have access to consistent supplies of safe and nutritious food to meet their needs for an active healthy life. Food must be available, affordable and useable.

Factors that affect food insecurity include:

- increasing birth rate
- changing diets in developed countries
- new pests and pathogens affecting farming
- environmental changes
- cost of agricultural inputs
- conflicts and war.

Figure 8.60 Conflicts and war destroy farmland

When food is insecure, people have a shorter life expectancy and suffer more ill health.

REMEMBER!

Learn the factors that cause food insecurity.

1 What is food security?

2 Identify four factors that affect food security.

How factors affect food security

Increasing birth rates mean that populations are growing faster. In some countries demand for food is increasing faster than the increased food production, which is threatening their food security. Changing diets in developed countries have resulted in foods being transported from countries where it is scarce, to other countries around the world. These crops are often grown at the expense of staple foods, for example, coffee.

Climate is a major threat to food security. Parts of Africa have suffered food insecurity and famine for many years, due to decreasing rainfall.

Agricultural inputs, for example, fertilisers and pesticides, cost money. Sometimes the costs to produce large yields are too high. This is not sustainable and crops are not grown.

In regions where there is insufficient food and water, conflicts may happen as people fight for survival.

Figure 8.61 Decreasing rains have caused crops to fail in Ethiopia

3 **Describe how climate and increasing birth rates affect food security.**

4 **How have changing diets affected food security?**

Food security and biodiversity

There are millions of plant species but we cultivate only a few for food. This reliance on a relatively few species makes food production vulnerable to:

- changes in climate and weather patterns
- new pests and pathogens
- natural disasters, for example, flooding and heat waves.

Biodiversity must be protected to ensure food security. Pollinators, especially bees, are vulnerable to environmental changes. In some areas in which monocultures are grown, nectar is only available for a short period. Sustainable methods must be found to feed all people on Earth while protecting the biodiversity on our planet.

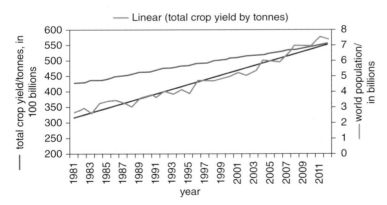

Figure 8.62 Graph showing world population and total crop yield

5 **Look at Figure 8.62. Interpret the data to evaluate food security in the future.**

6 **Why, if ample food is produced, do some regions suffer food insecurity?**

DID YOU KNOW?

Many families suffer food insecurity in developed countries.

Maintaining food security

Learning objectives:

- describe some intensive farming methods
- explain ethical issues related to intensive farming
- evaluate modern farming techniques
- describe methods to maintain sustainable fisheries.

KEY WORDS

factory farming
sustainable
  fisheries

Modern agricultural practices have increased food production.

Food production

Modern farming methods use machines and chemicals to produce large yields very quickly. Methods include:

- using fertilisers
- using insecticides and fungicides to kill pests, and herbicides to kill unwanted plants (weeds)
- growing plants in glasshouses
- using hydroponics to grow plants
- **factory farming** and fish grown in cages or tanks.

1 What is intensive farming?

2 How can intensive farming increase food production?

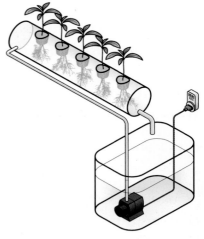

Figure 8.63 Growing plants without soil is called hydroponics

Exploring food production

A hydroponics system has advantages over growing plants in soil:

- mineral supplies are controlled and unused minerals are recycled. Production costs are lower with no pollution risk to waterways.
- the system is under cover so control of conditions and disease is improved.

Overfishing has depleted the Earth's oceans of fish stocks. Fish stocks can be increased and **sustainable fisheries** created for future generations through the use of:

- fishing quotas to conserve stocks at sustainable levels
- control of net size to ensure that young fish mature and breed, or some species may disappear from certain regions; for example, cod in the northwest Atlantic Ocean.

Fish in fish farms are fed high-protein food so they grow quickly. Obtaining our fish from fish farms allows wild fish stocks to recover.

3 Explain the advantages of using hydroponic systems.

4 How can fish stocks be maintained?

5 Discuss the advantages and disadvantages of modern fish farming.

Figure 8.64 The closeness of fish in fish farms makes them prone to disease

Efficiency and ethics

Intensive farming is very efficient. The efficiency is increased by:

- killing weeds, so reducing their competition with crops
- fewer animal pests to eat crops or cause disease in livestock
- feeding animals high protein foods to increase growth
- restricting the movement of animals so that energy used for movement is minimal and more can be used for growth
- controlling temperatures, which reduces the energy that an animal uses to keep cool or warm. The energy saved is used for growth.

Figure 8.65 Predators can be used to control pests in glasshouses

Mass production of animals for food causes suffering through:

- close confinement (increasing the risk of disease spreading)
- lack of movement
- the nature of the pens/cages
- poor transportation or slaughtering techniques.

Figure 8.66 Battery cows often have no room to move

6 Explain how the factory farming of animals increases energy transfer.

7 Evaluate modern farming techniques.

REMEMBER!

Learn about factory farming, fish farming and ethical issues.

Using biotechnology

Learning objectives:

- describe some uses of biotechnology
- explain the advantages of some uses of biotechnology
- evaluate some uses of biotechnology.

KEY WORDS

biotechnology
mycoprotein

Modern biotechnology can help to meet the needs of growing human populations.

Using biotechnology

Biotechnology is using living organisms to make a product to improve the quality of life. Biotechnology is used to produce:

- genetically modified (GM) crops
- human insulin
- **mycoprotein** (a protein-rich food).

GM crops can:

- be resistant to insect attack or to herbicides
- produce increased yields because of the characteristics chosen, such as larger fruits, disease resistance or herbicide resistance.

GM crops could provide more food or more nutritious foods, for example, golden rice. Rice is a staple food for Asian communities. Many people in Asia suffer from vitamin A deficiency, which causes night blindness. This is when the eyes are unable to adjust to dim light. Golden rice has been developed using genes from maize and a common soil bacterium which produce the β-carotene (which we say as 'beta-carotene') that makes vitamin A.

> **DID YOU KNOW?**
>
> The first commercially grown GM food crop was a tomato, which stayed firmer for longer.

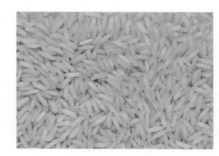

Figure 8.67 The more β-carotene that golden rice contains the more yellow-orange it is.

1. **How is biotechnology useful in meeting the needs of growing human populations?**

2. **Describe how golden rice can prevent night blindness.**

Human insulin

Blood sugar levels in the body are controlled by a hormone called insulin. People with diabetes cannot control their blood sugar levels. Diabetes is becoming increasingly common. Some people with diabetes need to inject insulin.

Genetically modified bacteria are now used to produce a constant supply of human insulin. The insulin is harvested and purified before being used to treat diabetes. However, some people think that genetic engineering is not ethical.

> **REMEMBER!**
>
> Learn about GM crops, insulin and mycoprotein production.

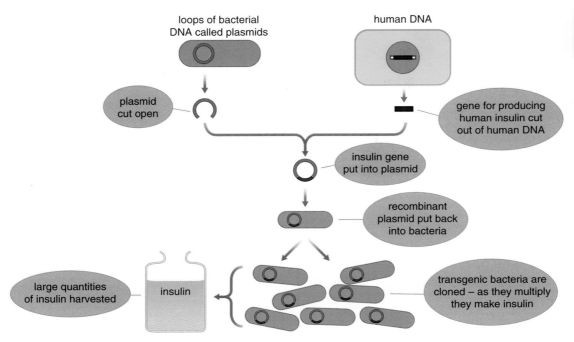

Figure 8.68 Steps in human insulin production

3 **Explain the advantages and disadvantages of using biotechnology to produce insulin.**

4 **Why are large amounts of insulin made?**

Mycoprotein

Modern biotechnology techniques allow large quantities of microorganisms to be cultured in vats, called fermenters, for food. The fungus *Fusarium* is used to make mycoprotein, a protein-rich, low-fat food (that is suitable for vegetarians). *Fusarium* grows and reproduces rapidly on glucose syrup (made from waste starch), in aerobic conditions. *Fusarium* can double its biomass every 5 hours. The biomass is harvested, purified and dried to make mycoprotein, but it is almost tasteless. A range of textures and flavourings can be added.

The fermenters have:

- constant oxygen supplies for *Fusarium* to respire
- water jackets to remove heat produced during respiration
- pH and temperature probes to monitor conditions and allow adjustments to be made quickly
- stirrers to spread heat, oxygen and syrup evenly through the vats and keep the fungus in suspension.

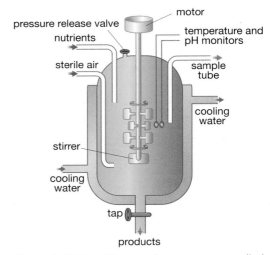

Figure 8.69 Conditions in the vats are controlled to maximise production

5 **How are conditions controlled in the fermenting vats?**

6 **Evaluate the use of biotechnology to produce food.**

MATHS SKILLS

Using graphs to show relationships

Learning objectives:

- to recognise direct proportionality in a graph
- to calculate reaction rates in linear graphs
- to use the gradient of a graph to calculate the rate.

KEY WORDS

directly
  proportional
gradient
intercept
linear
rate

It is sometimes useful to see how one variable relates to another, such as how much we exercise affects how fit we are. A graph is a good way of displaying the relationship.

Types of graph

When we plot the results from investigations we sometimes get graphs in which the points are approximately in a straight line, so we can draw a straight line of best fit. Even if the line does not go through the origin (0,0), it has a strong correlation that approximates a **linear** relationship.

If this straight line passes through the origin (0,0), the relationship is **directly proportional**. This means that if one variable doubles in size, the other variable does too. The point where the line crosses the y-axis is called the **intercept**.

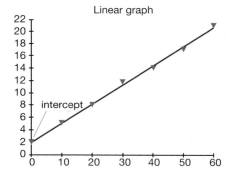

1. **Where does the line intercept the y-axis in the linear graph in Figure 8.70?**

2. **Where is the y-intercept for any line showing direct proportionality?**

Using a graph to work out the rate at which something is happening

Sometimes when we use data to plot a graph, one variable is a measurement of time. This is common in biology as we are often interested in the **rate** at which something is happening.

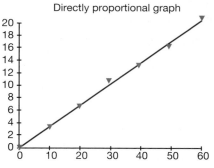

Figure 8.70 What is the difference between these two graphs?

Look at this graph showing how puppies gain weight over the first couple of years. They all gain weight more quickly when they are young and the weight gain levels off when they become adults. The rate is shown by the **gradient**. The gradient is the steepness of the line. In this case, the steeper the gradient the more weight the dogs are gaining each month. The graph also shows that the rate is greater for larger breeds. Knowing the gradient is useful.

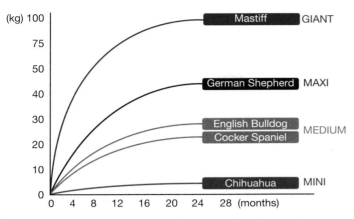

Figure 8.71 Rate of growth of different dog breeds.

3 **a** How much weight will a German shepherd puppy put on in the first 4 months?
   b On average, what would this be every month? (Your answer will be in kg/month – this is the rate.)
   c How much weight will the same dog add in the second 4 months?
   d What is the average rate during this period?

4 Why might dog breeders find a graph like this useful?

5 Why might people choosing a breed of dog find this graph useful?

Calculating gradients

Linear relationships, as we've seen, have straight lines. The gradient is constant. It means the rate of increase (or decrease) isn't changing over time.

The gradient is the change in the y-values divided by the change in the x-values.

For example, if we heat water, the longer we heat it for, the hotter it gets, up to 100°C.

In this experiment, from the end of the first minute to the end of the third minute the temperature went from 60°C to 80°C.

The change in y-values (the temperature) was $80 - 60 = 20$ and the change in x-values (the time) was $3 - 1 = 2$. The gradient of the graph can be dividing 20 by 2, which is 10, indicating that the rate at which the water is heated is 10°C per minute i.e. for every minute on the graph, the temperature rises by 10°C.

6 Look at the graph from the end of the third minute to the end of the fifth minute:
   a What is the change in y-values?
   b What is the change in x-values?
   c What is the rate?

7 Why would this trend not continue after the fifth minute?

> **DID YOU KNOW?**
>
> The independent variable always goes on the x-axis, so that's where time goes, as we can't control how quickly it passes.

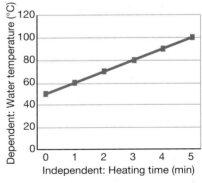

Figure 8.72 Graph showing how the temperature of water changes as it is heated.

> **KEY INFORMATION**
>
> Straight-line graphs show a linear relationship. If the line goes through the origin, (0,0), it is also directly proportional.

Check your progress

You should be able to:

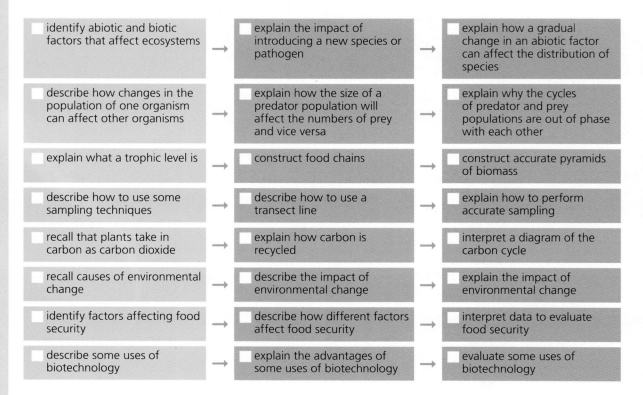

identify abiotic and biotic factors that affect ecosystems	explain the impact of introducing a new species or pathogen	explain how a gradual change in an abiotic factor can affect the distribution of species
describe how changes in the population of one organism can affect other organisms	explain how the size of a predator population will affect the numbers of prey and vice versa	explain why the cycles of predator and prey populations are out of phase with each other
explain what a trophic level is	construct food chains	construct accurate pyramids of biomass
describe how to use some sampling techniques	describe how to use a transect line	explain how to perform accurate sampling
recall that plants take in carbon as carbon dioxide	explain how carbon is recycled	interpret a diagram of the carbon cycle
recall causes of environmental change	describe the impact of environmental change	explain the impact of environmental change
identify factors affecting food security	describe how different factors affect food security	interpret data to evaluate food security
describe some uses of biotechnology	explain the advantages of some uses of biotechnology	evaluate some uses of biotechnology

Worked example

Some students investigated the average number of plant species growing in a habitat. They investigated different sized areas of the habitat. Look at the graph.

1. **Describe what the graph shows.**

   The number of different plants found increases from 0 to 18

   > Maximum number of plants found identified.
   >
   > More detailed responses include: as area sampled increases the number of species found increases; the ideal area to sample is 8 m².

2. **Amy and Laura decide to investigate how the number and type of plant species growing in a salt marsh change with distance from the sea.**

   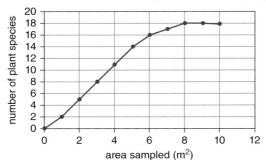

   What equipment should they use to do this?

   A quadrat

   > A more detailed answer would include a measuring instrument, e.g. tape measure.

3. **Describe how Laura and Amy should use this apparatus.**

   They make a straight line and place the quadrat every metre along it. Then count the different species and their numbers.

4. **Why will the results of Amy's and Laura's investigation be an estimate of the population sizes?**

   They have only sampled the habitat

   > Correct answer identified. A more detailed response would include that it is impossible to count every plant.

5. **Laura and Amy repeat the process in two other locations. Explain why.**

   To improve the accuracy of their results

   > This is a correct response.
   >
   > Overall, there is some good understanding of the topic shown but the answers need to be more detailed.

End of chapter questions

Getting started

1 What group of organisms break down dead organisms?   `1 Mark`

2 What are quadrats used for?   `1 Mark`

3 Give one advantage of intensive farming.   `1 Mark`

4 Why can badgers be primary and secondary consumers?   `1 Mark`

5 Give two factors needed for decay.   `2 Marks`

6 Explain why bluebells grow before the trees in a wood are in leaf.   `1 Mark`

7 Why do homes using biogas need to be well ventilated?   `1 Mark`

8 Why is unmixed plant material in a sealed bin in the shade slow to decay?   `2 Marks`

Going further

9 Why is it important to place quadrats in a random way rather than choose an area?   `1 Mark`

10 What is a trophic level?   `1 Mark`

11 The biomass of algae in a pond is 1000g. The biomass of tadpoles is 100g and the biomass of diving beetles is 10g. Draw a pyramid of biomass, to scale, for this food chain.   `2 Marks`

12 Ken has lots of moss growing in his lawn. He wants to estimate what percentage of his lawn is moss. Describe how Ken could do this, making sure the data he collects is random.   `4 Marks`

13 The graph shows human population growth and the number of extinctions over time.

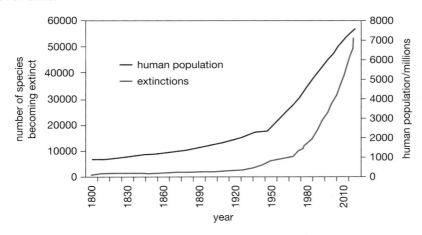

Describe what the graph shows.　　　　`2 Marks`

More challenging

14 Pyramids of biomass are used to show the mass of living material at different trophic levels in a foodchain. At each stage there is less biomass than the stage before. This is often only one-tenth of the previous biomass. Why does this biomass reduce?　　`2 Marks`

15 Explain how destruction of rainforests without replanting has helped increase the concentration of carbon dioxide in the atmosphere.　　`2 Marks`

16 Alistair notices that the pond in his garden has turned green and all the fish have died. He thinks it might have been because he used too much inorganic fertiliser in his garden. Explain how using inforganic fertiliser on his garden may have caused the fish to die in his pond.　　`6 Marks`

Most demanding

17 Explain why reliance on a few plant species could affect food security.

2 Marks

18 Jessica is carrying out an investigation into how many dandelions plants there are in her school playing field. The field is 20 metres by 30 metres. She selects 9 spots at random and samples each spot with a 1 m² quadrat. Jessica's results are shown below.

Sample number	Number of dandelion plants
1	5
2	4
3	2
4	4
5	3
6	1
7	3
8	2
9	4

Jessica calculates the mean number of dandelions to be 3.1.

a Calculate the median and mode values for the number of dandelions across the areas sampled.

1 Mark

b Use the data to estimate the number of daisy plants in the playing field.

1 Mark

19 **Sewage is accidentally leaked into a river.**

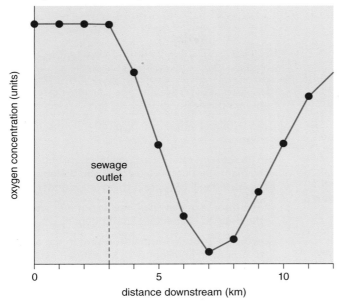

Evaluate the evidence to suggest that the sewage leak has
affected the environment.

6 Marks

Total: 40 Marks

Glossary

A

abiotic factor physical or non-living conditions that affect the distribution of a population in an ecosystem, such as light, temperature, soil pH

absorption the process by which soluble products of digestion move into the blood from the small intestine

abstinence method of contraception whereby the couple refrains from intercourse, particularly when an egg might be in the oviduct

accommodation ability of the eyes to change focus

acid rain rain water which is made more acidic by pollutant gases

active site the place on an enzyme where the substrate molecule binds

active transport in active transport, cells use energy to transport substances through cell membranes against a concentration gradient

adaptation features that organisms have to help them survive in their environment

ADH (antidiuretic hormone) a hormone released by the pituitary gland, which acts on the kidney in response to changes in the concentration of the blood plasma

adrenaline a hormone released quickly from the adrenal glands during a 'flight or fight' situation

adrenal medulla part of the adrenal gland that produces the hormone adrenaline

adult cell cloning a form of cloning where an embryo is produced from an adult body cell. The embryo is then implanted into a female animal or used for therapeutic reasons, (e.g. this process was used to created Dolly the sheep)

adult stem cells rare, unspecialised cells found in some tissues in adults that can differentiate only into the cell type where they are found, e.g. blood cells

aerobic respiration respiration that involves the use of oxygen

agar plate plastic dish, with a lid, containing a nutrient gel upon which bacteria are grown in a lab

alleles inherited characteristics are carried as pairs of alleles on pairs of chromosomes. Different forms of a gene are different alleles

alveolus (plural: alveoli) air sacs; the site of gaseous exchange in the lungs

amino acids small molecules from which proteins are built

amylase a digestive enzyme (carbohydrase) that breaks down starch

anaerobic respiration respiration without using oxygen

antibacterial chemicals chemicals produced by plants as a defence mechanism; the amount produced will increase if the plant is under attack

antibiotic e.g. penicillin; medicines that work inside the body to kill bacterial pathogens

antibody protein normally present in the body or produced in response to an antigen, which it neutralises, thus producing an immune response

antimicrobial resistance (AMR) an increasing problem in the twenty-first century whereby bacteria have evolved to develop resistance against antibiotics due to their overuse

antiretroviral drugs drugs used to treat HIV infections; they stop the virus entering the lymph nodes

antiseptic substance that prevents the growth of bacteria

antitoxins chemicals produced by white blood cells (lymphocytes) that neutralise toxins to make a safe chemical

antivirals drugs used to treat viral infections; they are specific to the virus but can only slow down viral development

aorta artery that carries oxygenated blood from the left ventricle to tissues around the body

apex predator carnivores with no predators, they are at the top of the food chain

aphids (greenfly) disease-causing small insects that reproduce rapidly; aphid infestation causes reduced growth, yellow leaves, wilting and death

archaea primitive bacteria that usually live in extreme conditions

artificial heart a temporary heart used to keep patients alive while they are waiting for a transplant

artificial pacemaker a device fitted under the skin that sends out electrical impulses to control the heartbeat

arteries blood vessels with thick elastic walls that carry oxygenated blood away from the heart under high pressure

aseptic technique measures taken to reduce contamination when preparing bacterial cultures, e.g. working with a flame

asexual reproduction reproduction involving only one parent

aspirin painkiller derived from willow bark

ATP molecule used to store energy in the body

atrium (plural: atria) the upper chambers of the heart that receive blood from the body or lungs

autoclave hot oven in which contaminated materials used in cell culture are sterilised

auxin a type of plant hormone produced at the tips of plant shoots and roots

B

bacteria single-celled microorganisms which can either be free-living organisms or parasites (they sometimes invade the body and cause disease)

bacterial growth curve bell-shaped curve that shows how bacteria in culture increase in number exponentially until a build-up of toxins in the culture causes them to die off

barrier methods of contraception; where a condom or diaphragm is used to prevent sperm reaching an egg

basal metabolic rate the level at which chemical reactions take place in the body at rest

behaviour animals adapt their movement and feeding patterns according to the conditions of their environment

Benedict's test a test for sugar in which Benedict's reagent turns orange-red in the presence of sugar

benign tumour slow-growing, harmless mass of cells that can usually be easily removed

bias when a data set is inaccurate or unrepresentative; it contains systematic errors

bile an alkaline substance made by the liver and stored in the gall bladder, which emulsifies fat to aid digestion

biotic factor factors caused by living organisms, such as food availability and competition between species

binary fission the process by which a bacterium divides and doubles its numbers, approximately every 20 minutes

binomial system the scientific way of naming an organism whereby each organism has a genus name and a species name; e.g. Homo sapiens

biodiversity range of different plant and animal species living in an ecosystem

biogas energy released when bacteria break down biomass, which can be used as a cheap fuel source

biomass the dry mass of organisms

Biuret reagent chemical indicator used to test for the presence of protein in a sample

blood sugar level the amount of glucose in the blood (controlled by the hormones, insulin and glucagon)

biological catalysts enzymes – molecules in the body that speed up chemical reactions

biotechnology using living organisms to make products to improve quality of life

body mass index (BMI) measure of someone's weight in relation to their height

breed new form of animal created with particular characteristics, created through selective breeding

bronchus (plural: bronchi) branches of the trachea – one going to each lung

C

calibrate the process of checking the accuracy a piece of equipment using standard data

camouflage a method of avoiding detection used by prey to avoid predators and by predators to surprise prey

capillaries small blood vessels that are one cell thick and permeable for diffusion of gases; join arteries to veins

capillary network tree of capillaries surrounding alveoli that are adapted for efficient gaseous exchange

carbohydrase enzyme that breaks down carbohydrates into simple sugars

carbohydrates one of the main groups of nutrients; e.g. glucose, starch

carbon cycle a natural cycle through which carbon moves by respiration, photosynthesis and combustion, in the form of carbon dioxide

carbon sink 'stores' of carbon dioxide in oceans or forests

carcinogen substance or virus that increases the risk of cancer, e.g. ionising radiation

carrier someone who is heterozygous for a disease-causing gene so that they do not have the condition but 'carry' the defective allele, which they could pass on to their children

causal mechanism where a direct link has been made between a risk factor and a disease; e.g. smoking and lung cancer

cell lines cells grown in a lab that are used to study a certain condition

cell membrane layer around a cell which helps to control substances entering and leaving the cell

cell cycle a series of three stages during which a cell divides

cellulose polymer of glucose; strengthening component of plant cell walls

central nervous system (CNS) collectively, the brain and spinal cord

cerebellum area at the base of the brain that coordinates muscular activity

cerebral cortex highly folded outer layer of the brain that controls consciousness, intelligence, memory and language

cervix the neck of the womb

Charles Darwin scientist who proposed the theory of evolution by natural selection

chlorophyll pigment found in plants which is used in photosynthesis (gives green plants their colour)

chloroplast a cell structure found in green plants that contains chlorophyll

chlorosis when a plant develops yellow leaves due to a lack of magnesium

chromosomes thread-like structures in the cell nucleus that contain DNA

cilia tiny hair-like structures found throughout the respiratory system that beat together to move mucus out of the lungs

classify to organise and present data in a logical order, e.g. in a table

clinical trial process during which a new drug is tested on patients and healthy individuals

clone in asexual reproduction, the offspring produced are identical to the parent (clones)

collision theory an explanation of enzyme action; enzymes cause more collisions between reactants thereby increasing the rate of reaction until reactants are used up

colony a spot containing millions of bacteria growing on an agar plate

combined contraceptive pill a hormonal method of contraception, taken orally, which contains synthetic forms of oestrogen and progesterone

communicable disease infectious condition caused by a virus, protist, bacteria or fungi; e.g. measles, HIV

community all the plants and animals living in an ecosystem, e.g. a garden

companion cells found next to phloem cells, companion cells control the activities of the phloem

compare to examine your data set next to a known set of results

competition when different species in an ecosystem compete for the same resources

complementary describes how each DNA strand in a double helix pairs with the opposite strand through base pairs (A-T and G-C)

complex diseases conditions that are caused by a combination of genetic and environmental factors

compost dead and decaying plant material

condom rubber sheath that covers the erect penis and prevents sperm reaching the egg; a barrier form of contraception

cones light-sensitive receptor cells in the retina that allow us to see in colour

conservation a way of protecting a species or environment

cooling the body loses heat through sweating

coordination centre region of the body that receives and processes information from receptors, e.g. brain, pancreas

cornea transparent region at the front of the eye through which light refracts when it enters the eye

coronary artery vessel that provides the heart muscle with oxygen and glucose for respiration

coronary heart disease a condition caused by build-up of fatty deposits in the coronary artery, leading to reduced blood flow and less oxygen and glucose reaching the heart tissue, which can cause a heart attack

correlation relationship between two sets of results

culture the growing of cells in a lab

culture medium the liquid (nutrient broth) or gel (agar) in which bacteria are grown in a lab

culture solutions liquids with known amounts of specific minerals in which plants are grown to investigate mineral deficiencies

cycle predator–prey relationships follow a cyclical pattern whereby the numbers of prey rise and fall over a period of years, and the numbers of predators follow the same pattern

cystic fibrosis disorder of cell membranes that is caused by a recessive allele

D

daughter cells cells produced during mitosis that are identical to the parent cell

deamination the removal of amino groups, as ammonia, from excess amino acids, in the liver

decomposer an organism that breaks down carbon in dead animals and plants

deforestation removal of large areas of trees to provide land for cattle or growing crops

dehydration result of the body losing too much water

demographic describes the structure of a population

denatured an enzyme is denatured if its shape changes so that the substrate cannot fit into the active site

depression a mental health condition

dialysis process by which the blood is filtered by a dialysis machine, in someone with kidney failure

diarrhoea a symptom of food poisoning; frequent passing of liquid stools

differentiation when cells gain certain features needed for their function; they become specialised

digitalis a heart drug extracted from foxgloves

directly proportional describes a relationship between two variables on a graph where the line of best fit goes through 0,0

disc-diffusion technique used to test effectiveness of antibiotics or disinfectants on plates of cultured bacteria using paper discs soaked in the test substance and measuring the size of clear area around the discs

disease resistance crop plants that are not affected by certain diseases and can therefore be cloned in large numbers for farmers

distribution how living organisms are dispersed/spread out over an ecosystem

DNA (deoxyribonucleic acid) found as chromosomes in the nucleus – its sequence determines how our bodies are made, and gives each one of us a unique genetic code

DNA bases four chemicals that are found in DNA, they make up the base sequence and are given the letters A, T, G and C

domain the level of classification group above kingdom, first suggested by Carl Woese

dominant an allele that is expressed when one or two copies are present; represented by a capital letter

dominant hand the hand a person uses most often for single-handed tasks, e.g. writing

donor refers to the person or organism where donated cells or tissue come from

dormant describes seeds that have not germinated; they can be sprayed with the plant hormone gibberellins, which induces them to germinate

dose amount of a drug given to a patient

double-blind trial a drugs trial where neither doctors nor patients know whether the patient has received the test drug or the placebo

double circulatory system where the blood is pumped to the lungs then returned to the heart before being pumped round the body

double helix the shape of DNA – two strands of nucleotides that wind around each other like a twisted ladder

dredging the process of digging up sea beds; an example of environmental change induced by humans

dual bar chart bar graph on which two data sets are plotted at each point

E

Ebola a deadly virus that kills 50–90% of people infected

ecosystem the interaction of a community of living organisms with the non-living parts of their surroundings

effector muscle or gland that brings about a response to a stimulus

efficacy how effective a drug is at preventing or curing the disease

embryonic stem cells unspecialised cells found in the early embryo that can differentiate into almost any cell type in the body

embryo screening the process whereby a few cells are taken from an embryo produced by IVF are checked for any defective genes

embryo transplants form of cloning where cells are taken from a developing embryo from a pregnant animal, separated and grown in culture then implanted into host mothers

emulsify the process by which fats are broken into smaller droplets by bile

endocrine gland part of the body that releases chemical messengers, or hormones, directly into the bloodstream

endocrine system a control system in the body that communicates using chemical messengers, or hormones, to produce slow but long-lasting responses

endothermic reaction chemical reaction which takes in heat

environmental change a shift in the external conditions caused by human activity or a natural event

environmental variation differences between individuals that arise during development; e.g. wearing glasses

enzymes biological catalysts that increase the speed of a chemical reaction

epidemic an outbreak of a disease that spreads quickly to many people

epidermal tissues layers of cells that cover the top surface of a leaf; they let light penetrate

epiphyte rainforest plant that is adapted to grow above ground level (e.g. on trees)

estimate a guess based on prior knowledge

ethanol product of anaerobic respiration in plant and yeast cells; used in the manufacture of alcoholic drinks

ethene a plant hormone released as a gas that speeds up fruit ripening

ethical whether a process is considered to be right or wrong, based on moral considerations

eugenics refers to the notion of improving a population by selecting certain desirable characteristics

eukaryotic cells cells that contain a true nucleus in the cytoplasm; e.g. plant and animal cells

eutrophication when waterways become too rich with nutrients (from fertilisers), allowing algae to grow wildly and subsequently decay, resulting in the oxygen being used up

evaporation the process by which the body loses heat through sweating

evolution the gradual change in organisms over millions of years caused by random mutations and natural selection

evolutionary tree a diagram created by scientists that shows how animals are believed to have evolved from common ancestors

exchange surfaces specialised areas with large surface area to volume ratios for efficient diffusion

exothermic reaction chemical reaction in which heat is given out

extinction the elimination of all members of a species; there have been five mass extinctions over geological time

extracellular digestion referring to when food is broken down outside an organism, e.g. bacteria secrete enzymes to breakdown dead material

extremophile organisms that can survive very extreme environments, e.g. high temperature or high salt

F

factory farming a form of intensive farming where animals are reared in sheds or cages under controlled conditions

family tree a diagram used to show how a condition is passed down through a family

fermentation anaerobic respiration in yeast cells; produces ethanol and carbon dioxide

fertiliser chemical containing minerals (nitrates, phosphates, potassium and magnesium) put on soil to improve plant growth

fertility drug combination of hormones given to a woman with low FSH levels to stimulate egg production

field of view the area of a sample visible under a microscope

filtrate dissolved substances that cross into the kidney after the blood has been filtered

flaccid floppy

follicle the sac left once an egg is released in the ovary; it produces the hormone progesterone

food security when all people have access to a consistent supply of food to meet their needs

fossil the remains of organisms that lived millions of years ago

fossil record an incomplete account of organisms that lived millions of years ago

fraction (in genetics) the number of people affected by a recessive condition expressed as a fraction could be 1/4

FSH (follicle stimulating hormone) a reproductive hormone that causes eggs to mature in the ovaries

functional adaptation an adaptation involving how the body operates, e.g. camels don't sweat to avoid water loss

fungicide chemical used to kill fungi

G

gametes the male and female sex cells (sperm and eggs)

gel derived from silicone, the material that contact lenses are made from

gene section of DNA that contains the instructions for a particular characteristic

gene expression whether genes are switched on or off, which is regulated by non-coding DNA

generator large tanks full of rotting organic matter, used for biogas production

gene theory dogma developed in the twentieth century that brought together the work of many scientists on inheritance, DNA structure and the genetic code

gene therapy medical procedure where a virus is used to 'carry' a gene into the nucleus of a cell (this is a new treatment for genetic diseases)

genetic code a sequence of three DNA bases that codes for a single amino acid

genetic cross a way of working out the probability of the genotype of the offspring produced from the parents' genotype for a particular characteristic, e.g. using a Punnett square

genetic engineering transfer of specific genes from one organism to another

genetic marker gene a gene inserted into a modified organism alongside the required gene; the marker gene is often resistant to an antibiotic, so that when cultures of cells are treated with that antibiotic, cells that do not contain the marker gene will be killed

genetic variation the product of meiosis, mutations and sexual reproduction, which all lead to changes in our genome

genome the entire genetic material of an organism

genome editing an experimental method of detecting and correcting defective genes in an individual

genomics the science of studying the human genome to increase our understanding of human DNA

genotype the alleles present for a particular gene make up the organism's genotype

geotropism a plant's growth response to gravity (also called gravitropism)

gibberellins plant hormones that speed up seed germination

gills a gas exchange surface found in fish and tadpoles

global warming the increase in the Earth's temperature due to increases in carbon dioxide levels

glucagon hormone released by the pancreas that, along with insulin, controls blood glucose levels

glucose tolerance test a way of testing for diabetes by measuring how the body responds to glucose

GM crops varieties of crops that have had their genomes modified by the insertion of genes from a plant or other organism

goblet cells cells in the epithelium that produce mucus to trap dust particles

golden rice rice that has been genetically modified to contain β - carotene, which converts to vitamin A

gonorrhoea a sexually transmitted disease caused by bacteria, resulting in pain when urinating and a thick discharge from the vagina or penis

gradient the steepness of a line plotted on a graph

graticule piece of glass or plastic onto which a scale has been drawn, which is used with a microscope

Gregor Mendel Austrian monk who established the science of genetics through his work on pea plants

guard cells cells surrounding the stomata that open and close the pore to control the exchange of gases and water loss

H

habitat the place where an organism lives in an ecosystem; e.g. the worm's habitat is the soil

haemocytometer device used for counting the number of cells, e.g. under a microscope

haemodialysis filtering of the blood by a machine in patients with kidney failure; takes four hours, three times a week

haemoglobin chemical found in red blood cells which binds to oxygen to transport it around the body

health a state of compete mental and physical wellbeing

heterozygous a person who has two different alleles for a characteristic, e.g. someone with blond hair may also carry the allele for red hair

HIV (human immunodeficiency virus) virus transmitted through sexual contact that attacks the immune system

homeostasis the regulation of internal conditions, such as temperature, in the body

homozygous a person who has two alleles that are the same for a characteristic, e.g. a blue-eyed person will have two 'blue' alleles for eye colour

hormone chemical messenger that acts on target organs in the body

hybrid infertile organism created through interbreeding organisms from two different species

hybridoma cells cells made from combining tumour cells with lymphocytes, to give cells that can produce antibodies and divide (for the production of monoclonal antibodies)

hydrocarbon a compound composed of the elements hydrogen and carbon

hydroponics growing plants in mineral solutions without the need for soil

hyperopia long-sightedness; where people can see objects far away but not close up

hypothesis a scientific explanation formed based on prior knowledge that can be tested in an experiment

I

immune system the body's defence against pathogens

immunity when the body is protected from a pathogen as it has already encountered it and can therefore produce antibodies against it, rapidly

inbreeding breeding closely related animals

incubation period of time a culture of bacteria is left at a certain temperature in order for the bacteria to grow

indicator species organisms used to measure the level of pollution in water or the air

inoculating loop length of wire with a ring at the end used for transferring bacteria to an agar plate

insulin hormone made by the pancreas that controls the level of glucose in the blood

interbreeding organisms of the same species can interbreed to give fertile offspring

intercept the point where the line on a graph crosses the y-axis

interdependence plants need animals for carbon dioxide and seed dispersal; animals need plants for food and oxygen; i.e. they depend on each other for survival

interspecific competition competition between two different species of organism

intraspecific competition competition among organisms from the same species

invasive species a cause of extinction as a result of a new species of organism being introduced into a population, which may bring disease, be a predator or compete with native organisms

inverse square law (in terms of light intensity and photosynthesis) the intensity of light is proportional to $1/d^2$ (where d = distance from the light source); for example, if the distance is doubled, the light intensity is quartered

in-vitro **fertilisation (IVF)** a fertility treatment, whereby an egg is fertilised with sperm outside the body, in a laboratory

iodine solution an orange solution used to test for starch (turns blue-black when starch is present)

iris part of the eye that has sets of muscles that control the size of the pupil and regulate the light reaching the retina

IUD (intrauterine device) a device containing hormones that is placed in the uterus to prevent a fertilised egg from implanting; a hormonal method of contraception

IVF cycle the process of IVF from stimulation of the ovaries to implantation of an embryo; if unsuccessful, another cycle can be attempted after about two months

K

kingdom the level of classification above phylum and below domain of living organisms

L

lactic acid product of anaerobic respiration in muscles

larva immature but active form of an insect that's created from an egg and turns into a pupa

laser surgery a technique used to correct eye defects by changing the shape of the cornea

LH (luteinising hormone) a menstrual cycle hormone that stimulates an egg to be released from an ovary

lichen slow-growing plant that grows up trees

lignin substance found in some plant cell walls that gives them strength and rigidity

limiting factor factors such as light, temperature and carbon dioxide, which affect the rate of photosynthesis

lipases enzymes that digest fats into fatty acids and glycerol

lock and key theory to explain how enzymes work; the substrate is the 'key' and the active site is the 'lock'

lumen the central part of a vessel

lymphatic system network of vessels containing a colourless fluid into which fatty acids and glycerol are absorbed

lymphocytes white blood cells that produce antibodies and antitoxins to destroy pathogens

M

magnetic resonance imaging (MRI) scanning technique that uses strong magnetic fields to produce detailed images of the body; used to diagnose nervous system disorders

magnification the factor by which an object is enlarged by a microscope; calculated as: size of image/size of real object

malaria condition characterised by recurring fever, which is spread by mosquitoes and can be fatal

malignant tumour abnormal cancerous mass of cells that grows quickly and can spread to other parts of the body

mathematical model an independent method used to show that Mendelian genetics is consistent with evolution

median the middle number in a set of values when they are arranged in order from lowest to highest

medulla region at the base of the brain that controls unconscious activities, such as breathing

mean average value calculated by adding up all the values in a data set then dividing by the number of values

meiosis cell division that results in gametes being produced, with half the number of chromosomes as the parent cell

melanism darkening of appearance – a form of colour variation in animals, e.g. melanic moths

Mendelian inheritance name given to single-gene disorders, such as cystic fibrosis

menstrual cycle monthly cycle in females which is controlled by reproductive hormones

mental health a feeling of well-being, having a positive frame of mind

meristem regions at tips of roots and shoots where cell division takes place

metabolism the sum of all reactions in a cell or the body

micrograph image captured using a microscope

microorganisms tiny organisms that can only be viewed with a microscope – also known as microbes

migration patterns historical movements of humans across the world, which have been deduced through studying genomes

mimicry mechanical defence mechanism where a plant uses features to trick animals into not feeding or not laying eggs

mineral deficiency when a plant cannot get enough of an essential mineral from the soil

mineral ions substances found in the soil that are essential for a plant's survival; e.g. plants need magnesium ions to make chlorophyll

missing links evidence for evolution found as intermediate forms between different organisms, as they evolved

mitochondria structures in a cell where respiration takes place

mitosis cell division that results in genetically identical diploid cells

mixed population plants bred from two different varieties for their desirable traits

mode the most frequently occurring number in a set of values

monoclonal antibodies antibodies made from cells that are cloned from one cell that are specific to one binding site on an antigen. They are used in medicine to treat cancer, and in research

monocultures where only one crop is grown in an area

motor neurone nerve cell carrying information from the central nervous system to muscles

MRSA bacteria that are resistant to the antibiotic, methicillin

mulch compost or decaying leaves put around rose bushes in the spring to prevent rose black spot spores reaching the stems

mutation where the DNA within cells has been altered (this happens in cancer)

mycoprotein a protein-rich food made from microorganisms in fermentation vats

myelin sheath a fatty insulating layer that surrounds neurones and speeds up nerve transmission

myopia short-sightedness; when people can see objects at short distances but not far away

N

natural selection process by which advantageous characteristics that can be passed on in genes become more common in a population over many generations

negative feedback a regulatory process in the body whereby changes in the body can be reversed once they have happened

nervous system a control system in the body that uses electrical impulses to communicate rapidly and precisely

neurone a nerve cell that is specialised to transmit electrical signals

nitrates nutrients found in fertilisers and the soil that plants need for protein synthesis, and therefore growth

non-coding DNA sections of DNA that do not code for genes but turn genes on or off instead; they regulate gene expression

non-communicable disease conditions caused by environmental or genetic factors that are not spread among people; e.g. cancer, cardiovascular disease

non-invasive describes brain mapping techniques, such as MRI scans, that can be carried out without the need for surgery

nuclear transfer type of cloning that involves taking a nucleus from a body cell and placing it into an egg cell

nucleic acid group of complex molecules found in living organisms including DNA and various types of RNA

nucleotide a sugar and phosphate group with a chemical base (A, T, C, G) attached to the sugar

nutrient broth liquid used for growing bacteria

O

oestrogen the main female reproductive hormone produced by the ovaries

opiates a group of painkillers found in poppies

optic nerve nerve found at the back of the eye that carries impulses from the retina to the brain

optimum the conditions, in terms of temperature and pH, at which an enzyme works best

optimum dose the amount of a new drug needed to be effective, determined in a trial by starting at low doses

order a level of classification below class but above family; e.g. butterflies and moths are orders of the insect family

order of magnitude in microscopy, the difference between sizes of cells, calculated in factors of 10

organ group of tissues that carries out a specific function

organ system arrangement of organs in the body according to function; e.g. respiratory system

osmosis the diffusion of water molecules through a partially permeable membrane, from a dilute solution to a concentrated solution

out of phase refers to the cycle of a predator lagging behind that of its prey

ovulation release of an egg from the ovary

oxygen debt the amount of oxygen that the body needs to breakdown lactic acid after muscles undergo anaerobic respiration

oxyhaemoglobin bright red substance formed when oxygen binds to haemoglobin in red blood cells; this is how oxygen is transported to tissues

P

pacemaker a group of cells in the right atrium that controls the heart rate

palaeontologists scientists who study fossils

palisade mesophyll tightly packed together cells found on the upper side of a leaf that carry out photosynthesis

pandemic when an outbreak of a disease becomes global

parasitism a relationship between two organisms where one benefits and the other is harmed, e.g. tapeworms in humans

partially permeable membrane a membrane that allows some small molecules to pass through but not larger molecules

pathogen harmful microorganism that invades the body and causes infectious disease

peatlands areas of peat bog formed in marshlands; peat is partially decomposed leaf matter that is used as fuel and cheap compost

peer review scientific findings are scrutinised by independent experts before they can be published

penicillin an antibiotic, isolated from Penicillum mould, which was discovered by Alexander Flemming

percentage a number or amount expressed per hundred

peripheral nervous system (PNS) network of nerves leading to and from the spinal cord

peristalsis the movement of food through the digestive system by muscle contraction

peritoneal dialysis a type of dialysis where fluid is pumped into the abdomen and waste products diffuse into it across a membrane in the body called the peritoneum

Petri dish plate, with a lid, used for growing bacteria in a lab

phagocyte a type of white blood cell that enters tissues and engulfs pathogens then ingests them

phagocytosis process by which a white blood cell (phagocyte) engulfs a pathogen

phenolphthalein indicator a pink solution (at pH 10) that turns colourless when fatty acids are added

phenotype the characteristic that is shown or expressed

phloem specialised transporting cells which form tubules in plants to carry sugars from leaves to other parts of the plant

phosphates nutrients found in soil and fertilisers that plants need for respiration and growth

photosynthesis process carried out by green plants where sunlight, carbon dioxide and water are used to produce glucose and oxygen

phototropism a plant's growth response to light

pituitary gland known as the 'master gland' as it controls other endocrine glands, such as the thyroid gland

placebo a treatment that does not contain a drug

placenta organ that forms during pregnancy that provides the foetus with oxygen and nutrients

plant cuttings in plant cloning, a leaf and stem are cut off a plant and then dipped in hormone powder to encourage rooting, a new plant then grows

plant organ system roots, stems and leaves, which collectively form a plant's transport system

plasma straw-coloured liquid part of blood

plasmid small ring of DNA found in prokaryotic cells

plasmolysis/plasmolysed the shrinking of a plant cell due to loss of water, the cell membrane pulls away from the cell wall

platelets cell fragments which help in blood clotting

polydactyl condition caused by a dominant allele in which the sufferer has extra fingers or toes

polymer a molecule, such as DNA, made up of repeating units

population the total number of one species in an ecosystem

potometer piece of equipment used to measure water uptake by plants

preclinical testing testing of a new drug in a lab using cells, tissues or live animals

prey animals that are eaten by a predator

primary consumer organism in a food chain that gets its energy from eating food (producers)

probability the likelihood of an event happening; e.g. there is a 1 in 4 chance of a couple who are both heterozygous for cystic fibrosis having a child with cystic fibrosis

producers organisms in a food chain that make food using sunlight

progesterone reproductive hormone that causes the lining of the uterus to be maintained

progestogen-only pill (or mini pill or POP) contraceptive pill that only contains progesterone (not oestrogen), which may be more suitable (than the combined pill) for women who are older or have high blood pressure

Prokaryota group of single-celled organisms including bacteria and Archaeans

prokaryotic cells single cells of bacteria and Archaeans with DNA found in a loop not enclosed in a nucleus

proportion (in genetics) the number of people affected by a condition expressed in relation to the total population

proteases digestive enzymes that break down proteins into amino acids

protist a type of single-celled organism, for example the pathogen that causes malaria

pruning snipping the tops off plant shoots in the spring, a treatment for rose black spot

pulmonary artery vessel that carries deoxygenated blood from the right ventricle to the lungs

pulmonary vein vessel that carries oxygenated blood from the lungs to the left atrium

Punnett square a grid used to determine possible outcomes of a genetic cross

pyramid of biomass table of dry weight of organisms at different trophic levels in an ecosystem, which forms a pyramid shape

Q

qualitative reagents chemicals used to test for the presence of a substance in a sample

R

radiometric dating method of determining the absolute age of rocks

range a measure of spread; the difference between the biggest and smallest values in a set

range bar line drawn on a graph to show the range of a set of values

rate the speed at which an event is occurring over time; can be calculated using the gradient on a graph

rate of photosynthesis is affected by temperature, light intensity, carbon dioxide concentration and amount of chlorophyll

ratio the relationship between two variables, expressed as 1:4, for example

reaction time the time it takes the body to respond to an event

receptors cells in the body that detect changes in the environment

recessive two copies of a recessive allele must be present for the characteristic to be expressed; represented by lowercase letters

red blood cells blood cells with a concave shape which are adapted to carry oxygen from the lungs to body cells

reflex action rapid automatic responses to a stimulus

reflex arc pathway taken by nerve impulses through the spinal cord during a reflex action

refraction the bending of light rays as they travel from one medium to another, e.g. as they enter the eye

regeneration restoration (of a habitat)

relay neurone neurone found in the spinal cord that transmits impulses from the sensory to the motor neurone

repeatability consistent data with a narrow range has a high repeatability; it is likely that the conclusions drawn from such a data set are valid

resolving power the ability of a microscope to distinguish between two points; the resolving power of electron microscopes is higher than that of light microscopes

respiration the process used by all organisms to release the energy they need from food

retina area at the back of the eye where light-sensitive receptor cells are found

ribosome structures in a cell where protein synthesis takes place

risk factor a lifestyle or genetic factor that increases the chance of developing a disease

rods light-sensitive receptor cells in the retina that allow us to see in dim light

root hair cells specialised cells in plant roots that are adapted for efficient uptake of water by osmosis and mineral ions by active transport

rose black spot fungal disease that affects plant growth; can be treated with fungicide or by destroying the leaves

run-off when fertilisers from a farmer's field are washed off into rivers or lakes (can lead to eutrophication)

rusts type of plant disease caused by fungi

S

Salmonella a type of bacteria that causes food poisoning

sampling techniques methods used to choose a small, representative number of individuals from a large population

scale bar a line drawn on a micrograph used to measure the actual size of the object

scanning electron microscope (SEM) works by bouncing electrons off the surface of a specimen that has had an ultrathin coating of a heavy metal, usually gold, applied. Used to view surface shape of cells or small organisms

sclera tough outer layer of the eye

secondary consumer organism in a food chain that gets its energy from eating primary consumers

secondary sex characteristics features that develop during puberty, as a result of sex hormones being released, such as a deep voice and hair growth in boys, and breast development in girls

secondary tumour abnormal growth of cells that forms from a malignant tumour elsewhere in the body

secrete to produce a substance; e.g. decomposers secrete enzymes to help break down food

selective breeding process of breeding organisms with the desired characteristics (also known as artificial selection)

selective reabsorption in the kidney, useful substances, such as glucose and amino acids, are absorbed back into the blood after filtration

selective weedkiller preparations of synthetic plant hormones that only kill weeds with broad leaves, for example, and leave grasses and crops unharmed

self-supporting ecosystem all ecosystems can support themselves; i.e. everything organisms need for growth and survival is present

sensory neurone nerve cell carrying information from receptors to central nervous system

sewage waste water and excrement, a form of pollution

sex chromosomes a pair of chromosomes that determines gender; XX in females, XY in males

sex determination whether a fertilised egg develops into a male (XY) or female (XX) depends on the 23rd pair of chromosomes

sexual reproduction form of reproduction involving two parents, which introduces variation

sexually transmitted disease (STD) disease spread through sexual contact; e.g. HIV

sieve plates the end walls of phloem cells that contain many small pores

species basic category of biological classification, composed of individuals that resemble one another, can breed among themselves, but cannot breed with members of another species

specialised when cells or tissues become adapted to carry out their specific function

speciation formation of a new species over a long period of time, through separation of a population

spermicide a cream that is toxic to sperm; it is used alongside other contraceptive techniques, e.g. condoms

sphere shape that has the smallest surface area compared with its volume

spiracles tiny holes along the side of an insect's body used in gas exchange

spongy mesophyll layer of cells found in the middle of a leaf with an irregular shape and large air spaces between them for diffusion of gases

spores small seed-like particles released by a fungus to help it spread

stacked bar chart bar graph where two data sets are stacked on top of each other for each group/bar

standard form way of writing very large or small numbers using powers of ten; e.g. $1.0 \times 10-3$

starch a complex carbohydrate found in animals and plants

statins drugs that stop the liver producing so much cholesterol

stem cells unspecialised body cells (found in bone marrow) that can develop into other, specialised, cells that the body needs, e.g. blood cells

stents a treatment for heart disease; a catheter with a balloon attached is inserted to open up a narrowed coronary artery

sterilise the process of removing contaminated material (e.g. from an agar plate)

stomata (singular stoma) small holes in the surface of leaves which allow gases in and out of leaves

structural adaptation a physical adaption to a body part of an animal, e.g. elephants have larger ears to maximise heat loss

surface area to volume ratio the relationship between the surface area of an organism or structure and its volume; SA:V ratios are large in single-celled organisms for efficient diffusion

suspensory ligaments ligaments in the eye that are attached to the ciliary muscle and hold the lens in place

sustainable long-lasting; something that can be maintained for future generations

sustainable fisheries measures to improve fish stocks, which were declining due to over-fishing, to include controls on net size and fishing quotas

synapse the gap between two neurones

systematic error mistakes made during data collection that give false outcomes

T

target organ the site where a hormone has its effect

testosterone male sex hormone produced by the testes that stimulates sperm production

therapeutic cloning the process of creating stem cells with the same genes as the patient, through nuclear transfer

thermoregulatory centre a region in the brain that detects changes in body temperature

three-domain system developed by Carl Woese, a classification system where organisms are divided into Archaea, Bacteria and Eukaryota (animals, fungi and plants)

thyroid gland gland at the base of the neck that makes the hormone thyroxine

thyroxine hormone produced by the thyroid gland that increases the body's metabolism

tissue group of cells that work together, with a particular function

tissue culture process that uses small sections of tissue to clone plants

tobacco mosaic virus (TMV) pathogen that causes mosaic pattern of discolouration on leaves and affects growth of plant

trachea (lungs) windpipe, through which air travels to the lungs

trachea (insects) small tubes in insects containing water, through which gases diffuse into body cells

transect line across an area to sample organisms

translocation the movement of sugars through a plant

transmission electron microscope (TEM) uses an electron beam to view thin sections of cells at high resolution

transpiration the movement of water up through a plant and its loss from the leaves

trend a pattern in a data set

trophic level feeding positions in a food chain

tropism the response of a plant by growing towards or away from a stimulus

tundra regions in the Arctic where few plants grow as the subsoil is permanently frozen

turgid plant cells which are full of water with their walls bowed out and pushing against neighbouring cells

type 1 diabetes a condition where the pancreas cannot produce enough, or any, insulin

type 2 diabetes a condition where the body cells no longer respond to insulin produced by the pancreas

U

umbilical cord joins the foetus to the mother during pregnancy and is a source of stem cells

uncertainty a measure of the range about the mean, calculated by the range divided by two

urea waste products formed by breakdown of excess proteins, excreted in urine

V

vaccination injection of a small quantity of inactive pathogen to protect us from developing the disease caused by the pathogen

vaccine preparation of an inactive or dead form of a pathogen given by injection or nasal spray

valid refers to results that are accurate, representative and repeatable

valves flaps of tissue that prevent the backflow of blood in the heart and in veins

variation the differences between individuals brought about by both genetic and environmental influences

vascular bundle (veins) group of xylem and phloem cells that transport water and glucose around the plant

vasoconstriction in cold conditions, the diameter of small blood vessels near the surface of the body decreases, which reduces blood flow

vasodilation in warm conditions, the diameter of small blood vessels near the surface of the body increases, which increases blood flow

vector a carrier, usually a plasmid, used to transfer a gene into an organism to be genetically modified; also, mosquitoes are vectors – they spread malaria but don't cause it themselves

veins blood vessels with thin walls and valves to prevent backflow, that carry blood at low pressure back to the heart

vena cava vein that carries deoxygenated blood from the body to the right atrium

ventilate the movement of air into and out of the lungs

ventricles the lower chambers of the heart that pump blood around the body (left) or back to the lungs (right)

W

Wallace (Alfred Russel) scientist that proposed the theory of evolution through natural selection, independently from Darwin

warning colouration bright colours or patterns found on animals, used to deter predators

X

X-chromosomes sex chromosome present in males (XY) and females (XX)

xylem cells specialised for transporting water through a plant; xylem cells have thick walls, no cytoplasm and are dead, their end walls break down and they form a continuous tube

Y

Y-chromosomes sex chromosome found only in males

yield the amount (of crop) produced

Z

zones of inhibition areas of no bacterial growth on an agar plate around an antibiotic disc

zygote a fertilised egg cell

Index